THE SCIENTIFIC BASIS OF FILTRATION

edited by

K. J. IVES

Professor of civil engineering
Department of Civil and Municipal Engineering
University College London, U.K.

SPRINGER-SCIENCE+BUSINESS MEDIA, B.V. 1975

Proceedings of the NATO Advanced Study Institute
on The Scientific Basis of Filtration, held at
Cambridge, U.K., July 2-20, 1973

ISBN 978-94-015-3987-6   ISBN 978-94-015-3985-2 (eBook)
DOI 10.1007/978-94-015-3985-2

THE SCIENTIFIC BASIS OF FILTRATION

NATO ADVANCED STUDY INSTITUTES SERIES

Proceedings of the Advanced Study Institute Programme, which aims at the dissemination of advanced knowledge and the formation of contacts among scientists from different countries.

The series is published by an international board of publishers in conjunction with NATO Scientific Affairs Division

A	Life Sciences	Plenum Publishing Corporation
B	Physics	London and New York
C	Mathematical and Physical Sciences	D. Reidel Publishing Company Dordrecht and Boston
D	Behavioural and Social Sciences	Sijthoff International Publishing Company Leyden
E	Applied Sciences	Noordhoff International Publishing Leyden

Series E: Applied Sciences
Volume 2. The Scientific Basis of Filtration

TABLE OF CONTENTS

VI

INTRODUCTION

K.J. Ives

Professor of Public Health Engineering
University College London

The filtration of liquids is a process which is widely used in industry. In some cases the object of the process is to clarify the liquid, as in the filtration of beer, or sugar solutions; in other cases the object is to recover solids suspended in the liquid, for example, ores in mineral processing, or the production of pigments such as titanium dioxide.

In addition to the many industrial applications of liquid filtration, a considerable investment lies in municipal applications for the filtration of drinking water, wastewater and the sludges arising from water and sewage treatment. Indeed, in terms of volume filtered, it is likely that the municipal application exceeds that of industry.

The difference between clarification of the liquid and recovery of the solids is often expressed in the division of the processes into deep bed filtration and cake filtration. There are exceptions to this division, and some applications are not clearly in one field or the other. An example is the diatomite filtration of drinking water used in small municipal supplies for emergency circumstances and for military purposes. In this application the slurry (diatomite body feed) is the filtering agent which forms a cake, clarified liquid is required, and the filtration of small impurities in the water may be taking place in the depth of the diatomite layer.

Generally, deep beds are of loose granular media, such as sand, whereas cake filtration takes place on some primary medium such as a cloth, with the liquid impelled by a vacuum pressure difference or a positive pressure difference. In all these cases Darcy's Law applies, and frequently the Kozeny-Carman equation is a useful mathematical model of the process. Consequently, some basic hydrodynamic laws should apply to the liquid flow through the porous media, and some physical concepts of structure should be available to describe the media. Furthermore, the particles suspended in the liquid interact with each other and with the filter media in a manner involving colloid and interfacial forces. Therefore, filtration, which has evolved principally as an empirical or ad hoc technology, should have some scientific basis evolved from physical and chemical considerations.

It is this scientific basis which formed the lecture content of the NATO Advanced Study Institute held in Cambridge, England, in July 1973. These lectures have been compiled into the present volume, and therefore it represents a collection of styles of presentation and individual emphasis. There has been no attempt to suppress these individualities by a concept of editorial conformity. Each author is an expert of international standing, so the structure of each lecture reflects this expertise which no editor should have the temerity to question.

However, the overall structure of the book does reflect editorial opinion. Chapters 2 to 5 are fundamental enough to apply in any area of liquid filtration. However, after this the unification cannot be maintained and the deep bed and cake filtration differences emerge, so that Chapters 6 to 12 relate to deep bed filtration and 13 to 17 deal with cake filtration. In each of these major sections there is a progression from fundamental aspects to applications, but in both cases stops short of engineering design and the development of the hardware of machinery, containers, pumps, etc.

It is assumed that the reader is acquainted with the basic sciences of physics, chemistry, fluid mechanics and mathematics, for these are all used to present the scientific basis of filtration.

PARTICLE CHARACTERISATION

D. C. Freshwater

Loughborough University of Technology

1. INTRODUCTION

The behaviour of a system of particles is determined by the 'size' of the particles, their spatial position relative to one another and the nature and properties of the surrounding fluid. The reason for putting the word 'size' in inverted commas will become clear later. Suffice it to say at this point that size has no very precise meaning and in this context includes both shape and distribution. Consider for a moment a system of mono-sized plate-like crystals evenly dispersed at low concentration in a thin liquid of less density than the particles. Such a system will have a very high porosity and a negligible resistance to flow of the fluid. However, if the particles are large enough so that gravitational forces outweigh diffusional ones (> 1 μm say), the particles will settle into a closely packed bed probably preferentially oriented with the plates lying on top of one another. In this spatial arrangement the particle system will be almost impermeable and may have a very low porosity. Thus the behaviour depends not only on the size but the spatial position of the particles. Again a system of spherical particles of a wide range of sizes would behave in both circumstances (dispersed and settled) in a quite different way.

The aim of the particle technologist is to (a) describe the particle system precisely and (b) to use this description to predict its behaviour in a given geometry (shape of container and flow rate). Just as one can predict the heat transfer coefficient for a given fluid flowing at a known rate in a circular pipe of a particular diameter, so one would like to be able to predict say the permeability of a bed of particles of known size to a given fluid. Anyone with the slightest familiarity with filtration theory (or indeed almost all solid/fluid processes) will know that this aim is far from being realised in the present state of our knowledge. The so-called basic filtration relationships are largely empirical, and cake resistances for example must be determined experimentally for each individual particle/fluid system. The reasons for this lie in our inability to describe in precise quantitative terms a system of particles and our lack of a quantitative description of flow of a fluid through a packed bed. Nevertheless we do know quite a lot about the qualitative relationships which determine the behaviour of particle systems in terms of the particles themselves. Thus the first step in tackling any filtration problem is, or should be, an examination of the particles themselves and attempting to characterize them. Therefore in what follows I shall first of all consider ways of describing particles both as individual particles and as systems. Then I shall go on to discuss the more important principles involved in measuring these properties and finally discuss how these measurements may be used.

2. DESCRIPTIONS OF PARTICLES

2.1 Single Particles

Particles occurring in real systems are almost always irregular in shape and of a range of sizes. In general therefore one has to describe an irregular shape by some simple parameter or set of parameters of size and similarly to describe the range of size. Consider first the problem of the individual particle.

It is possible to define the size of a particle in two fundamentally different ways. The first of these is in terms of an equivalent sphere and the second as a statistical diameter. The methods of making the measurements are the same but the manner in which these measurements are interpreted and applied are very different.

Although historically not the first, the so-called 'equivalent-sphere' approach is the most widely used. In this the particle is replaced by a sphere which is equivalent to the original particle either in a geometric or in a dynamic sense. As an example, an irregular particle viewed from above as in a microscope can be represented by its projected two-dimensional image. This can be assigned a size in terms of its length in some arbitrary direction which is then considered as the diameter of a sphere equivalent, in this respect, to the particle. More commonly, the projected image is viewed as an area and the particle is characterised by a circle of equivalent area, this circle being the projected two-dimensional version of a sphere. Again, a sphere might be imagined equivalent in volume to the particle. Geometric identities are not the only ones which can be used, particles settling in a fluid under the influence of gravity are often given a size equal to the sphere of the same density which falls at the same speed, i.e. a dynamic or Stokes' size. The following equivalent sphere sizes may be defined:-

(i) The same projected area as the particle when viewed in a direction perpendicular to the plane of greatest stability-symbol d_a

(ii) The same projected perimeter as the particle when viewed from above (d_p)

(iii) The same volume as the particle (d_v)

(iv) The same surface area as the particle (d_s)

(v) The same free falling velocity as the particle (d_f) or if the particle obeys Stokes' Law, d_{st}

(vi) Correspond to a square aperture of side A through which the particle will just pass d_A

Immediately one of the main disadvantages of representation of particle size by an equivalent sphere is evident. It is that the same particle can be represented by different sized spheres, depending upon which equivalence is used. It is important to appreciate this because various practical methods of particle size determination measure different equivalent diameters, e.g. the microscope method gives (i) above, sedimentation (v), sieve analysis (vi) and so on. Thus it is desirable to choose a method appropriate for the purpose for which the measurements are to be used. It is also important when one wishes to compare a particle size determined by one method with that measured by another.

The great advantage of the equivalent sphere representation is that it results in a single parameter description of a particle of irregular shape which, because of the mathematical properties of a sphere serves to describe diameter, area, volume and surface. One has to ask eventually whether this simplicity is a worthwhile exchange for all the information about the irregular particle that this method throws away.

The second method of characterising a single particle is by a statistical diameter. This is the linear dimension of the particle measured in some specified way. Two such statistical diameters are well known, the Feret diameter and the Martin diameter (see Fig. 1). Thus the Feret diameter is the distance between two parallels which just touch the particle in the direction in which it is being scanned. Another linear statistical measurement is the Martin diameter. These measurements are not meaningful when applied to a single particle and only gain significance when applied to describe large collections of particles.

A third statistical description of much more recent origin is the random chord method. In this the particle is described by the distribution of a series of parallel chords drawn in a random direction across the projected image of the particle. The chords could be extended into two dimensions, thus producing a distribution of cross-sectional areas. These two distributions together with a knowledge of the volume of the particle completely and uniquely describe the particle. This method promises to provide a powerful analytical description of particle systems but as yet is at the development stage.[1]

2.2 Systems of particles

In practice we are always concerned with systems containing very large numbers of particles. These particles are not identical in size, indeed in many filtration processes, it is not uncommon to find a hundredfold variation between the smallest and largest particles. The particles may not all be identical in shape (although it is generally assumed that they are) and very little work has ever been done to determine the effect of different shapes [2] on the analysis of particle systems. Particles in a large system may also be of different species in which case it is impossible to characterize the system by any sedimentation or elutriation method. Even less work has been done on this problem therefore, having drawn attention to these difficulties, I shall in what follows assume that one is dealing with particles all of the same species* and same geometric shape.

Clearly for a system containing a large number of particles of different sizes the only complete description is a distribution curve showing number of particles against some individual particle parameter, i.e. a frequency distribution curve. Fig. 2 shows three types of frequency distribution curves commonly used in describing real systems of particles. There are at least eight other distribution functions which have been proposed and/or used by various workers to describe particle distribution. The aim is always to present all the information in terms of a few parameters — in the case of the normal distribution, for example, the mean and standard deviation. For a more detailed treatment,

---

* If the particles are of the same species they all have the same density and therefore particle weight is directly proportional to geometric particle size cubed.

reference should be made to Herdan[3] and Beke.[4] The desirability
of being able to represent particle distributions by curves of
this type which obey well known mathematical laws is obvious.
This desire for pattern and conformity is, however, a dangerous
one as it tends to mask real and significant differences between
particle populations.

Although the frequency distribution curve is the fundamental
one representing size data of the system, it is not the most convenient method of expressing in graphical form the characteristics
of a powder. For this purpose it is better to use the cumulative
curve which is in effect the integral of the frequency distribution curve, and is in general constructed more readily from the
data (Fig.3). These may be plotted to arithmetic or logarithmic scales so as to approximate the straight lines. Again,
reference should be made to one of the standard works cited [3,4]
(see also Appendix A).

3. MEASURING TECHNIQUES

There are many different measuring techniques which have been
used but the number of physical principles involved is relatively
few. A large number of the techniques are summarised in Table 1,
although there are some which do not fit this classification.

Table 1

Technique	Common Examples	Approximate size range commonly covered.	Diameter which can be measured.
Field Scanning Techniques	Microscope Counting	1μ – 100μ	Length area
	Electron Microscope	$.001\mu$ – 5μ	statistical diameters
Stream Scanning Techniques	Coulter Counter Light Scattering Photometer	1μ – 100μ $.5\mu$ – 50μ	volume
Classification Techniques	Sieving Air Classifiers	$> 10\mu$ 2μ – 50μ	Sieve diameter Stokes diameter
Sedimentation Techniques	Pipettes Balances Photosedimentometers Centrifugal Methods	1μ – 60μ $.05\mu$ – 10μ	Stokes diameter

It is not possible within the compass of these notes to
describe these methods in detail, but some techniques of especial
use in filtration technology will be considered. The first of
these is microscope counting.

3.1 Microscope Counting

Examination by microscope of particles to be filtered is or
should be an automatic step in the ordered solution of a filtration problem because visual inspection of the particles can provide valuable information as to their filtration behaviour. Thus
direct observation will immediately show if the particles are
gelatinous flocculated or discrete or contain a significant proportion of small particles (< 10 μm) and so on. Some idea of the
size range of particles may be obtained by using an eyepiece which
has a scale engraved on it. This scale is calibrated for the
degree of magnification being used by viewing a similarly engraved
slide placed on the microscope stage. The method of accurate
sizing used is an extension of this in that an eyepiece graticule
is used on which a series of circles of different sizes have been
micro-photographed. The size of the particle is assigned as being
between that of two adjacent reference circles (see Fig. 4). The
subject of particle counting and sizing by microscopy is very com-
prehensively covered in BS.3406 part 4 (1963).

This method, however, suffers from two limitations:

(a) the large amount of time required to count a comparatively
     small number of particles, and

(b) the unreliability of manual counts.

In recent years a number of techniques have been evolved to
make counting faster and easier. They range from the semi-automatic in which the measurements are made mechanically but with the
operator retaining his ability to discriminate, e.g. between
touching or re-entrant particles, to the completely automatic in
which discrimination is preset into the machine and the results
delivered as a computer print out. The range of methods has been
described (Scarlett[5]). It should be observed that the British
Standard method automatically counts the particles as equivalent
spheres (area) whilst most of the semi-automatic or automatic
methods measure a statistical diameter.

However, the major importance of the microscope in filtration
technology is to obtain some idea of the likely nature of the
problem, (a) by visual examination of the particles themselves and
(b) by obtaining an approximate idea of the size range.

3.2   Sedimentation

The range of techniques used to measure particle size by sedimentation is very wide and therefore I shall begin this section with an outline of the basic principles which apply to them all.   The size of particle that is measured by sedimentation is, as has been noted above, a dynamic size and is usually expressed as the diameter of a sphere which would fall at the same velocity as the particle under the conditions to which the Stokes' equation applies.   It will be remembered that this relationship only applies to individual spherical particles falling at their terminal velocity in a vessel of infinite cross-sectional area.   The diameter of such a sphere is given by the relationship

$$d_{st} = 10^4 \sqrt{\frac{18\mu}{(\rho_s - \rho_1)g} \cdot \frac{h}{t}}$$

where d_{st} is in microns, viscosity μ in poises, height of fall h in centimetres, time in t seconds and the densities ρ_s of the solid, and ρ_1 of liquid, in g/cm^3.   In a given apparatus the viscosity, densities and height of fall are usually maintained at constant values, hence the diameter measured is proportional to the square root of 1/time of fall.   Table 2 shows the size of sphere of three different densities which will fall in two kinds of liquids in times ranging from 1 second to 12 hours and explains amongst other things why there is a wide range of techniques of sedimentation available.

It must be remembered that Stokes' equation is only followed in the laminar flow region when the particle Reynolds number, $(\frac{d_p u \rho}{\mu})$ is less than 0.2.   Thus for particles and liquids of the density considered in Table 2 there is a maximum diameter of particle which can be measured accurately by sedimentation and this is shown in Table 3.

Table 2   Settling Time of Particles in Different Fluids

density of particle	= 2g/cc		= 3g/cc		= 10 g/cc	
Time	Water	Glycerin/water	Water	Glycerin/water	Water	Glycerin
1 sec	600	1400	420	970	200	450
100 sec	60	140	42	97	20	45
10^4 sec 3 hr	6	14	4	9.7	2	4.5
12 hr	3	7	2	4.85	1	2.75

Table 3   Particle Size which just obeys the Re $<$ 0.2 Criterion

---

	p = 2	p = 3	p = 10
Water	71µ	56µ	34µ
Water/Glycerin	223µ	136µ	104.5µ

$$d^3 = \frac{3.6\, \eta^2}{(p - p_o)p_o g}$$

The British Standard on sedimentation (BS.3406 Part 2 1963) therefore recommends that particles whose size is to be determined by
this means should first be sieved on a 200 mesh sieve (75 micron)
before being analysed.   Even so this will leave some particles
which fall in the region beyond the strict application of streamline flow.   It is possible to correct for this effect[6] but in
practice it is usual simply to assume the Stokes' equation applies.

Just as there is a maximum particle size beyond which sedimentation should not be used, so there is a minimum particle size
for gravity sedimentation.   The limit in this case is due to the
particle motion caused by molecular bombardment (Brownian movement).   For particles of a density of 2 g/cm^3 settling in water,
this will become noticeable at 5 microns and BS.3406 recommends
that the minimum particle size for gravity sedimentation measurements should be 3 microns.   Thus if we are going to use the
Stokes' approximation and avoid problems of trying to describe
particle shape, we are limited by using sedimentation in a liquid
to an upper size of 60 microns and a lower one of 3 microns.   The
lower limit may be decreased by using a centrifugal force to cause
the particles to settle, but this will also reduce the upper size
limit as well.   The probable range of a centrifugal sedimentometer
is approximately 10 to 0.05 microns.

It should be noted that since the Stokes' equation is strictly
applicable only to a single particle in a vessel of infinite cross-
section, sedimentation analysis should be carried out with very
dilute suspensions.   For example, the German standard recommends
a weight/volume concentration of 0.02 %.   Again, thermal convec-
tion currents in the liquid can if not minimised cause serious
errors in the analysis, and very close temperature control is
necessary preferably within a range of $\pm$ 0.1°C.

In sedimentation, as in other methods of analysis, one must
decide whether one wants to measure the particles as they exist in

the system, or in a more discrete form; thus if the particles in
the system are flocs, one is interested in the size of the flocs
and not of the size of the individual particles going to make up
the flocs.   Care should be taken, therefore, in the use of dis-
persing agents which may break up these flocs.   On the other hand,
if one wishes to measure the fundamental particles in the system
which may exist as aggregates, then one must apply sometimes quite
severe dispersive action before carrying out the analysis.   The
existence or otherwise of aggregates is another important reason
for using the microscope for a preliminary examination.

    In gravity sedimentation in a liquid there are two main types
of methods.   The first of these is the cumulative method in which
the sediment collected at the base of the container is measured.
These methods include sedimentation balances, liquid columns with
sediment extract and radiometric methods.   The other main group
may be classified as incremental methods in which the concentration of powder at a fixed depth is measured as a function of time.
Examples of these types are pipette methods and photo-sedimento-
meters.   A brief outline of the theory applying to the principal
methods of each group is given in Appendix B.

3.3   Stream Scanning methods

    Stream scanning methods can be applied to particles which are
dispersed in a fluid.   The suspension is passed through a small
sensing zone and the particles are counted and sized individually.
These methods count a large number of particles yielding a number
: size analysis which is useful for many purposes and could, in
some cases, be adapted for on-stream monitoring.   Several physical
principles have been used to detect particles, but the one that is
of most interest in filtration technology is the Coulter principle.

3.3.1 Description of the Coulter Principle.   The Coulter Counter
determines the number and size of particles suspended in an electrolyte.   This is done by forcing the suspension to flow through
a small aperture submerged in the electrolyte and having an elec-
trode on either side.

    As a particle passes through the aperture, it changes the
resistance between the electrodes.   This produces a voltage pulse
of short duration whose magnitude is proportional to particle size.
The voltage pulses are amplified and fed to a threshold circuit
having an adjustable threshold level.   If this level is reached or
exceeded by a pulse, the pulse is counted.   The threshold level
is indicated on an oscilloscope screen by a brightening of the
pulse segments above the threshold, facilitating the selection of
appropriate counting levels.   The threshold level can be shown to

be proportional to particle size.   Thus the instrument will count all particles above a pre-selected threshold or particle size.

The current between the electrodes is maintained at a constant level so that any change in the cell resistance results in a change of voltage across it.   The presence of a particle of a different resistivity in the orifice will thus result in a change of resistance and this will be proportional to the volume of the particle.   The theory of this is elaborated in Appendix C.   From this it can be seen that the response of the Coulter Counter is not linear with particle volume.   The non-linearity depends upon particle shape, but is not significant provided that the longest particle measured is not greater than 40% of the orifice diameter.

A problem peculiar to all stream scanning methods is that of coincidence, i.e., the simultaneous arrival in the sensing zone of two or more particles.   This has been the subject of considerable discussion and research in respect of the Coulter Counter (see reference 11).   The results of this may be summarised by saying that coincidence cannot be eliminated but its effect can be minimized by counting sufficiently large numbers of particles.

3.4   Sieving

Many industrial filtration operations are concerned with relatively large particles which cannot conveniently be examined by methods so far discussed.   Nevertheless, it is desirable in considering the approach to the design of a suitable filtration system to have some knowledge of the particle size.   For such particles, sieving is probably the best method of characterisation. The British Standard fine series of sieves have apertures ranging from about 50 to 500 microns and there are similar (although not exactly the same) continental and American standards (see Appendix D).   Details of the techniques of sieve analysis are given in B.S.1796 (1952) and it is not intended to go into detail here on this, but rather to draw attention to what it is that a sieve analysis measures and some of the considerations to follow from this.

When a system of particles is presented to the surface of a sieve, what determines whether a particle will pass through this sieve and hence be counted as below that nominal sieve size or not, is whether it has some dimension on a major axis less than the maximum size of aperture on the sieve.   This in turn depends on the shape of the particle.   Thus, for example, an elongated and flaky particle could pass through a sieve opening across the diagonal (= 1.414 times the nominal sieve aperture).   On the other hand, an elongated particle but of a more nearly rectangular cross-section could only pass through the square aperture so that sieve analysis perhaps more than any other form of characterisation is peculiarly sensitive to the shape of the particles which are being measured. The extensive researches of the late Professor Heywood on this subject are well summarised in references (7,8) when he showed that the size of particle which a sieve most nearly measured was its breadth and that it was necessary to use shape factors in terms of a volume coefficient to convert the size measured by a sieve into an equivalent spherical size. The relationships he derived and the shape factors for a number of materials are given in Appendix E.

Apart from this effect, sieving is also the least accurate method of characterisation described because whether or not a particle passes through the sieve depends not only on its size but also on the probability of its reaching the aperture in a preferred direction. Sieving is a probabilistic process and therefore there cannot be a sharp cut between the different sizes.

4. USE OF MEASUREMENTS

Each of the methods described above measures a different property of the particles being examined and there are other methods of size analysis which measure yet other properties, e.g. Air permeametry which measures the surface area of the particles. It is absolutely essential when carrying out a size analysis to understand which property the method that is being used is measuring and to report it as such. Unfortunately, it is not uncommon to find two different methods have been used to cover a wider size range than is possible by one technique and although the different methods measure different properties, no attempt has been made to bring either method back to a common base. Since relationships are available to enable the particle size measured by one technique to be related to that by another, let us consider how this may be done in terms of equivalent spherical sizes. The surface area of a particle is proportional to the square of some characteristic dimension and the volume to the cube of this dimension. Proportionality constants will depend on which dimension is chosen to characterise the particle and this is the projected area diameter in the following discussion. Thus, surface of the particle equals $f_a d_a^2 = \pi d_s^2$ and volume of the particle equals

$k_a d_a^3 = \dfrac{\pi d_v^3}{6}$ where f_a is the surface coefficient and k_a is

the volume coefficient. The volume coefficient k_a is much easier to determine than the corresponding surface coefficient f_a. The former may be calculated knowing the number, mean size, weight and density of the particles composing a fraction graded between close size limits. f_a, however, must be determined by geometrical

analogy from measurements made on larger particles. If the particle size is measured by any method, it may be related to the projected area diameter by noting that the volume of the particle equals $k_a d_a^3 = k_x d_x^3$ where the suffix x denotes the method of measurement is unspecified. Similarly, $f_a d_a^2 = f_x d_x^2$.
Table 4 shows

Table 4

Equivalent Diameter d_x	Ratio $\dfrac{d_x}{d_a}$	Ratio $\dfrac{d_a}{d_x}$
d_a	1.00	1.00
d_s	0.91	1.10
d_v	0.78	1.28
d_{st}	0.69	1.45
A	0.71	1.41

typical values of the ratios of diameters measured by different procedures when the actual shape coefficients k_a and f_a are 0.25 and 2.60, respectively. (Note this corresponds to an angular particle).

  If we want to apply this to a system of particles, then we must use similar transforms for each of the size increments.

  It is often desired to characterise a system of particles by a mean size and this may be done assuming that the particles all have the same density and shape. We can express the properties of total number, length, surface area, volume or weight in the following ways:

Number in Group	$x^0 dN$	Total for System	$\sum x^0 dN = N$
Length " "	$x\, dN$	" " "	$\sum x\, dN = L$
Surface " "	$x^2 dN$	" " "	$\sum x^2 dN = S$
Volume " "	$x^3 dN$	" " "	$\sum x^3 dN = V$

Suppose the particles have been sized by a microscope count, the distribution can be expressed by means of a simple histogram in which the area of each rectangular block represents the number of particles in the respective groups. The height of each block would be equal to dN/dx where dx is the difference between the respective size limits of each group. If we now wish to express the distribution for property represented by a higher order of x, each element of the number frequency curve must be multiplied by x, x^2 or x^3, respectively. If the law relating dN/dx and x is known, this summation may be made mathematically, but this is generally not the case and the summation has to be made by tabulation. One important point to note is that in carrying out this kind of conversion, the higher the order of x, the more important effect the larger particles have on determining the mean, as shown in Fig. 5.

The mean diameter for a system of particles is thus obtained by dividing the sum of the moments about the $x = 0$ axis of the constituent elements by the sum of the areas of the elements. The most important point to note, however, is that if a system of non-uniformly sized particles is to be represented by a system of uniformly sized particles having the same shape and density, the two systems can be regarded as equivalent in respect of two, but only two, size dependent properties. This restriction is illustrated in Fig. 6 which shows an imaginary system of 20 spheres of varying diameters. This non-uniform system can be represented by various uniform systems but equivalence can only be obtained for two properties, therefore the mean size calculated for the system will differ according to the properties that have been selected. For more details of this procedure and its use, reference should be made to Heywood.[9] The relationships may be expressed conveniently in terms of moments of distribution and a comprehensive system for doing this has been worked out by Rumpf and his colleagues.[10] (see Appendix F)

This, however, shows up again one major disadvantage of the equivalent sphere approach or indeed any method which attempts to represent a particular system in terms of a single parameter. In fact, this limitation is prohibitive as far as ever being able to explain the behaviour of particulate systems in terms of sets of equivalent spheres; thus much of the elegant work that has been carried out on the determining the packing of unequal size spherical particles is only at best an interesting academic exercise and cannot really lead to a quantitative understanding of the behaviour of a packed bed.

What then should we measure? The answer to this is that we do not yet know precisely what properties need to be measured in order quantitatively to describe particulate systems. What we are interested in as filtration engineers, for example, is the behaviour of a particulate system in terms of macro properties such as porosity and permeability and indeed the kind of laboratory tests we carry out on our particulate systems are usually ones which measure this property together perhaps with a specific cake resistance, and which can then be applied in existing filtration equations. However, these relationships are only valid for the set of particles on which they have been determined, and we cannot as yet predict how these macro properties will change as the micro properties of the particle system such as size and size distribution change. The real problem is therefore the discovering of the relationship between these macro properties and the micro properties of the system. Until we understand what these are (and it is likely that they are related to factors more than just those of geometric or even dynamic size), then we are not going to be able to deal quantitatively or predictively with filtration from an examination of the primary properties of the particle system. Existing methods of particle size analysis therefore are useful in two ways. Firstly, as already explained, in giving some idea of the type of size distribution and type of particles which will enable us by experience to make useful qualitative estimations about the kind of filtration problem we are likely to have, and secondly from the point of view of control purposes. Thus, if we have designed a filtration unit to work successfully for a given particulate system, we know that if the particle size and distribution changes significantly, then the filtration problem will be different, may be more severe and perhaps the filtration equipment will not work satisfactorily. Therefore conventional size analysis is a useful control check.

REFERENCES

1. Scarlett, B. Proceedings of 1970 Conference on Particle Size Analysis. Society of Analytical Chemistry (London 1972)

2. Bird, K.E. University of London, Faculty Mech. Eng., M. Phil. Thesis, 1966.

3. Herdan, G. Small Particle Statistics, Butterworth (London 1960).

4. Beke, B. Principles of Comminution, McLaren (London 1964).

5. Scarlett, B. Particle Characterization and its Application, New York & Chicago. Loughborough Univ. of Technology (Loughborough 1972).

6. Lloyd, P. J. Particle Characterization and its Application. New York & Chicago. Loughborough Univ. of Technology (Loughborough 1972).

7. Heywood, H. Proc. Inst. Mech. Eng. 140, 257 (1938).

8. Heywood, H. Trans. Inst. Min. Metall. 55, 373 (1945-6).

9. Heywood, H. Jl. Pharmacy & Pharmacology Supplement, 15, 56T (1963).

10. Rumpf, H., Debbas, G., & Schönert, K.   Chem. Ing. Tech. 39 (1967).

11. Lloyd, P.J., Scarlett, B., Sinclair, I.   Proceedings of 1970 Conference on Particle Size Analysis. Society of Analytical Chemistry, London, 276, 1972.

APPENDIX A

STATISTICAL FUNCTIONS

1. Normal Distribution

This is given as:-

$$N = \frac{\Sigma N}{s\sqrt{\pi}} \quad \exp\left[-\frac{1}{2}\left(\frac{d - \bar{d}}{s}\right)^2\right]$$

where

$\bar{d}$ = arithmetic mean diameter = $\dfrac{\Sigma(N\,d)}{\Sigma N}$

s = arithmetic standard deviation = $\left[\dfrac{\Sigma N(d - \bar{d})^2}{\Sigma N}\right]^{\frac{1}{2}}$

d = particle diameter

N = number of particles with diameter d.

The function is closely obeyed by many biologically occurring systems, e.g. diameter of pollen grains, red blood cells, starch grains, etc.

2. Log Normal Distribution

Given by the expression:-

$$N = \frac{\Sigma N}{\log s\sqrt{\pi}} \quad \exp\left[-\frac{1}{2}\left(\frac{\log d - \log \bar{d}}{\log s}\right)^2\right]$$

where

$\log \bar{d}$ = geometric mean diameter = $\dfrac{\Sigma(N \log d)}{\Sigma N}$

$\log s$ = geometric standard deviation = $\left[\dfrac{\Sigma\left[N(\log d - \log d)^2\right]}{\Sigma N}\right]^{\frac{1}{2}}$

The function is closely obeyed by many powder systems.

There are two points of importance common to both these statistical functions which are worthy of mention.

a) Absolute agreement of the experimental curve with that obtained from the equation is not necessary for valid use of either function. For instance, a true normal distribution is symmetrical about the mean, but deviation from this, i.e. skew, can be assessed quantitatively and a° skew parameter used to describe the system. In the same way, extremes of spread can be defined by a Kurtosis factor.

b) It is possible to linearise both normal and log normal distribution functions using the probit transformation. In the equation for a normal curve, the term:-

$$\frac{d - \bar{d}}{s} \quad (= u)$$

occurs. u is referred to as the normal equivalent deviate, and is the deviation of a particle diameter from the mean expressed in terms of the standard deviation. u is zero at the mean (cumulative frequency 50%) negative for smaller, and positive for larger particles. Values of u can be calculated at various levels of the cumulative frequency curve, +5 is added to make all values positive, and these probits are used as the proportion axis of a proportion VS size plot. Tables for converting per cent to probits are available, but it is convenient to use special probability graph paper.

3. Rosin-Rammler Distribution (Rosin and Rammler (1933)

Written as:-

$$P = 100 \exp (- bd^{n})$$

P = percentage of particles greater than diameter d
b = constant
n = constant

Linearization of the curve is possible by writing the expression as:-

$$\log (\log \frac{100}{P}) = \log b + \log \log e + n \log d$$

A plot of P on log-log reciprocal VS d on log axes gives a line of slope n and intercept (b log e)

Although much used for describing the products of breakage

treatments, especially coal, it is a strictly empirical equation depending on curve fitting and graphical interpretation. Another criticism is its insensitivity and the fact that it is difficult to derive meaningful parameters therefrom.

APPENDIX B — SEDIMENTATION THEORY

1.   Cumulative Methods

Consider the sedimentation of an initially uniform suspension.   After a selected time t, the weight settled will consist of two parts.   The first consists of the particles of Stokes' diameter equal to or greater than that given by equation on page 7 All these particles will have fallen from the surface of the liquid to the bottom.   The second part consists of those particles of smaller diameter which will have fallen from intermediate heights.

If the weight of particles per micron range in interval between D and D + dD is W(D), the total weight that has settled can be described by

$$W = \int_{D1}^{D_{max}} W(D) \; dD + \int_{D_{min}}^{D1} \frac{vt}{h} \; W(D) \; dD \qquad (2)$$

where v is the velocity of fall of particles of size D and h is the height of the suspension.   D min and D max are smallest and largest sizes in the distribution.

$$\text{Differentiating (2)} \quad \frac{dW}{dt} = \int_{Dmin}^{D1} \frac{v}{h} \; W(D) \; dD$$

$$\text{or} \qquad t \; \frac{dW}{dt} = \int_{Dmin}^{D1} \frac{vt}{h} \; W(D) \; dD \qquad (3)$$

combining (2) and (3)

$$\int_{D1}^{Dmax} W(D) \; dD = W - t \; \frac{dW}{dt} \qquad (4)$$

where W is the weight of particles greater than D_1.

The term $t \; \frac{dW}{dt}$ can be evaluated by plotting the weight of the sediment W versus time and drawing tangents to the curve at selected times t.

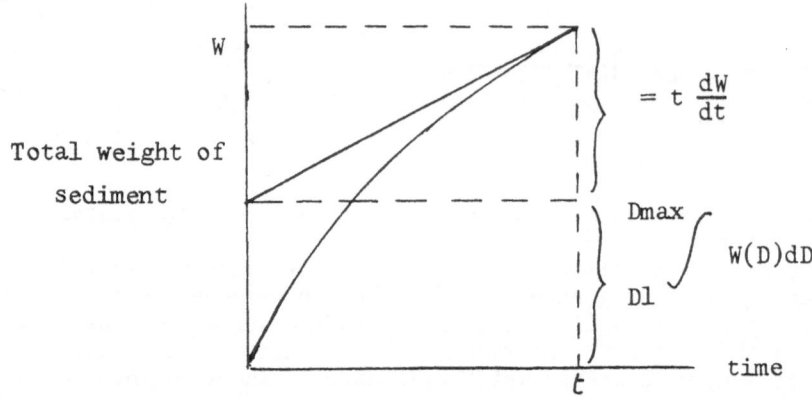

The drawing of tangents to experimental curves is very difficult
to do accurately.   However, equation (4) can be written

$$\int_{D_1}^{D\ max} W(D)\ dD = W - \frac{dW}{d\ \log_e t}$$

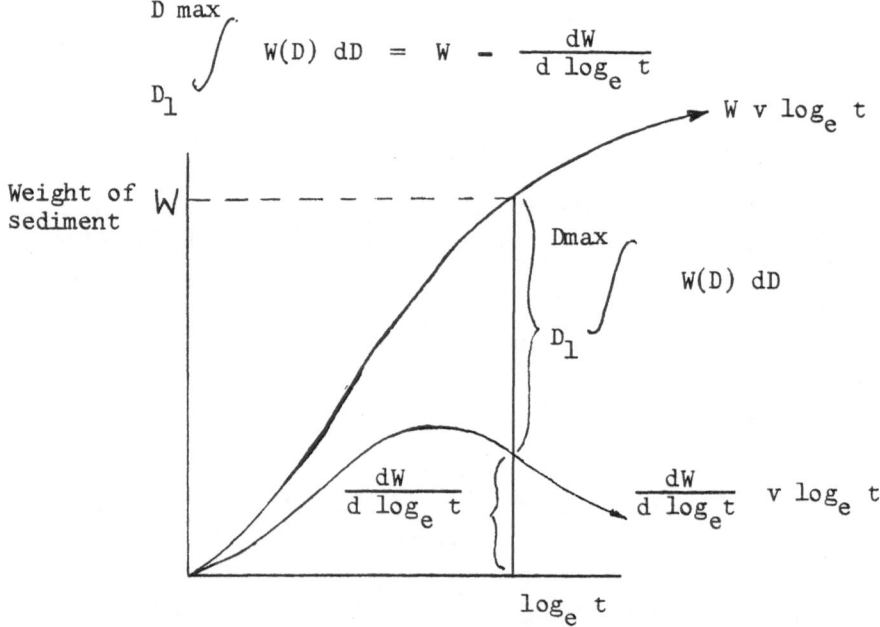

This method is claimed to be more accurate than the previous
since the tangents can be drawn more precisely at low values of ·
t.   A method which avoids the drawing of any graphs is obtained
by using a finite difference method and is recommended in British
Standard BS 3406 (2).   Suppose that equation (4) is written in
terms of fractional weights and finite differences

$$Q_1 = W_1 - t_1\ \frac{(W_2 - W_1)}{t_2 - t_1}$$

provided the ratio of t_2 to t_1 is not too large, Q_1 corresponds to the fractional weight oversize corresponding to an approximate time of $(t_1 t_2)^{\frac{1}{2}}$

in general
$$\frac{Q_1}{Q_{Tot}} = P_R - t_R \frac{(P_{R+1} - P_R)}{t_{R+1} - t_R}$$

where $P_R = \dfrac{W_R}{W_{Tot}}$ and $W_{Tot} = Q_{Tot}$

or
$$\frac{Q_1}{Q_{Tot}} = \frac{kP_R - P_{R+1}}{k - 1} \quad \text{where } k = t_{R+1}/t_R$$

For ease of calculation it is recommended that k is kept equal to 2

whence
$$\frac{Q_1}{Q_{Tot}} = 2 P_R - P_{R+1}$$

2. Incremental Methods

The technique consists of measuring the concentration of particles versus time at a given height h below the surface. Consider a small volume at a time t after the initially homogenous suspension has been allowed to sediment. The concentration of particles of size such that they have already fallen the distance h will be zero, but the concentration of all other particles will be the same as the initial concentration, since although some will have left the volume, their place will have been taken by an equal number from above. Thus the concentration measured is always that equivalent to the weight of the particles under the size given by the Stokes' diameter

$$\text{concentration at time } t = \frac{\int_0^{D_t} W(D)dD}{\int_0^{Dmax} W(D)dD}$$

$$= \text{weight less than } D_t$$

The advantage of these techniques is that the calculation is very simple. However, it must be remembered that the above theory requires that the concentration is measured over an infinitesimal layer and in most cases, since the initial concentration must be low to prevent particle interaction, the sampling region is a large region.

The main methods used are sample removal by pipette, density measurement by hydrometer or diver and by the adsorption of light in the photosedimentometers.

APPENDIX C.   COULTER COUNTER THEORY

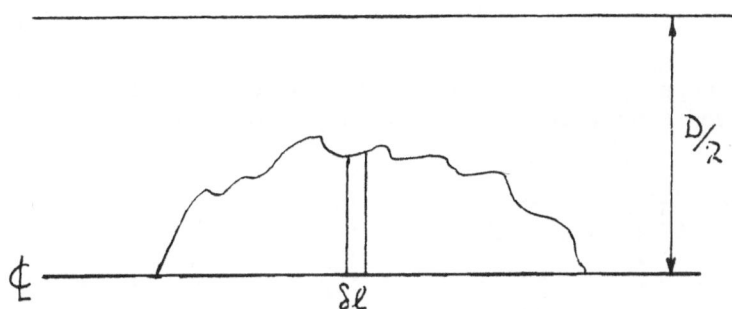

Consider an orifice of cross-sectional area A, diameter D. Also the liquid has a resistivity ρ_o and the solid particle ρ. The elemental resistance of the disc of thickness $\delta\ell$ will be

$$dR = \frac{\rho\,\rho_o\,dl}{\rho\,(A-a) + \rho_o a} \qquad \text{where a is the area of the elemental strip.}$$

$$= \frac{\rho\,\rho_o\,dl}{\rho A + a(\rho_o - \rho)}$$

Thus the change in the elemental resistance is

$$\delta\,(dR) = \frac{\rho\rho_o\,dl}{\rho A + a(\rho_o - \rho)} - \frac{\rho_o\,dl}{A}$$

$$= \frac{a\,\rho_o(\rho - \rho_o)\,dl}{A^2(\rho - \frac{a}{A}(\rho - \rho_o))}$$

$$= \frac{a\,\rho_o(1 - \rho_o/\rho)\,dl}{A^2(1 - \frac{a}{A}(1 - \rho_o/\rho))}$$

$$= \frac{a\,\rho_o\,k\,dl}{A^2(1 - \frac{ka}{A})} \qquad \text{where } k = 1 - \rho_o/\rho$$

Thus the total resistance change is

$$\Delta R = \frac{\rho_o \, k}{A^2} \int \frac{a \, dl}{(1 - {}^{ka}/A)} \qquad (1)$$

for a rod shaped particle lying along the axis, this equation can be solved simply because a is not a function of ℓ .

so $$\Delta R = \frac{\rho_o \, k \, v}{A^2 \, (1 - k \, {}^a/A)}$$

Batch[1] solved equation (1) for spherical particles and cone shaped particles by expanding the expression under the integral sign.   Gregg and Steidley[2] obtained exact solutions by solving the integral explicitly before expansion.   The resulting equations for change in resistance are summarised below:

TABLE 1

Particle Shape	ΔR. Batch	ΔR. Gregg & Steidley
rod length $\gg$ dia.	$\frac{\rho_o kv}{A^2} \left(1 + \frac{ka'}{A} + \frac{k \, a'^2}{A^2} + ..\right)$	$\frac{\rho_o \, v}{\pi^2 D^4} \; \frac{1}{(1 - K^2)}$
rod length $\sim$ dia.	———	$\frac{\rho_o v}{\pi^2 D^4}(t\pi r_1^2 + 6\pi r_1^3)(\frac{1}{1-K^2})$
sphere	$\frac{\rho_o kv}{A^2} \left[1 + {}^4/5 \, \frac{ka'}{A} + \frac{24}{35} \frac{k^2 a'^2}{A^2} + ..\right]$	$\frac{2 \, \rho_o}{\pi D} \left[\frac{\tan^{-1} k(1-K^2)^{\frac{1}{2}}}{(1 - K^2)^{\frac{1}{2}}} - K\right]$
cone	$\frac{\rho_o kv}{A^2} \left[1 + {}^3/5 \, \frac{ka'}{A} + {}^3/7 \, \frac{k^2 a'^2}{A^2} + ..\right]$	———

a' = maximum cross-sectional area of particle

* Gregg & Steidley assume ρ_o/ρ to be negligible, $K = \frac{d}{D}$

$$= \frac{\text{particle diameter}}{\text{orifice diameter}}$$

The equation for the sphere approximates to

$$\Delta R = \frac{\rho_o v}{A^2} \left[\frac{1 + 0.3\,K^2 + 0.13\,K^4}{\sqrt{1 - K^2}} + \ldots \right] \qquad (2)$$

Thus for any shaped particle provided a/A or $\frac{d}{D} = K$ is small the change in resistance is directly proportional to the volume of the particle. Allen[3] has shown that this assumption is valid up to particle/tube diameter ratios of 20% and the assumption of the linear response up to 40% ratio can lead to an error in particle diameter of 7.8% for rod shape particles and 3.6% for spherical shaped particles.

The value of the term $(\frac{1 + 0.3\,K^2 + 0.13\,K^4}{\sqrt{1 - K^2}})$, the error

term obtained by assuming linear response is given in the following table.

d/D	0.1	0.2	0.3	0.4	0.5
$\frac{\Delta R.\ A^2}{V}$	1.008	1.033	1.077	1.145	1.26
error in volume %	0.8	3.3	7.7	14.5	26.0
error in diameter %	0.3	1.1	2.9	4.8	8.7

Thus it is seen that the response of the Coulter Counter is not linear with particle volume. The non-linearity is dependent on particle shape. However, the error is not significant if the largest particle measured is of the order of 40% of the tube diameter. Particles larger than this can be measured provided the "error" in diameter is acceptable — although one is not likely to use the counter for particles greater than 40% of the orifice tube diameter because of the dangers of tube blocking. The errors can be recognized in this method and are not greater than those found in other methods of analysis, whether "recognized" or not.

The effect of particle shape has further been investigated by Gregg & Steidley in some interesting experiments, using a larger model orifice in an electrolytic tank. Wax models of cylinders and discs were pulled through the orifice and the changes in resistance noted. Justification for the type of equation given in Table 1 for thin discs was obtained experimentally.

The consequence of the resistivity of the particle has been ignored in the foregoing discussion. It is seen from Batch's equations (Table 1) that if $k = (1 - \rho_o/\rho)$ is less than 1, the magnitude of the errors will be decreased. However, it has been found in practice that a majority of powders, whether conducting or not, behave in the electrolyte as particles with infinite resistance (see Berg[4]). Thus the assumption that $K = 1$ can generally be assumed to be good.

The foregoing theory contains an implicit assumption that the electric field remains parallel to the axis of the aperture and that it is not disturbed by the presence of the particles. What would be the effect of taking this into account has not been investigated.

REFERENCES

1. Batch, B.A.   J. Inst. Fuel, 37, 455 (1964).

2. Gregg, H & Steidley, D.C.   Biophysics J., 3, 393 (1965).

3. Allen, T.   Proceedings of 1966 Conference on Particle Size Analysis.   S.A.C. (London) 1970.

4. Berg, R.H.   A.S.T.M.   Spec. Tech. Bull. 234. (1958)

APPENDIX D. COMPARISON OF B.S., A.S.T.M. AND TYLER SIEVE SCALES

British Standard B.S. 410		Tyler Equivalents		U.S. Standard A.S.T.M. - E11	
Sieve Number	Aperture (microns)	Sieve Number	Aperture (microns)	Sieve Number	Aperture (microns)
		$3\frac{1}{2}$	5613	*$3\frac{1}{2}$	5660
		4	4699	4	4760
*4	4000	5	3962	*5	4000
5	3350	6	3327	6	3360
*6	2800	7	2794	*7	2830
7	2400	8	2362	8	2380
*8	2000	9	1981	*10	2000
10	1680	10	1651	12	1680
*12	1400	12	1397	*14	1410
14	1200	14	1168	16	1190
*16	1000	16	991	*18	1000
18	850	20	833	20	841
*22	710	24	701	*25	707
25	600	28	589	30	595
*30	500	32	495	*35	500
36	420	35	417	40	420
*44	355	42	351	*45	354
52	300	48	295	50	297
*60	250	60	246	*60	250
72	210	65	208	70	210
*85	180	80	175	*80	177
100	150	100	147	100	149
*120	125	115	124	*120	125
150	105	150	104	140	105
*170	90	170	88	*170	88
200	75	200	74	200	74
*240	63	250	61	*230	63
300	53	270	53	270	53
*350	45	325	43	*325	44
		400	38	400	37

Sieve numbers marked with an asterisk (*) correspond to those proposed as an International Standard (I.S.O.) Scale; it is recommended that these sieves should be used for test data that are intended for International publication.

APPENDIX E. RELATIONSHIP BETWEEN PARTICLE SIZE AND SIEVE APERTURE

Material	Size	Volume coefficient k_a	$\dfrac{d_a}{A}$	$\dfrac{d_v}{A}$
	mesh			
Copper shot	10	0.520	1.05	1.05
Sand	10	0.258	1.40	1.11
"	18	0.261	1.39	1.10
Sillimanite	10	0.232	1.50	1.14
"	100	—	1.48	—
"	300	—	1.45	—
Coal (average)	10	0.227	1.48	1.12
"	100	—	1.46	—
"	300	—	1.40	—
Blast Furnace slag	10	0.193	1.48	1.06
" " "	100	—	1.40	—
" " "	300	—	1.41	—
Limestone	10	0.164	1.56	1.06
Plumbago	10	0.161	1.61	1.09
Talc	10	0.163	1.76	1.20
"	100	—	1.60	—
"	300	—	1.65	—
Gypsum	10	0.128	1.56	0.98
"	100	—	1.53	—
"	300	—	1.61	—
Flake Graphite	18	0.023	1.69	0.60
Mica	10	0.003	1.68	0.30
Stone Chippings:—	inches			
Limestone	$\frac{1}{2}$	0.315	1.19	1.00
Ragstone	$\frac{1}{2}$	0.288	1.30	1.06
Quartzite	$\frac{1}{2}$	0.318	1.24	1.05
Basalt	$\frac{1}{2}$	0.314	1.20	1.01
"	$\frac{1}{2}$	0.315	1.18	1.00
Granite	$\frac{1}{2}$	0.320	1.18	1.00
"	$\frac{1}{2}$	0.292	1.23	1.01
"	$\frac{1}{2}$	0.298	1.21	1.00
Gravel	$\frac{1}{2}$	0.271	1.29	1.04
Blast Furnace Slag	$\frac{5}{8}$	0.297	1.24	1.03
Coals:—				
Coal A	2	0.287	1.30	1.06
" B	$2\frac{1}{2}$	0.288	1.32	1.08
" C	2	0.278	1.47	1.19
" D	2	0.326	1.26	1.08
" E	$1\frac{1}{2}$	0.285	1.26	1.03
Anthracite	1	0.296	1.36	1.12

APPENDIX F

Transformation of size by method of moments

Let particle size be x where x depends on method of determination.

Distribution of size $-$ cumulative $N_r(x)$ or differential $n_r(x)$

$$\int_{x_{min}}^{x} n_r(x)dx = N_r(x) \qquad (1)$$

N_r is normalised so that $N_r(x_{min}) = 0$ & $N_r(x_{max}) = 1$

A series of subscripts r is used to denote

    $r = 0$ number distribution
    $r = 1$ length distribution
    $r = 2$ area distribution
    $r = 3$ volume or weight distribution

The moment of a distribution is defined as

$$M_{k,r} = \int_{x_{min}}^{x_{max}} x^k n_r(x)dx \qquad (2)$$

So that $n_r(x) = \dfrac{x^r n_o(x)}{\int_{x_{min}}^{x_{max}} x^r n_o(x)dx} = \dfrac{x^r n_o(x)}{M_{r,o}} \qquad (3)$

Combining equations (2) and (3)

$$M_{k,r} = \dfrac{\int_{x_{min}}^{x_{max}} x^k x^r n_o(x)dx}{M_{r,o}}$$

$$= \dfrac{M_{k+r,\,o}}{M_{r,o}} \qquad (4)$$

From this it can be shown that

$$M_{-1,3} = \frac{M_{2,0}}{M_{3,0}} = \frac{M_{1,1}}{M_{2,1}} = \frac{M_{0,2}}{M_{1,2}} = \frac{1}{M_{1,2}}$$

Since $M_{0,2} = 1$ (by normalisation).

These relationships can be used to convert any size distribution determined by one method into the corresponding distribution that would result if another method measuring a different property had been used (ignoring shape variation and assuming constant density).

Means of distributions, standard deviations and coefficients of variation may similarly be derived.

Thus :-

$$(\bar{x}_{k,0})^k = \frac{\int_{x_{min}}^{x_{max}} x^k n_0(x)dx}{\int_{x_{min}}^{x_{max}} n_0(x)\ dx} = \frac{M_{k,0}}{1}$$

Thus $\bar{x}_{k,0} = \sqrt[k]{M_{k,0}}$

and $\bar{x}_{k,r} = \sqrt[k]{M_{k,r}}$

Similarly, it can be shown that

$$(\sigma_0)^2 = M_{2,0} - (M_{1,0})^2$$

or, more generally $(\sigma_r)^2 = M_{2,r} - (M_{1,r})^2$

Likewise the coefficient of variation C is given by

$$C^2 = \frac{(\sigma_r)^2}{(\bar{x}_{1,r})^2} = \frac{\bar{x}_{1,r+1}}{\bar{x}_{1,r}} - 1$$

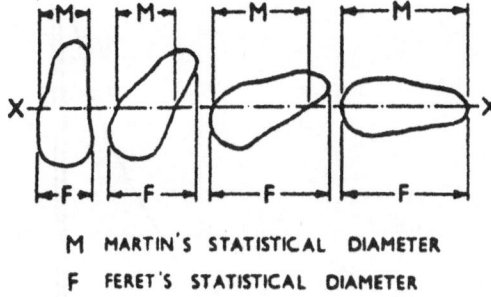

M MARTIN'S STATISTICAL DIAMETER

F FERET'S STATISTICAL DIAMETER

Figure 1. Comparison of Statistical Diameters

34

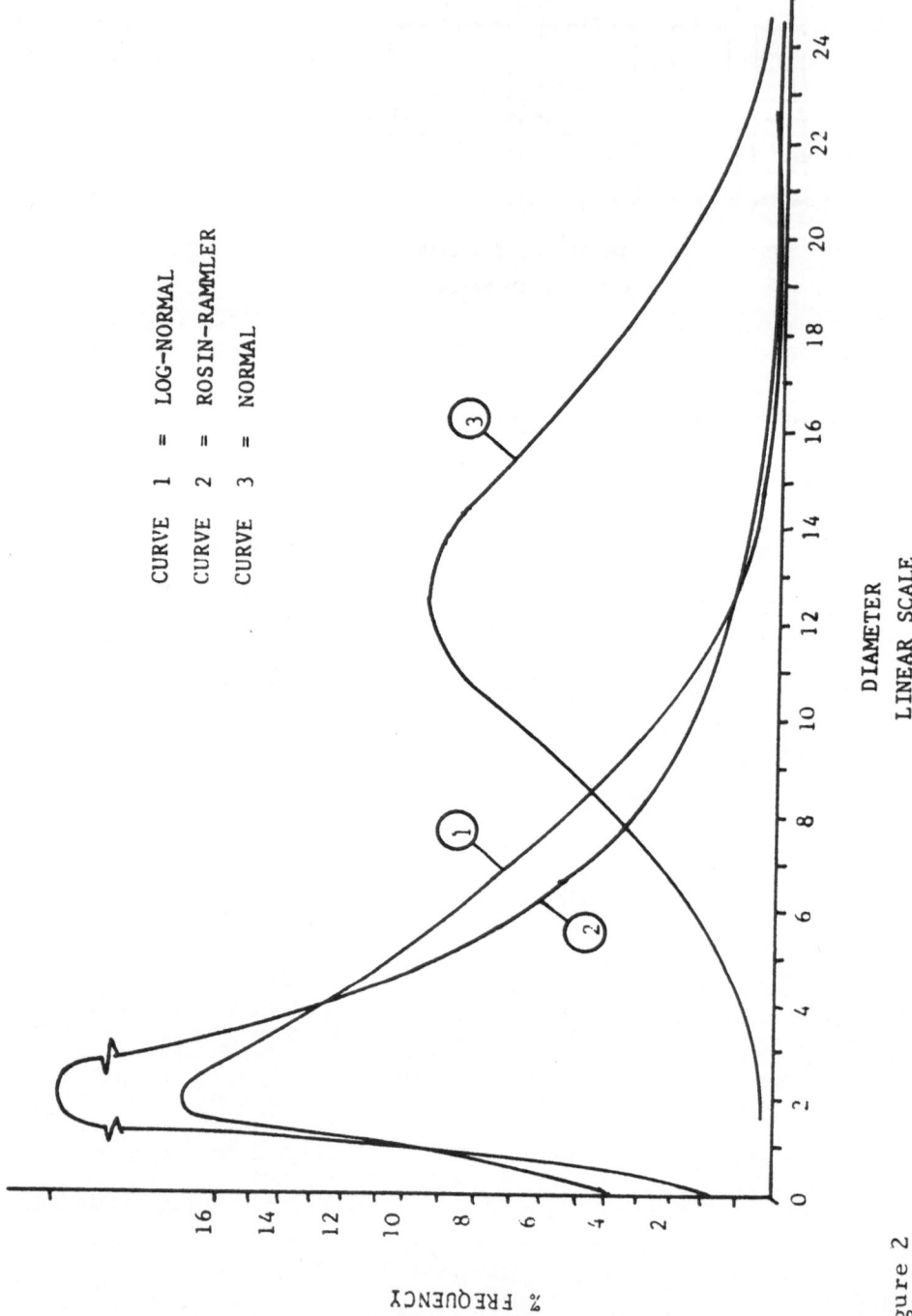

CURVE 1 = LOG–NORMAL

CURVE 2 = ROSIN–RAMMLER

CURVE 3 = NORMAL

DIAMETER
LINEAR SCALE

% FREQUENCY

Figure 2

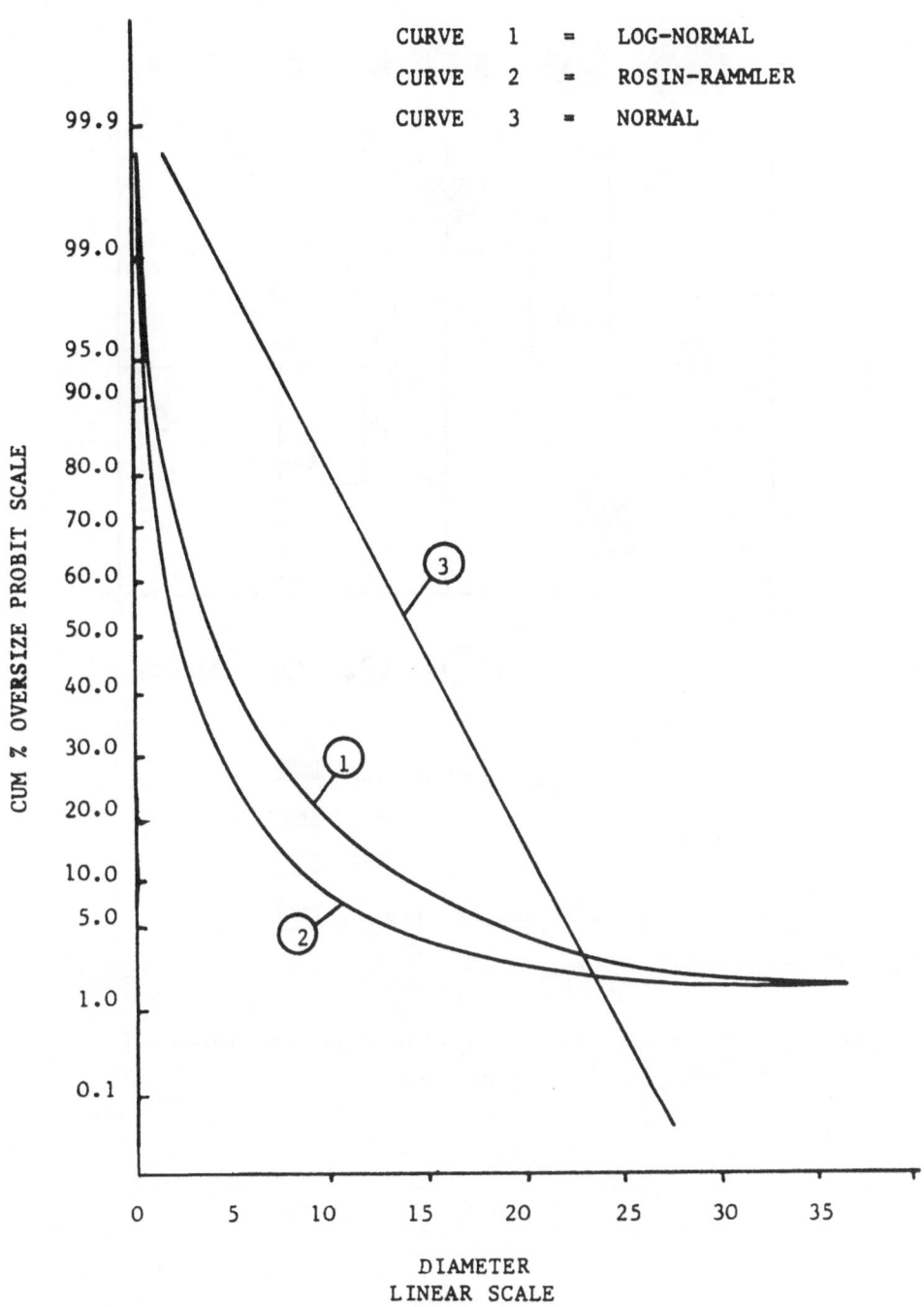

Figure 3

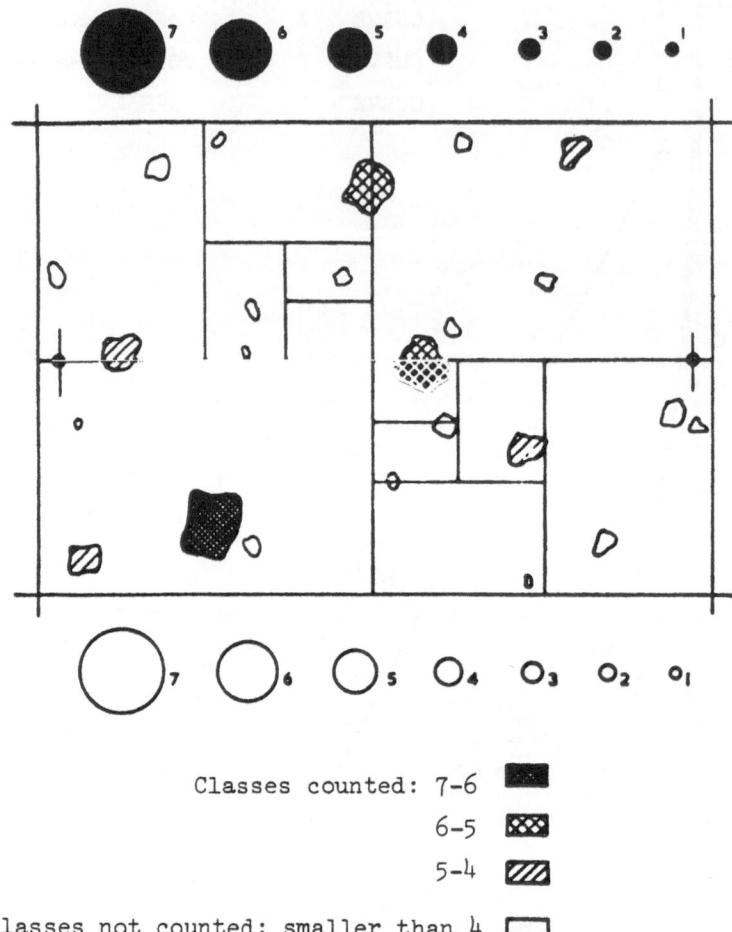

Figure 4. Field of particles at a concentration suitable for
counting by B.S. graticule

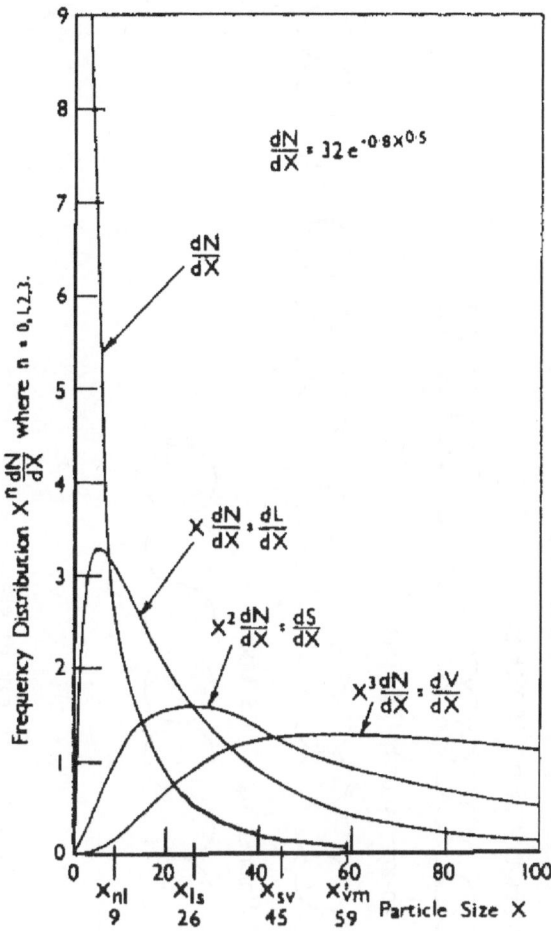

Figure 5. Frequency distributions and mean particle sizes

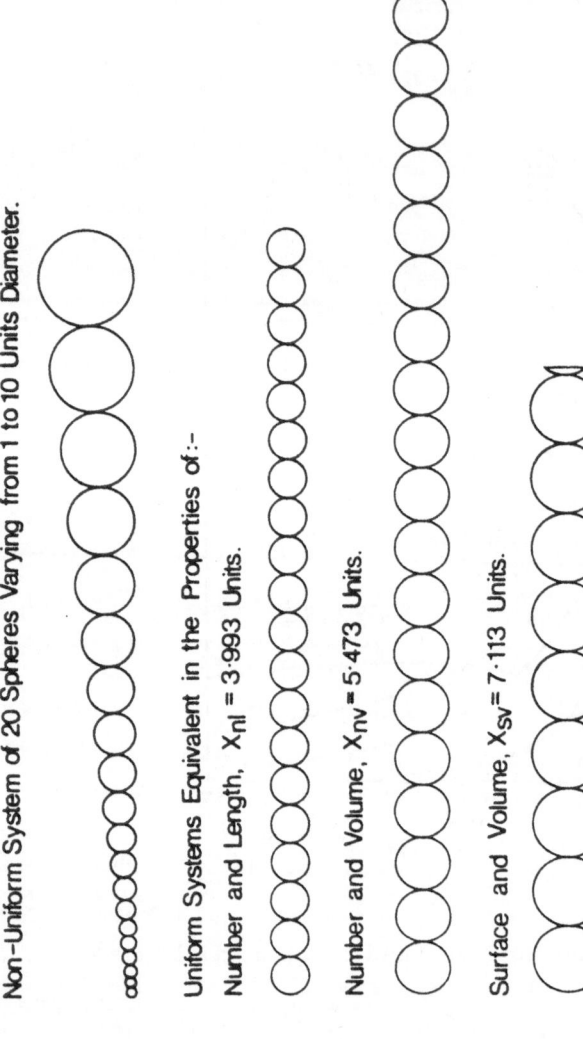

Non-Uniform System of 20 Spheres Varying from 1 to 10 Units Diameter.

Uniform Systems Equivalent in the Properties of :-

Number and Length, $X_{nl} = 3.993$ Units.

Number and Volume, $X_{nv} = 5.473$ Units.

Surface and Volume, $X_{sv} = 7.113$ Units.

Figure 6. Equivalent system of spheres

DEFINITIONS OF EFFICIENCY OF FILTRATION PROCESSES

D. C. Freshwater

Loughborough University of Technology

INTRODUCTION

The efficiency of filtration is a quantity that is not simple to define and may often not be easy to measure. In the case of solid/liquid separation there is very little published literature on the definition or measurement of the efficiency of the process. For example a standard literature survey, Solid/Liquid Separation by Oliver and Doyle published by the H.M.S.O. contains no direct reference to the subject.

Again, whilst there is an extensive literature in the gas filtration field on the efficiency of the process, various definitions are used, none of which is faultless. It seems appropriate therefore to first outline some of the criteria which have been proposed for defining the efficiency of filtration processes and then to discuss how far these are satisfied by practice in various practical situations.

1. MATERIAL BALANCE EFFICIENCY

The object of filtration is the separation of particles (usually solid but sometimes liquid) from a fluid stream by passing the mixture over a porous membrane or through a porous bed.

Therefore an overall filtration efficiency can be defined as

$$\frac{\text{Mass of solids removed}}{\text{Mass of solids in feed}} \quad \text{per unit volume}$$

This may be rearranged in a number of simple algebraic variations to take account of the concentration range of feed or the effectiveness of the filter.

Whilst this is unimpeachable as defining the overall mass performance of the filter it is very uninformative in a number of ways. For example it does not say anything about the relative particle sizes or size distributions of the captured material and that which is not removed, yet this may be the most important characteristic that determines whether the filter will be satisfactory for a particular purpose. This definition therefore is not to be recommended.

2.   ENERGY BALANCE EFFICIENCY

It has been suggested that an alternative definition of efficiency would be

$$\frac{\text{work done at filter media (i.e. volume x pressure drop)}}{\text{total energy supplied to the filtration apparatus}}$$

This is not a basic relationship since it depends on the individual machine and its engineering as well as on the system. Its usefulness is therefore very doubtful.

3.   THERMODYNAMIC EFFICIENCY

A more basic energy efficiency, exactly analogous to the overall material balance efficiency would be to relate the energy expended across the filter media, i.e. volume x pressure drop, to the energy of mixing of the solid/fluid system. This would be the thermodynamic efficiency and would be different for different size distributions of the same material and also for equivalent size distributions of different materials. Unfortunately we do not know sufficient about the thermodynamics of heterogenous systems to be able to calculate mixing energies for feeds encountered in filtration problems. Therefore this definition is useful at present only in a qualitative manner although its potential might be remembered.

4.   POINT EFFICIENCY

We come then to the third basic method of describing the efficiency of filters which I shall call point efficiency. This simply expresses the mass balance efficiency as a series of values corresponding to each particle size in the feed mixture. Thus one obtains an overall retention efficiency curve which is the point efficiency plotted against the particle size. (The so-called 'Tromp' curve). The suggestion that this kind of definition should be used for filtration separation (just as it is for air classifiers) has come from several persons including J. Murkes and M. Wells. This definition is strictly based on material balance considerations and has the great advantage of describing the performance in terms of the particle size distribution of the material to be removed. Hitherto it has not been easy to determine but a technique recently developed has overcome this difficulty (see Appendix I).

It is recommended that because of its strict derivation and its descriptiveness of the filter performance over a wide range of particle sizes, the point efficiency definition be used as the standard basis for filter efficiency.

5. PROCESS CONDITIONS

So far the discussion has been concerned with factors which characterise the performance of the filtration apparatus. However any apparatus is also affected by the process conditions under which it operates and any definition of filtration must include these conditions within its description. We may say that these factors are as follows:-

(i)   The nature of the solids

(ii)   The particle size distribution

(iii)   The concentration of the suspension

(iv)   The physico-chemical properties of the fluid

(v)   The flow rate of the fluid

(vi)   The temperature

Generally it will be sufficient simply to specify factors (i), (iv), (v) and (vi) ((iv) being defined by specifying the fluid composition or species).

Factor (ii) is covered as we have seen by the point efficiency definition.

Factor (iii) is rather more difficult to classify in a simple way and affects the practical application of these definitions as will be seen in the next section.

However it can be stated at this stage that any description of filter performance that does not also give details of conditions listed under factors (i) to (vi) is of little significance.

6.   SOME PRACTICAL CONSIDERATIONS

This note began by referring to solid/liquid and solid/gas filtration as two separate operations.   In practice, from the point of view of filter performance it is probably more useful to classify filtration operations rather on the basis of suspension characteristics.   Thus we may consider those where the suspension is dilute and complete or near complete recovery of suspended material is the aim so as to leave a particle-free liquid and those where the suspension is concentrated and the purpose of the separation is to recover the suspended material but slight contamination of the filtrate is not objectionable.

Examples of the first class of operations are clarification water treatment and high efficiency gas filtration.   In the second class of operations one can cite cake filtration, bag filtration and cyclone separation.

It is seen immediately from these descriptions how it is necessary to rearrange the material balance efficiency equation to produce meaningful numbers in each of the two classes of operations.   However, the point efficiency definition can be applied to both classes with equal ease.   Let us, however, consider some practical cases and current practice and see how far these meet the needs of providing meaningful descriptions of the performance of filters.

6.1   Air Filtration

The capture efficiency of an isolated fibre or even a bank of fibres in a regular geometric array may be predicted with considerable accuracy on hydrodynamic considerations and allowing for the combined effects of inertia and diffusion.   This is the class of filtration in which theoretical analysis and prediction has proceeded furthest.   However, three factors prevent this approach being applied to practical gas filters.   These are (i) effect of other capture mechanisms, positive ones such as electrostatic forces and negative ones such as "bounce off";   (ii) lack of geometric regularity in the fibre arrangement both as regards diameter and array;   (iii) the effect of blocking of the filter as its time of service proceeds.   In practice, therefore, the most common efficiency quoted is the material balance efficiency.

This is the one given in B.S.2831, for the A S H R A E test and many other tests in use throughout the world.   It is also the basis of B.S.3928 for high efficiency air filters and also the U.S. D O P test.   Although these standards often state that the particle size distribution (or even particle shape as well) must be quoted, these tests still suffer from the basic deficiencies of the "material balance" definition discussed earlier.

Point efficiency is often used in describing the performance of cyclones and electrostatic precipitators but is not as yet in general use for air filters. Some alternative efficiency definitions sometimes encountered in air filtration work are: (a) number efficiency, (b) surface area efficiency, (c) decontamination factor, (d) decontamination index. (a) and (b) are of virtually no practical importance, whilst (c) and (d) are simply algebraic rearrangements of the material balance efficiency.

So far no satisfactory method has been found of dealing with the time dependency effect in the filter.

6.2 Clarification of Liquids

Although there are a number of British Standard Specifications relating to the testing of Air Filters there is only one in existence for liquid filtration and this refers only to sintered glass discs used for laboratory filtration.

In the special case of some importance − that of hydraulic fluids, special tests and definitions have grown up which constitute a jargon of their own within this technology. Most of these are when analysed material balance efficiencies and would be better replaced by "point efficiency" standards. Some of these tests will be found in the references given.[1,2]

Generally clarification is carried out either for improving the visual appearance of liquids, e.g. soft drinks, lubricating oil or for bacterial control, e.g. beer, potable water. In such cases simple visual inspection may be used. If quantitative measurements are to be made recourse is usually had to nephelometry, or more rarely light scattering methods. Essentially these techniques are concerned only with measuring the amount of material transmitted through the filter, rather than with the efficiency as defined in this paper. Few if any studies have been made of point efficiencies in this class of filtration (due to difficulties of accurate sampling and measurement presumably). However, the information about the filtration process that is provided by this analysis as described in Appendix I is such as to recommend this as an important area of investigation.

6.3 Dense Suspensions or Cake Filtration

In clarification we are concerned solely with the solids in the process stream downstream of the filter and tend to ignore the solids collected by the filter. In cake filtration the exact opposite is the case – we tend to ignore the solids in the downstream fluid or filtrate and devote attention wholly to the solids collected by the filter, or the cake. Just as for clarification material balance efficiency has little meaning, so for cake filtration it is not of any significance. The difference between the two systems is that clarification is invariably a deep bed filter process whilst cake filtration is a thin membrane filter process at least at the commencement of the operation. Thus clarification has its highest efficiency at the start of the process, whilst cake filtration has its lowest efficiency at the start. Any measurement of point efficiency on cake filtration should therefore be related to the time of filtration and is likely to give more information about the filtration process when it is carried out at the beginning of the operation.

However, practical considerations of pressure drop and liquid retention by the cake are of considerably greater economic importance in industrial cake filtration than either material balance or point efficiency. It may be worth considering, therefore, whether in cake filtration some efficiency definition which takes into account these factors might not be more useful. An attempt to do this has been made by M. Rosch to which reference has already been made.[1] The dangers inherent in using this "raw" definition of energy efficiency have already been touched upon. The concept of describing filter performance in terms of residual moisture content of filter cake is interesting and further study should be made of this. A powder of a given particle size distribution can pack in an infinite number of different ways. Generally the highest porosity will give the greatest ease of filtering, but not the lowest residual moisture content. Recent work on free moisture content and bed structure shows that the relationship is a complex one but since the packing, for a given size distribution, will be a function of the machine geometry, this description of performance has some significance and is worth pursuing further.

7. OTHER CONSIDERATIONS

7.1 Experimental

So far we have ignored or adverted only briefly to any experimental problems in making measurements necessary to determine any of the efficiencies that have been described. There can be little doubt that the widespread use of the material balance efficiency in many types of filtration operation is due to the relative ease of its determination experimentally. However, this is not a good reason for its continued use in view of its lack of precision in describing the performance of the filtration in various conditions. Nevertheless it is well known that the determi-

-nation of particle size distributions required for point efficiency calculations is more difficult. This difficulty arises partly from the particle size measurements themselves and partly from the problem of sampling. Particle size measurement in the region of 1 micron and upwards now presents no especial problem. Thus if the point efficiency definition is to be developed further work must be carried out on the techniques of measurement of particle size in the sub-micron region. Even more important is the problem of sampling and sampling error. We know, largely from the work of C. N. Davies, what to expect from sampling of particles in gas streams in the region of 2 microns and upwards.

Reliable information on the factors affecting sampling of particles in gas streams in the sub-micron region is almost non-existent. Again predictive or quantitative relationships on the sampling of particles in liquid streams is not available at all in the literature. Much work needs to be done on this before we can apply either material balance or point efficiency techniques to filtration.

7.2   Time Dependency

Most filtration operations are batchwise rather than continuous. On the other hand, most investigations of filter performance consider static rather than dynamic conditions. There is as yet no satisfactory method of predicting the time-dependent behaviour of any class of filtration process, and this effect is not included in any efficiency definition. This is an unsatisfactory state of affairs and work should be carried out on this important problem.

8.   CONCLUSIONS

8.1   Definition of Filtration Efficiency

It is concluded that the best absolute definition of filtration efficiency is a point efficiency leading to a grade efficiency curve.

In special cases of industrial process filtration, another type of performance measure might be developed in terms of liquid retention of filtered solid.

8.2   Process Conditions

It is essential to specify process conditions (see section 5 above) when quoting filtration performances.

8.3 Experimental

Much more work is needed on the quantitative relationships between sampling conditions and particle size distribution, especially for liquid dispersed systems.

8.4 Other Problems

The outstanding problem on which further research work is required is the effect of time of operation on filter performance.

9. ACKNOWLEDGEMENTS

In preparing this paper I have been greatly helped by notes written by M. Rosch, M. J. Murkes, Dr. Guyer, Prof. Alt, R. M. Wells, R. G. Dorman and C. R. G. Treasure, and I should like to acknowledge their contribution.

REFERENCES

1. Specification for bulk filtration and water separation equipment for aviation fuels, Air B.P., Specification No. FB/S/F/1.

2. Testing technique for submicron liquid filters, Hodge, T.W., & Smith, A.J., U.K. A.E.A. report, A E R E E/M29.

APPENDIX   I

C. R. G. Treasure

Loughborough University of Technology

1.   INTRODUCTION

In the past, the testing of filters has amounted to little
more than the determination of the fractional amount of the feed
solids which has been retained by the filter with perhaps some
attempt to estimate the maximum size of particle which would be
found downstream of the filter.   This type of testing has
involved close definition of the conditions of test and, in particular, specification of the test solids by nature and size
distribution.   The results of these tests are useful for the
comparison of filters only when the filters are used under similar
conditions.   For practical use the results have only indicative
value since the conditions of use and the solids to be filtered
will in general differ widely from those of the test.   More
informative data of the filter performance can be obtained by
determining the actual point efficiency, that is, the fraction of
each size of particle which is retained.   This can be expressed
as a retention efficiency curve (point efficiency plotted against
particle size), the parameters of which can be investigated as
functions of test variables.

2.   DETERMINATION OF THE RETENTION EFFICIENCY CURVE

Given the mass fraction of the solids retained by the filter
and the size distributions (usually cumulative fractional weight
oversize or undersize) of the feed and the solids passing or retained, the two commonest methods of determining the retention
curve are either to construct a histogram of retention efficiency
by considering various size ranges or to calculate the point

efficiencies from the mass-frequency distributions. Both methods
are straightforward but rely on having accurate size distributions,
particularly at the ends of the size distributions. In the histogram method rather wide size ranges have to be considered at the
extremes of the ranges in order to obtain accuracy in the calculated mean efficiencies, and in the direct determination method
it is more difficult to determine the values of the mass-frequency
distributions at extreme sizes so that the calculated point effi-
ciencies.are in greater error.

A suggested alternative method is to use "mono sized" fractions of powders so that only the fractional weight retained has
to be determined and the point efficiency is found directly.
However, objections to this method arise from the difficulty of
obtaining fractions of small size range in quantity, that the
system is highly idealised, and any effect due to a wide size
distribution of the feed solids is eliminated.

For these reasons, therefore, the method which is proposed is
an adaptation of a technique recently developed in these laboratories for the analysis of the behaviour of classifiers. Essentially this consists of plotting the size distributions of the
material retained or passing the filter against the feed size distribution, and analysing the curves obtained, as explained below.
A further advantage of the method is that it allows considerably
more information about events occurring inside the filter to be
obtained than by the previously described methods which present
only an overall picture.

3. THEORY OF METHOD

In the simple case, any two of the three size distributions
which are obtained from the test are related by the retention
efficiency curve. In Fig. 1, the diagonal OA would be obtained
if no filtration occurred; if the filtration were such that all
material greater than some size given by $(1-\psi) = Q(D)$ were retained, then the lines OB AC would be obtained.

ψ then equals the fraction: $\dfrac{\text{mass of solids retained}}{\text{mass of food solids}}$

for cumulative undersize distributions.

In practice, curves such as OFA, OGA, will be found.

For the general case, we must also consider the situations
when:

(a) the filter is faulty, that is, the medium has broken down at
     some point or by-passing occurred so that a fraction of solids

with the feed size distribution passes with the filtrate;

(b) a fraction of the feed solids is held up, either due to the design of the system or by the action of the cake which has formed. In effect, this material has not been filtered by the medium.

Let D = Particle size

$Q(D)$ = Fractional weight less than D in Feed.

$R(D)$ = " " " " D in retained solids.

$F(D)$ = " " " " D in solids passing filter.

f = Fraction of feed passing unfiltered to filtrate.

r = " " " retained (not filtered by medium).

ψ = $\dfrac{\text{Total weight of retained solids}}{\text{Total weight of feed solids}}$

ψ_f = $\dfrac{\text{Weight of solids retained by the filtering action}}{\text{Weight of feed solids subject to filtering}}$

$\eta(D)$ = Overall point retention efficiency.

$\eta_f(D)$ = Point retention efficiency due to filtering action of medium.

The distribution of the solids will then be as shown in Fig. 2, the quantities $Q(D)$, $F(D)$, $R(D)$, ψ, being experimentally determined.

For any size D, we have the mass balance:

$$\psi \frac{dR}{dD} = \eta_f (1-f-r) \frac{dQ}{dD} + r \frac{dQ}{dD}$$

or $$\psi \frac{dR}{dQ} = \eta_f (1-f-r) + r \tag{1}$$

Similarly:

$$(1-\psi) \frac{dF}{dQ} = (1- \eta_f) (1-f-r) + f \tag{2}$$

Now, for $D \to 0$, $\eta_f \to 0$,

hence $\quad \dfrac{dR}{dQ} = r/\psi$ \hfill (3)

and $\quad \dfrac{dF}{dQ} = \dfrac{1-r}{1-\psi}$ \hfill (4)

For $D \to \infty$, $\eta_f \to 1$,

hence $\quad \dfrac{dR}{dQ} = \dfrac{(1-f)}{\psi}$ \hfill (5)

and $\quad \dfrac{dF}{dQ} = \dfrac{f}{1-\psi}$ \hfill (6)

Sufficient data are therefore available to determine f, r, ψ. However, as ψ can be experimentally determined, a check of f and r is available.

Fig. 3 illustrates a plot for the general case.

A further check of the tangents is available by considering their intersections, H, K, on Fig. 3. It is easily shown that

$$Q_H = Q_K = \frac{1 - \psi - f}{1 - f - r} \qquad (7)$$

We have also, from Fig. 2, that

$$\psi = \psi_f (1-f-r) + r$$

i.e. $\psi_f = \dfrac{\psi - r}{1-f-r}$ \hfill (8)

and $1 - \psi_f = \dfrac{1 - \psi - f}{1-f-r} = Q_H = Q_K$ \hfill (9)

To determine the point efficiency due to the medium, η_f, equation (1) or (2) can now be used with the values of

$\dfrac{dR}{dQ}$ or $\dfrac{dF}{dQ}$ determined at the appropriate points.

If the overall point efficiencies are required, these are determined also from equations (1) and (2), putting r=f=0.

Fig. 1   Size Distribution Plots

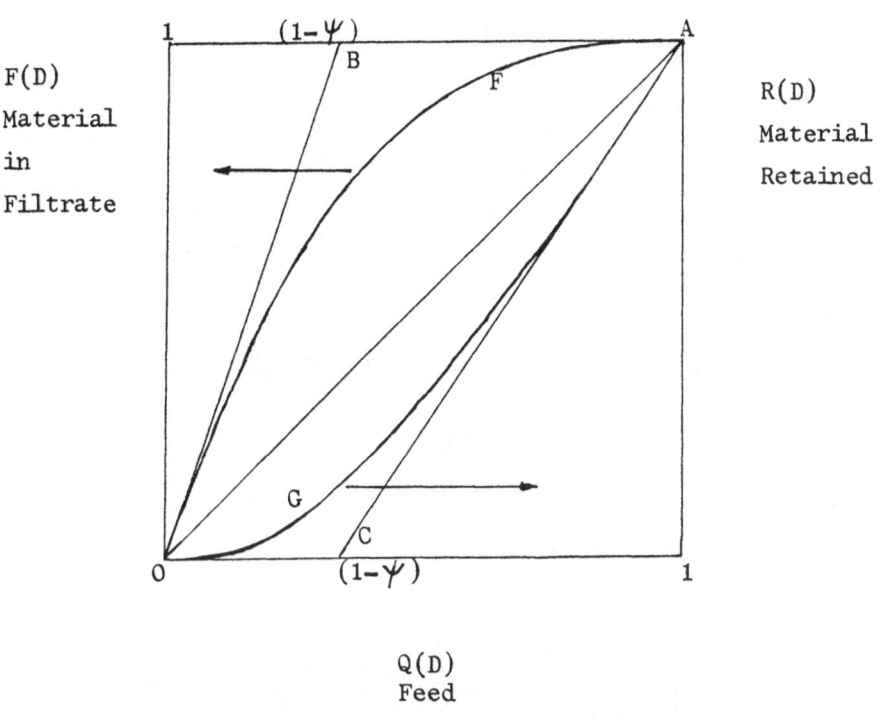

Q(D)
Feed

Fig. 2   Distribution of solids in filtration

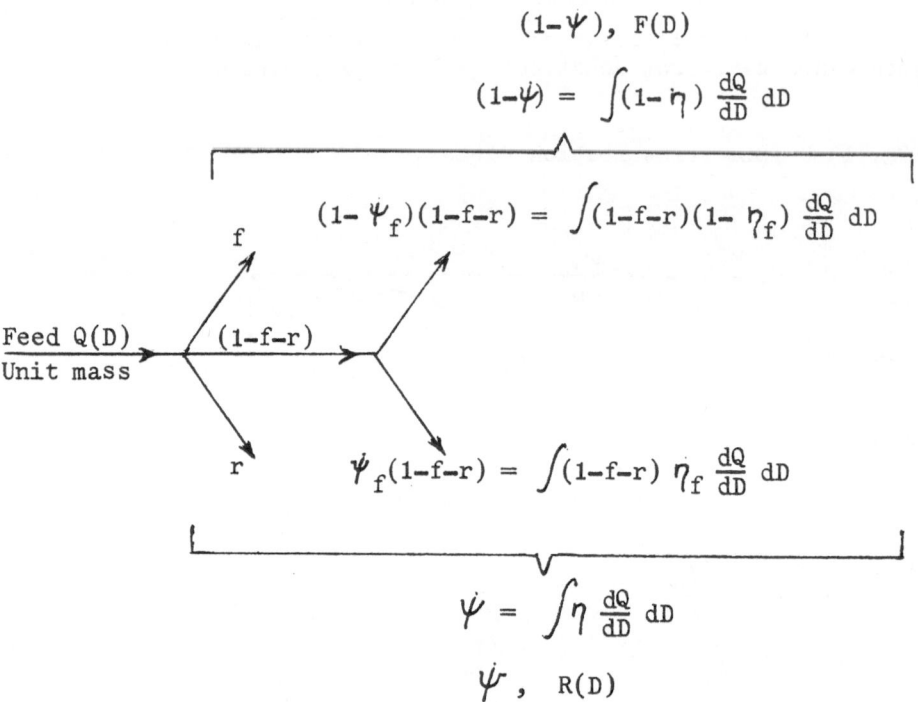

$$(1-\psi),\ F(D)$$

$$(1-\psi) = \int (1-\eta)\ \frac{dQ}{dD}\ dD$$

$$(1-\psi_f)(1-f-r) = \int (1-f-r)(1-\eta_f)\ \frac{dQ}{dD}\ dD$$

f

$$\frac{\text{Feed } Q(D)}{\text{Unit mass}} \quad (1-f-r)$$

r

$$\psi_f(1-f-r) = \int (1-f-r)\ \eta_f\ \frac{dQ}{dD}\ dD$$

$$\dot\psi = \int \eta\ \frac{dQ}{dD}\ dD$$

$$\psi,\ R(D)$$

Fig. 3   Plot for general case

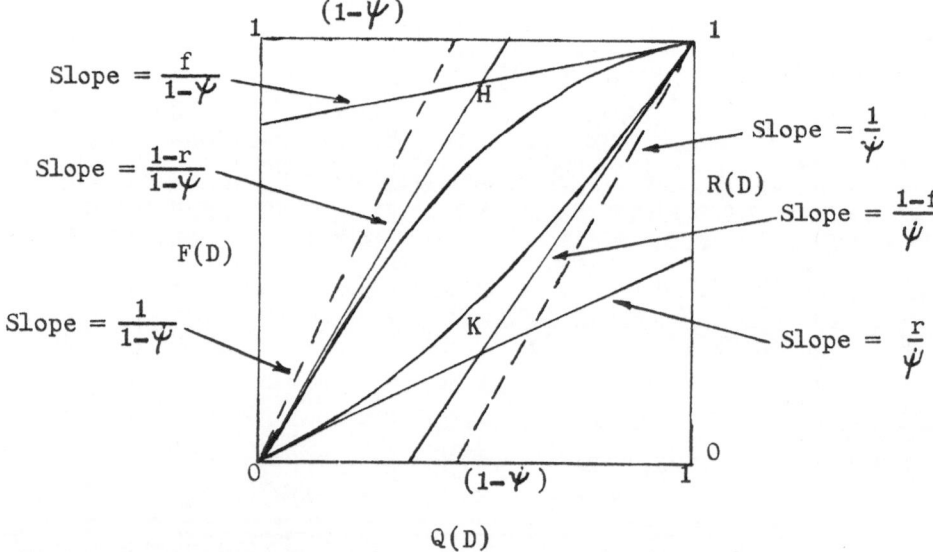

$$(1-\psi)$$

Slope $= \dfrac{f}{1-\psi}$

Slope $= \dfrac{1-r}{1-\psi}$

F(D)

Slope $= \dfrac{1}{1-\psi}$

H

K

Slope $= \dfrac{1}{\psi}$

R(D)

Slope $= \dfrac{1-f}{\psi}$

Slope $= \dfrac{r}{\psi}$

$$(1-\psi)$$

Q(D)

INTERFACIAL PHENOMENA

John Gregory

University College London

Symbols (S.I. units used throughout)

A_{12}	Hamaker constant for materials 1 and 2 in vacuo (J)
A_{132}	Hamaker constant for materials 1 and 2 in medium 3 (J)
a	Sphere radius (m) or radius of tube (m)
C	Integration constant in eq.(6)
c_f	Critical flocculation concentration (mol l^{-1})
c_i	Concentration of ion i (mol l^{-1})
d	Separation between particles or plates (m)
E	Field strength (V m^{-1})
E_s	Streaming potential (V)
e	Elementary charge (= 1.60×10^{-19} C)
h	Planck's constant (= 6.63×10^{-34} J s)
I_s	Streaming current (A)
K	Conductivity (V^{-1}A m^{-1})
k	Boltzmann's constant (= 1.38×10^{-23} J K^{-1})
l	Length of tube (m)
n	Number of ions per unit volume (m^{-3})
n_1, n_2, n_3	Limiting refractive index of materials 1, 2 and 3
P	Force per unit area between plates (N m^{-2})
p	Pressure difference across tube or plug (N m^{-2})
R	Distance between centres of spheres or molecules (m)
r	Distance from centre of sphere (m)
T	Absolute temperature (K)
U	Electrophoretic mobility ($m^2 s^{-1} v^{-1}$)
V_A	van der Waals attraction energy (J)(or J m^{-2} for plates)
V_R	Electrical repulsion energy (J)(or J m^{-2} for plates)
v_e	Electrophoretic velocity (m s^{-1})
v_{eo}	Electro-osmotic velocity (m s^{-1})
W	Stability ratio

X	Dimensionless distance ($= \varkappa x$)
x	Distance from interface (m)
y	Dimensionless potential ($= ze\psi/kT$) subscripts as for ψ
z	Valence of ion
α	Polarizability (m^{-3})
γ	$= \tanh(y_o/4)$
δ	Stern layer thickness (m)
ε	Permittivity ($C^2 N^{-1} m^{-2}$)
ζ	Electrokinetic or zeta potential (V)
η	Viscosity ($N s m^{-2}$)
$\varkappa$	Debye-Huckel parameter (m^{-1}) Eq.(5)
μ_1, μ_2	Dipole moments of molecules 1 and 2 (C m)
ν_1, ν_2	Characteristic frequencies of materials 1 and 2 (s^{-1})
σ_o	Surface charge density ($C m^{-2}$)
ϕ	Overall potential difference between phases (V)
χ	Chi potential (V)
ψ	Electric potential (V)
ψ_o	Surface potential (V)
ψ_δ	Stern plane potential (V)
ψ_1, ψ_2	Effective surface potentials for interaction (V)
ψ_m	Minimum potential between plates (V)

1. INTRODUCTION

For a granular filter to be effective in removing small particles, one condition has to be satisfied before all others - the particles must adhere to the filter grains. The question of whether a particle will stick or not is determined primarily by the surface characteristics of the particles and filter grains, which have nothing to do with the way in which particles are transported to the grains. The concept of transport and attachment as separate steps is now well established in filtration theory[1] and only the latter aspect will be discussed here.

As well as adhering directly to filter grains, particles should also adhere to existing deposits - otherwise the filter would cease to operate as soon as all the grains were covered with a monolayer of particles. This means, in practice, that suspensions to be filtered should be <u>colloidally unstable</u>, i.e. the particles should not repel each other sufficiently to prevent contact.

The principles governing colloid stability will be discussed mainly from the standpoint of the now classical Derjaguin-Landau-Verwey-Overbeek[2] (DLVO) theory, which is primarily concerned with the interaction of similar particles. Extension of the theory to cover the interaction of unlike particles is usually quite straightforward, as will be shown. The emphasis will be

on aqueous systems, although the broad principles will apply to
particles in non-aqueous media also (see Lyklema[4] for a detailed
review of non-aqueous colloid stability).

The DLVO theory views the stability of colloids as arising
from electric charge on the particles and a consequent electrical
repulsion between them. Opposing this repulsion are attractive
forces of the universal van der Waals type. If the repulsion
forces can be reduced sufficiently so that van der Waals
attraction becomes dominant, then particles colliding with each
other will stick together to form more-or-less permanent
aggregates. This aggregation process is known either as
COAGULATION or FLOCCULATION. Some workers use these terms inter-
changeably, but others imply some mechanistic distinction by their
choice of term. Unfortunately different distinctions have been
drawn, La Mer[5] using "flocculation" to denote "polymer bridging"
whereas, in water-treatment parlance, "flocculation" often implies
the aggregation of particles under the influence of velocity
gradients as opposed to "coagulation" by Brownian motion. In view
of this confusion, the term "flocculation" will be used here,
irrespective of mechanism.

The DVLO theory does not give a complete account of colloid
stability and, indeed, it is still vehemently opposed by a few
colloid scientists[6], who prefer to regard stability and
flocculation in terms of more specific chemical properties.
Nevertheless, it is widely thought to be correct as far as it
goes. Difficulties in testing theoretical predictions stem from
problems in determining the relevant electric potentials of the
particles and from uncertainties over the magnitude of the
van der Waals attraction. The theory is only concerned with
HYDROPHOBIC particles, such as insoluble oxides, clays etc. The
stability of many colloids is, at least partly, determined by the
presence of adsorbed, hydrated layers which prevent contact of
the particles. Such "protected" particles have much in common
with HYDROPHILIC colloids such as gums, starches etc., which are
essentially soluble macrocolecular substances. It is well known
that many hydrophilic colloids can coat hydrophobic particles,
giving extra stability.

In amounts much lower than those needed to give protection,
certain polymeric additives can actually promote flocculation by
a process which is thought to be the formation of "bridges"
between particles. This topic will be dealt with in the lecture
on Pre-treatment.

Since the most direct method of reducing colloid stability
is to lower the effective electrical repulsion between particles,
most attention will be devoted to a survey of electrical phenomena

at interfaces, electrokinetic effects and the electrical interaction between particles. Van der Waals attraction and the effect
of adsorbed layers will be treated more briefly.

2. ELECTRICAL PHENOMENA AT INTERFACES

2a The nature of interfacial potentials

       Between two phases in contact there often exists an
electrical potential difference. Guggenheim[7] long ago pointed
out that such potentials cannot be measured and, strictly speaking,
have no physical significance. Nevertheless, the concept of
interfacial potentials is of great value in understanding the
interaction between colloidal particles.

       The overall potential difference (GALVANI potential) between
the phases, ϕ , can be regarded as made up from two contributions:

$$\phi = \psi_o + \chi \qquad\qquad (1)$$

ψ_o is the VOLTA potential (often referred to simply as the
"surface potential") and arises from the unequal division of
charge between the two phases. For a solid-liquid system, this
is the potential that is "seen" from the liquid side of the phase
boundary and which is relevant to particle interaction.
 χ, usually known as the CHI potential, is a result of the
orientation of electric dipoles (i.e. polar or polarized molecules)
at the interface. In aqueous systems the orientation of polar
water molecules contributes significantly to χ .

       In figure 1 these concepts are illustrated for the case of
a solid in contact with an aqueous solution. The variation of
electric potential is shown from a point within the solid through
the interfacial region to a point far into the solution. As
drawn, the solid has a negative potential relative to the solution
and the sign of χ is positive, as would be the case if dipoles
were oriented with their positive ends towards the solid. In
this case the surface potential, ψ_o , is more negative than ϕ .
The variation of potential in solution is that expected for a
Gouy-Chapman type of double layer (see 2c) and it is assumed
that the potential does not vary within the solid.

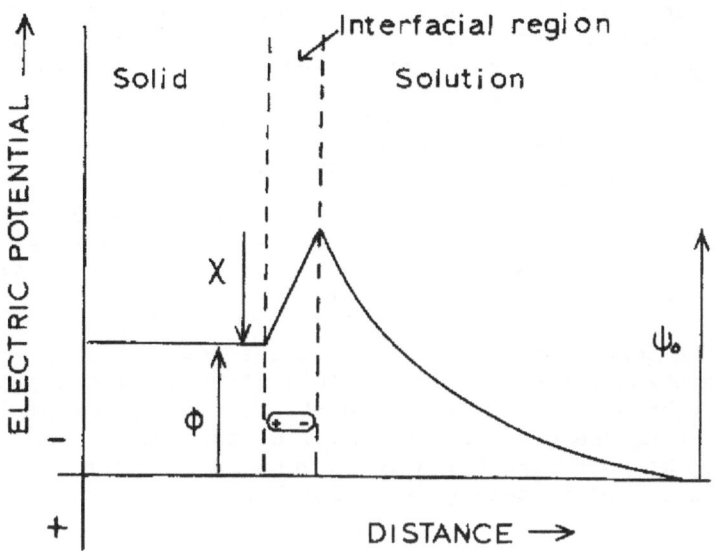

FIGURE 1 Potentials at a solid-solution interface

2b The origin of interfacial potentials

     Interfacial potentials can arise in at least five possible
ways:
i)   Unequal dissolution of constituent ions
ii)   Ionization of surface groups
iii) Isomorphous substitution
iv)   Specific adsorption of ions
v)   Dipole orientation

     In most practical cases two or more of these mechanisms
operate simultaneously, but it is convenient to consider each
separately.

i) Many crystalline solids have only limited solubility in water.
In contact with water these materials often acquire a charge
owing to a greater tendency of one or other of the constituent
ions to "escape" into the aqueous phase. Of the many possible
examples silver halides have received by far the most attention
and will be briefly discussed here, although the same principles
apply to other ionic solids such as calcium carbonate.
     The solubility of silver iodide in water is only about 10^{-8}
mol l^{-1} and so the solubility product $[Ag^+][I^-] = 10^{-16}$. By

increasing the concentration of one of the ions (e.g. by adding
KI solution) the concentration of the other ion must fall to
maintain the constant solubility product value. In this way
the charge on the silver iodide particles can be varied and
there is a characteristic concentration of silver ions (and hence
also of iodide ions) at which the charge on the particles is
zero. This is the POINT OF ZERO CHARGE (pzc) and, for silver
iodide, occurs when $[Ag^+] = 3.2 \times 10^{-6}$ mol l^{-1} and, hence,
$[I^-] = 3.1 \times 10^{-11}$ mol l^{-1}. These values show that silver ions
enter the aqueous phase more readily than iodide and that in
pure water, where $[Ag^+] = [I^-] = 10^{-8}$ mol l^{-1}, silver iodide
particles are negatively charged.

Since the constituent ions are interchangeable between
solid and solution phases, it is possible to apply thermodynamic
reasoning to the system and this leads to the well-known NERNST
equation relating the electric potential of the solid to the
concentration of POTENTIAL DETERMINING IONS in solution (i.e.
the constituent ions in this case). For low concentrations in
solution:

$$\phi = \text{constant} + \frac{kT}{z_i e} \ln c_i \qquad (2)$$

where k is Boltzmann's constant, T the absolute temperature, z_i
and c_i the valence and concentration of the potential-determining
ion respectively and e is the elementary charge.
    Differentiating with respect to $\log_{10} c_i$ gives:

$$\frac{d\phi}{d\log_{10} c_i} = 2.303 \frac{kT}{z_i e} \left(= \frac{59.2}{z_i} \text{ mV at } 25^\circ \text{ C} \right) \qquad (3)$$

Thus for singly charged potential determining ions, the
potential of the solid should change by about 59 mV for a ten-fold
change in concentration. The Nernst equation often describes
quite well the response of specific ion electrodes to changes in
solution concentration.

ii) Certain materials have chemical groups at their surface
which may ionize on contact with water. Well-known examples
are the synthetic ion-exchange resins, which consist of an
insoluble, polymeric matrix to which are bound ionizable groups,
such as the carboxylic group (-COOH) in the case of weak-acid
resins. In water the carboxyl groups can ionize:

$$R\text{-COOH} \rightleftarrows R\text{-COO} + H^+$$

This surface ionization depends on the pH of the solution - at low pH the high concentration of H^+ in solution suppresses the ionization, and the carboxyl groups will be mainly uncharged. Strong acid groups, such as $-SO_3H$ ionize over a wider range of pH.

Many naturally occuring particles of biological origin have a surface charge which can arise from acidic or basic groups. In proteins, both types of group are present, leading to the possibility of positive or negative surfaces and a point of zero charge at some characteristic pH value. In the most simplified form, this may be represented as:

$$R \begin{matrix} COOH \\ NH_3^+ \end{matrix} \xrightleftharpoons{-H^+} R \begin{matrix} COO^- \\ NH_3^+ \end{matrix} \xrightleftharpoons{-H^+} R \begin{matrix} COO^- \\ NH_2 \end{matrix}$$

At low pH, such a surface has a positive charge and as the pH is raised the surface charge becomes negative. Most biological particles, including many bacteria and algae, show a pzc in the acid region of pH and so, under most conditions, they are negatively charged.

Many oxides in water have surface hydroxyl groups, which may be acidic or basic in character, and so the surface can become negatively or positively charged, depending on the pH of the solution. The surface of a metal oxide, of the general formula MO_2 (e.g. SiO_2, TiO_2 etc.), may be represented in the hydrated form as:

$$M \begin{matrix} OH_2^+ \\ OH \end{matrix} \xrightleftharpoons{-H^+} M \begin{matrix} OH_2^+ \\ O^- \end{matrix} \xrightleftharpoons{-H^+} M \begin{matrix} OH \\ O^- \end{matrix}$$

Since H^+ and OH^- are not constituent ions of the oxide, the applicability of a Nernst-type equation is uncertain. Levine and Smith[9] have discussed this point in some detail.

The pzc of an oxide depends greatly on the nature of the metal and also, to some extent, on the origin of the oxide and its previous treatment. Parks[10,11] has given extensive information on the surface chemistry of oxides. A few pzc values from his compilations are presented below:

Oxide:	SiO_2	Fe_2O_3	Al_2O_3	MgO
Point of zero charge:	2.2	6.7	8.0	12.4

From these data it follows that silica particles should be
negatively charged for all pH values above 2.2 and that magnesium
oxide should be positive below pH 12.4.   Iron and aluminium
oxides should both show charge reversal at around neutral pH.
Hydrous oxides such as $Al(OH)_3$ often show pzc values not greatly
different from the corresponding oxide, but experimental data on
hydrous oxides are highly variable.

iii) Certain materials, notably clay minerals, have crystal
structures in which some cations may have been replaced by cations
of similar size but lower charge, e.g. Al^{3+} may be replaced by Mg^{2+}.
This is known as ISOMORPHOUS SUBSTITUTION and leaves the crystal
with an excess negative charge. This charge is balanced by
oppositely charged ions (COUNTERIONS), such as Na^+, which cannot
enter the crystal lattice and which are, to some extent, free to
migrate away from the clay particles when the latter are immersed
in water. For this reason, clay particles in water appear to have
a negative charge. This subject is treated extensively by
van Olphen[12].

iv) Another mechanism whereby a particle can acquire charge is
the preferential adsorption of certain ions from solution.
Surface active agents (surfactants) are strongly adsorbed from
aqueous solution on to a wide range of materials because of the
hydrophobic nature of the "tail" of the surfactant molecule.
Most surfactants are either anionic or cationic and so the particles
become similarly charged. For this reason, surfactants are widely
used for the preparation of stable dispersions.

    In the absence of surfactants, particles can still become
charged by preferential adsorption of simpler ions. Most common
anions are less strongly hydrated in solution than cations and,
for this reason, they are more strongly adsorbed, giving the
particles a negative charge. Even air bubbles in water are
negatively charged for this reason.

v) Uncharged molecules which carry a permanent dipole (e.g. water),
may adsorb at an interface with a preferential orientation, giving
a potential X . If the overall potential $\emptyset$ remains constant there
must be a corresponding change in ψ_o , which, according to eq(1),
is equal to $-X$. For molecules with their positive ends towards
the solid the surface potential, ψ_o , becomes more negative.

2c The electrical double layer

    An interfacial potential is accompanied by a characteristic
distribution of charge between the phases. Since the system has
overall electrical neutrality, the net charge must be zero.

The charge on a particle, together with the equal and opposite charge in solution comprise an ELECTRICAL DOUBLE LAYER.

A number of attempts have been made to account for the charge distribution in an electrical double layer, the simplest of which is the GOUY-CHAPMAN theory[13,14]. This is based on a number of simplifying assumptions:

i)   An infinite, flat, impenetrable interface
ii) The ions in solution are point charges, able to approach right up to the interface
iii) The solvent is a uniform medium, whose properties are independent of the distance from the interface

The main problem is to find the variation of potential in solution, ψ , with distance from the interface x. This depends on the distribution of excess ionic charge in solution (the DIFFUSE part of the double layer). The local concentrations of anions and cations can be expressed in terms of the local potential by means of the Boltzman expression and the difference between these two concentrations gives the charge density in solution at that point. Finally, the Poisson equation relating charge density and potential is employed. Full details are given by Verwey and Overbeek[5]. This treatment gives the following result for a flat double layer and for symmetrical electrolytes (i.e. those such as NaCl and $MgSO_4$ where the cation and anion have the same valence):

$$d^2y/dX^2 = \sinh y \tag{4}$$

In eq(4) the potential ψ at a distance x from the interface is expressed in dimensionless form, $y = ze\psi/kT$, (at $25^{\circ}C$, $\psi = 25.6y/z$ mV). The distance is also expressed in dimensionless form: $X = \varkappa x$, where:

$$\varkappa^2 = 2e^2nz^2/\varepsilon kT \tag{5}$$

where n is the number of cations (or anions) per unit volume of solution and ε is the (absolute) permittivity of the solution.
$\varkappa$ is the Debye-Huckel parameter and has the dimensions of reciprocal length.

Integration of eq(4) yields:

$$(dy/dX)^2 = 2 \cosh y - C \tag{6}$$

For an isolated double layer we have the boundary condition $dy/dX = 0$ at $X = \infty$, i.e. the potential gradient is zero at a very large distance from the interface (see figure 1).
Eq(6) then becomes:

$$dy/dX = - \sqrt{2\cosh y - 2} = - 2\sinh (y/2) \tag{7}$$

62

(The negative root is taken since the potential must fall with distance from the interface).

With the condition that $y=y_o (=ze\psi_o/kT)$ at $X=0$, integration of eq(7) gives:

$$y = 2 \ln\left[\frac{1+\gamma\exp(-X)}{1-\gamma\exp(-X)}\right] \qquad (8)$$

where $\gamma = \tanh(y_o/4)$

There are two useful approximate forms of eq (8). If the surface potential is small ($y_o < 1$):

$$y = y_o \exp(-X) \qquad (9)$$

For large X and any value of y_o:

$$y = 4\gamma \exp(-X) \qquad (10)$$

Eq(9) gives reasonable results for y_o up to 2 and eq(10) is a good approximation when $X > 1$.

The approximate exponential decay of potential from the interface means that, at $X=1$, the potential has fallen to $1/2.712$ of the surface value. At this plane $x = 1/\varkappa$ and this value is often regarded as the "thickness" of the double layer. As seen from eq(5), $1/\varkappa$ will increase as the concentration of ions in solution decreases and will be greatly dependent on the charge of the ions. Some values of $1/\varkappa$ for various solutions are given below:

Solution	$1/\varkappa$ (nm)
10^{-4}M NaCl	31
10^{-4}M MgSO$_4$	15
London tap water	3.6
Sea water	0.4

The surface charge density, σ_o, at an interface can be derived from the fact that the charge on the solid must be exactly balanced by the diffuse layer charge. The result is.

$$\sigma_o = (8n\varepsilon kT)^{\frac{1}{2}}\sinh(y_o/2) \qquad (11)$$

For low surface potentials ($y_o < 1$), this becomes:

$$\sigma_o = \varepsilon\varkappa\psi_o \qquad (12)$$

It is worth noting that eq(12) is the relationship between charge and potential for a parallel plate condenser when the distance between the plates is $1/\varkappa$.

So far, only a flat interface has been considered. In practice, many particles of interest may be treated as spheres and it would be useful to know how the potential around a charged sphere in aqueous solution varies with distance from the sphere.

An exact treatment of a spherical double layer is only possible by numerical techniques. Results have been tabulated by Loeb et al[15]. When the surface potential of the sphere is small, an analytical solution is possible:

$$y = \frac{y_o a}{r} \exp\left[\varkappa(a-r)\right] \qquad (13)$$

where a is the radius of the sphere and y is the potential in solution at a distance r from the centre of the sphere.

The surface charge density on the sphere is:

$$\sigma_o = \frac{\varepsilon \psi_o (1 + \varkappa a)}{a} \qquad (14)$$

When the sphere is very large ($\varkappa a \gg 1$) eqs (13) and (14) become equivalent to the corresponding expressions for a flat interface, eqs(9) and (12).

The Gouy-Chapman treatment of the double layer is open to a number of criticisms and does not adequately account for the charge-potential behaviour in real systems. However, a fairly simple extension to the basic theory can give significant improvement.

The main difficulty with the Gouy-Chapman theory is the assumption that ions in solution are point charges. In the STERN-GRAHAME[16,17] picture of the double layer, the finite size of real ions is recognised and this sets a lower limit to the distance to which an ion can approach the interface. The possibility of specific adsorption of certain ions is also taken into account, whereas, in the Gouy-Chapman approach, only electrostatic interactions are considered.

The resulting model of the double layer is one with a layer of fairly tightly held counterions (the STERN LAYER) adjacent to the interface. Across this layer the potential falls rather sharply from the surface value, ψ_o , to a value ψ_δ . δ is the thickness of the Stern layer and is of the order of 0.5nm (5Å). Outside the Stern layer the potential can be assumed to fall from ψ_δ according to Gouy-Chapman theory, i.e. as eq(8), but with y_o replaced by y_δ ($=ze\,\psi_\delta/kT$). In fact, since y_δ may be considerably smaller than y_o, it is more likely that the approximate expression, eq(9), will be adequate in the diffuse region.

Figure 2a is a representation of a Stern-Grahame type of double layer with moderate counterion adsorption in the Stern layer. With increasing ionic strength more counterions are located in the Stern layer and the value of ψ_δ diminishes. When the adsorption of counterions is very strong, it is possible for ψ_δ to be of opposite sign to ψ_o, as in figure 2b. This is an example of charge reversal and, for negative surfaces, can occur in the presence of cationic surfactants and polyelectrolytes and also with certain hydrolyzed metal ions such as the hydrolysis products of Al^{3+} and Fe^{3+}.

There are several other refinements to double layer theory such as the discrete ion effect[18], but these will not be discussed here. For details, the recent book of Sparnaay[19] is recommended.

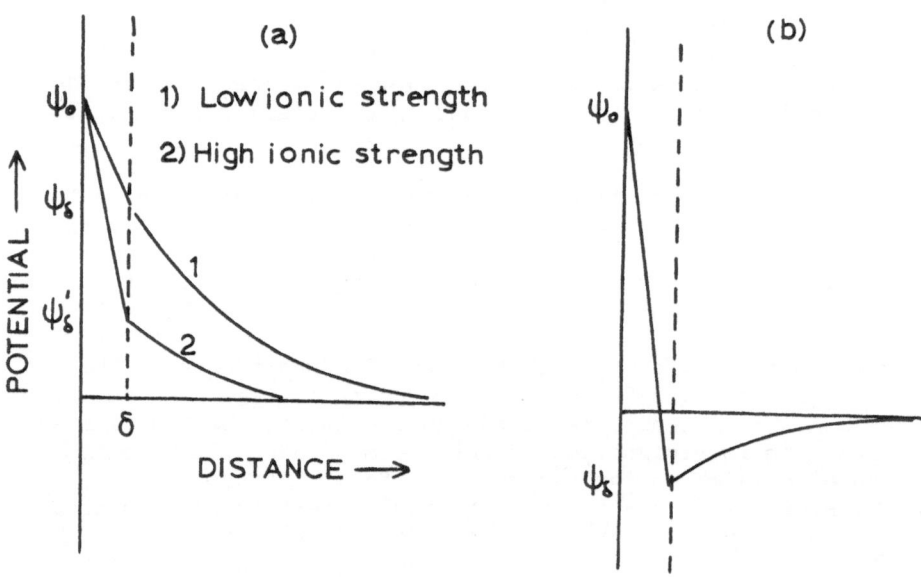

FIGURE 2   The Stern-Grahame double layer   a) showing the effect of ionic strength on ψ_δ . b) the effect of strongly adsorbed, charge-reversing counterions.

2d Electrokinetic phenomena

Electrokinetic effects arise when there is relative movement at the boundary between two phases (one a liquid) because at least some of the charge in the diffuse part of the double layer is mobile and can move with the liquid. Electrokinetic data are usually interpreted in terms of the potential at the boundary between the fixed and mobile parts of the double layer, i.e. at the "slipping plane". This potential is known as the ZETA POTENTIAL, ζ . It is often assumed that the slipping plane lies just outside the Stern layer and that $\zeta \sim \psi_{\delta}$. There is virtually no experimental evidence to support this assumption and it seems that in some cases, ζ may be considerably less than ψ_{δ} . Smith[20] has discussed some of the problems of interpreting zeta potentials and gives references to earlier work.

Four distinct electrokinetic phenomena are known, depending on whether relative motion between the phases is caused by electrical or mechanical means and whether discrete particles or porous media are concerned.

A STREAMING POTENTIAL is established when solution is forced through a porous plug (or tube) of material which acquires charge in contact with the solution.

A SEDIMENTATION POTENTIAL is set up when charged particles fall through a solution.

ELECTRO-OSMOSIS is the flow of liquid through a porous plug (or tube) under the influence of an applied electric field.

ELECTROPHORESIS is the phenomenon of migration of charged particles in an electric field.

It has been theoretically shown[21] that the same physical quantity (the zeta potential) is involved in all of these phenomena. A brief account of the underlying principles of the methods will be given here but no attempt will be made to describe the various experimental techniques, which have been well covered by Shaw[22].

In the case of streaming potentials it is convenient, initially, to consider flow through a uniform tube, radius a, length l, and with applied pressure difference p. It will be assumed that the radius of the tube is very much greater than the thickness of the double layer, i.e. $\varkappa a \gg 1$. For all aqueous solutions of interest, this condition would be satisfied for a $> 1 \, \mu$m. Using Poisseuille's equation for tube flow and the Poisson expression for charge density, it can be shown (e.g. Shaw[22]) that the flow of liquid causes a streaming current,

66

I_s, to be established, given by:

$$I_s = \frac{\pi \epsilon p a^2}{\eta l} \int_0^a x \frac{d^2\psi}{dx^2} dx \tag{15}$$

where η is the viscosity of the liquid and x is the distance from the tube wall. Assuming that the slipping plane is at the tube wall, $\psi = \zeta$ at x=0. Also, since $\varkappa a \gg 1$, $d\psi/dx=0$ at x=a. With these boundary conditions, the integration in eq(15) can be carried out, giving

$$I_s = \pi \epsilon p a^2 \zeta / \eta l \tag{16}$$

The streaming current is caused by the transport of counter-ions by the liquid flow. Under steady-state conditions, this is just balanced by a conduction current in the opposite direction, which is related to the streaming potential, E_s, by Ohm's law:

$$E_s = I_s l / K \pi a^2 \tag{17}$$

where K is the conductivity of the solution.

Combining eqs(16) and (17):

$$E_s = \epsilon p \zeta / \eta K \tag{18}$$

Eq(18) contains no terms for the dimensions of the tube and can, in principle, be used for porous media of arbitrary geometry, but this would only be justified if there were no significant surface conductance. In order to apply a correction for surface conductance, the geometry of the system should be known, as in the case of uniform tubes. In practice, eq(18) can be used for porous media if the value of K is taken as the effective conductivity, measured across the porous plug. This probably leads to little error if the surface conductance is not too great.

If the streaming potential is measured for a number of applied pressures, a plot of E_s vs p should be a straight line, the slope of which can be used to calculate ζ.

Although streaming potential is generally thought of as an experimental technique for determining the zeta potential of granular media, other aspects may be important. Flow in pipelines, for instance, may cause appreciable (and possibly dangerous) streaming potentials. Also, it is conceivable that streaming potentials set up across granular filters might affect filtration performance, although such an effect has not been convincingly demonstrated in practice.

When the pores are very small, the diffuse part of the double layer may occupy a significant proportion of the void volume. The rejection of salts by hyperfiltration (reverse osmosis) membranes has been explained[23] along these lines.

In the case of electro-osmosis it is possible to show, by similar reasoning to that above, that the velocity of liquid in a tube, caused by an applied field of strength E is:

$$v_{eo} = E\varepsilon\zeta/\eta \qquad (19)$$

This only applies when the radius of the tube, or the average pore size of porous media, is much greater than the double layer thickness.

Eq(19) also gives the electrophoretic velocity, v_e, of a particle under the influence of an applied field, when the particle radius is large ($\varkappa a \gg 1$).

Usually, electrophoretic velocity per unit field strength, the ELECTROPHORETIC MOBILITY, U, is quoted. This is given by:

$$U = v_e/E = \varepsilon\zeta/\eta \qquad (20)$$

Inserting values of ε and η for water at 25°:

$$\zeta \text{ (mV)} = 12.8 \text{ U} \qquad (21)$$

in which the units of U are $\mu m\ s^{-1}/V\ cm^{-1}$ (or $10^{-8} m^2\ s^{-1}\ V^{-1}$)

In fairly dilute electrolyte solutions, eq(20) is only accurate for particles of about $1\ \mu m$ diameter and greater (so that $\varkappa a > 100$). For many colloids of interest, $10 < \varkappa a < 100$, and, in this region, quite significant corrections have to be applied. These have been computed and tabulated by Wiersema et al[24] for a variety of conditions.

Because of these corrections, and the difficulty of interpreting zeta potentials, it is quite common for experimenters to quote results simply as electrophoretic mobilities, rather than converting to zeta potentials. However, when a rough value of ζ is required, eq(21) can be used in most cases if the mobility is not too high (or ζ not greater than 50-60mV).

As with the streaming potential technique, particle electrophoresis is usually considered as an experimental tool. However, electrophoretic painting and forced-flow electrophoresis in water purification[25] are examples of practical applications.

3. INTERACTION BETWEEN PARTICLES

3a Electrical Interaction

When two charged particles in water approach each other the diffuse parts of their double layers overlap and, if the charges are of the same sign, a repulsion is experienced between them. A thorough theoretical treatment[2] involves the derivation of interaction energy from the increase in free energy as the diffuse layers overlap. The resulting expressions can only be solved numerically and tabulated values of double layer interaction energies are available[5,26,27]. When the surface potentials are small, or when the distance between the particles is fairly large, approximate analytical solutions are possible. In such cases, the interaction energy can be simply derived from expressions for the force between double layers and this approach will be adopted here. The interaction between parallel flat plates will be considered first and, from this, it will be shown how expressions for sphere-sphere and sphere-flat plate interactions can be derived.

Unless the plates are very close, only the diffuse layers will be involved and the problem can be treated as the interaction of two Gouy-Chapman type double layers, but with ψ_o replaced by ψ_δ . This has the advantage that the assumption of low surface potential is more likely to be correct. Also, information on ψ_δ may be available (at least approximately) from electrokinetic data. In this section, since the interaction of unlike particles will be considered, and to avoid cumbersome symbols, ψ_1 , and ψ_2 will be used for the "surface potentials" of the plates, with the understanding that these quantities refer to the potentials at the Stern plane and not to the true surface potentials.

The potential distribution in the solution between two charged plates is shown schematically in figure 3. In this case, the plate potentials are of similar sign and a minimum potential ψ_m , occurs between the plates.

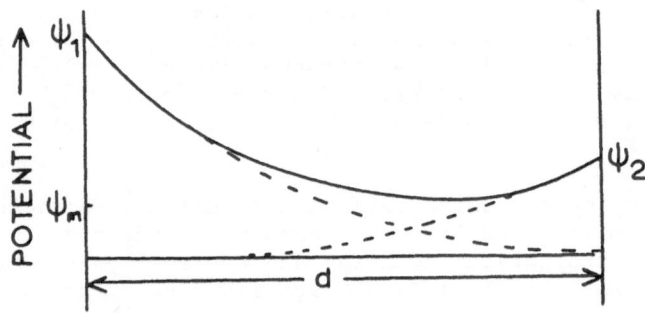

FIGURE 3 Potential distribution between two plates.

The form of the potential distribution is given, in principal, by the fundamental differential equation, eq(6), but the problem is, of course, to find the value of C, the integration constant. As shown by Derjaguin and Landau[2], C is directly related to the force by:

$$P = nkT(C-2) \qquad\qquad (22)$$

where P is the force per unit area between the plates.

If there is a minimum potential between the plates, it follows from eq(6) that $C=2\cosh y_m$ ($y_m = ze\psi_m/kT$) and hence:

$$P = 2nkT(\cosh y_m - 1) \qquad\qquad (23)$$

Eq(23) can be derived in a number of ways (see Verwey and Overbeek[3]) and it is especially interesting that the force does not depend directly on the plate potentials, but on the minimum potential between them.

If there is no minimum between the plates, $C<2$ and the force becomes an attraction. This would clearly be the case if the potentials were of opposite sign, but Derjaguin[28] showed that potentials of different magnitude but similar sign would also give an attractive force if the plates approached closer than a certain critical distance. It is clear from figure 3 that the minimum potential would disappear on close approach of the plates, provided that the plate potentials remain constant. Bierman[29] pointed out that, in such a case, the surface charge densities of the plates would become infinite on contact. For plates of equal potential, $\psi_1 = \psi_2$, the condition of constant potential implies that the surface charge densities would tend to zero on close approach of the plates. In this case, the minimum potential would be maintained and would reach a limiting value $\psi_m = \psi_1$

Most of the existing theoretical work on the interaction of charged particles is based on the assumption that the potential of the particles remains constant during approach. This is a consequence of the heavy emphasis on silver halides in colloid science, because the potential of such particles should depend only on the concentration of potential-determining ions in solution, according to eq(2).

It now appears[30] that, during the very rapid encounters between particles in a colloidal dispersion, electrochemical equilibrium could not be maintained and so the potential need not remain constant. In view of this, and the fact that many colloids of interest (e.g. clays) have a fixed charge, it might be better to assume constant surface charge density on the approaching particles. In this case, the surface potentials must increase as the particles approach, becoming (at least, in theory) infinite on contact. The consequences of assuming constant surface charge density have been considered by Frens[30,31]. At fairly large separation of the particles ($\varkappa d \gg 1$), results for the constant charge and constant potential conditions are not greatly different, but, at close approach, the constant charge assumption leads to much higher interaction energies. This is of great significance in understanding certain phenomena such as the re-dispersion of flocculated colloids.

Here, we will derive approximate expressions based on the constant potential assumptions, which are adequate in many practical situations. The corresponding expressions for the constant charge case will be quoted, if available.

Returning to the force expression, eq(23), it is necessary to find the value of the minimum potential, y_m. If the plate potentials are small, the expression of Hogg et al[32] can be used for the potential, y, at a distance x from the left-hand plate (figure 3):

$$y = y_1 \cosh \varkappa x + \left(\frac{y_2 - y_1 \cosh \varkappa d}{\sinh \varkappa d} \right) \sinh \varkappa x \qquad (24)$$

The minimum potential can be found by differentiating this expression:

$$y_m = \left[y_1^2 - \left(\frac{y_1 \cosh \varkappa d - y_2}{\sinh \varkappa d} \right)^2 \right]^{1/2} \qquad (25)$$

Provided that $y_m < 1$, the approximation $(\cosh y_m - 1) \sim y_m^2/2$ can be made and the force between the plates becomes, from eq(23):

$$P \sim nkT y_m^2 \qquad (26)$$

The energy of interaction per unit area, V_R, is the work necessary to bring the plates from infinite separation to a distance, d:

$$V_R = - \int_\infty^d P \, dd \qquad (27)$$

With eqs(25) and (26) the integration can be carried out directly giving the energy of interaction between two plates with fairly low potentials, which remain constant during approach of the plates:

$$V_R^\psi = \frac{nkT}{\varkappa} \left[(y_1^2 + y_2^2)(1 - \coth \varkappa d) + 2 y_1 y_2 \operatorname{cosch} \varkappa d \right] \qquad (28)$$

For equal plate potentials, y_1, eq(28) reduces to:

$$V_R^{\psi} = \frac{2nkT}{\varkappa} y_1^2 \left[1 - \tanh (\varkappa d/2) \right] \tag{29}$$

These approximations give reasonable agreement with "exact" results[26], for surface potentials up to $y_1 = 2$ (i.e. up to 50mV for 1-1 electrolytes).

For the <u>constant charge</u> condition a similar approach leads to the following expressions, in which y_1 and y_2 now refer to the potentials of the <u>isolated</u> plates:
For unequal potentials:

$$V_R^{\sigma} = \frac{nkT}{\varkappa} \left[(y_1^2 + y_2^2)(\coth \varkappa d - 1) + 2y_1 y_2 \cosh \varkappa d \right] \tag{30}$$

and for equal potentials:

$$V_R^{\sigma} = \frac{2nkT}{\varkappa} y_1^2 \left[\coth(\varkappa d/2) - 1 \right] \tag{31}$$

In the constant charge case, the assumptions of low plate potentials and the approximate force expression, eq(26), are much more dubious than in the constant potential case, since, at close approach, the potentials can reach very high values. Comparison with exact results for constant charge interaction[27] shows that eq(31) can give results which are 20 times too high at small separations[33] (see figure 4).

An alternative treatment[33] of the interaction between plates at constant charge gives, for equal plate potentials:

$$V_R^{\sigma} = \frac{2nkT}{\varkappa} \left[2y_1 \ln \left(\frac{B + y_1 \coth \varkappa d/2}{1 + y_1} \right) \right.$$
$$\left. - \ln \left(y_1^2 + \cosh \varkappa d + B \sinh \varkappa d \right) + \varkappa d \right] \tag{32}$$

where $B = \sqrt{1 + y_1^2 \cosh^2(\varkappa d/2)}$

For unequal plate potentials which are not greatly different eq(32) can be used if y_1 is replaced by $\sqrt{y_1 y_2}$, but for widely different potentials no convenient approximation is available in the constant charge case. Exact calculations have been performed by Bell and Peterson[34].

When the plates are not too close, the potential at some point between them can be approximated as the sum of the potentials which would be produced at that point by each plate in the absence of the other (see figure 3). This is sometimes called the Linear Superposition Approximation (LSA). With this assumption, and making use of eq(10):

$$y = 4 \left[Y_1 \exp(-\varkappa x) + Y_2 \exp(\varkappa x) \exp(-\varkappa d) \right] \tag{33}$$

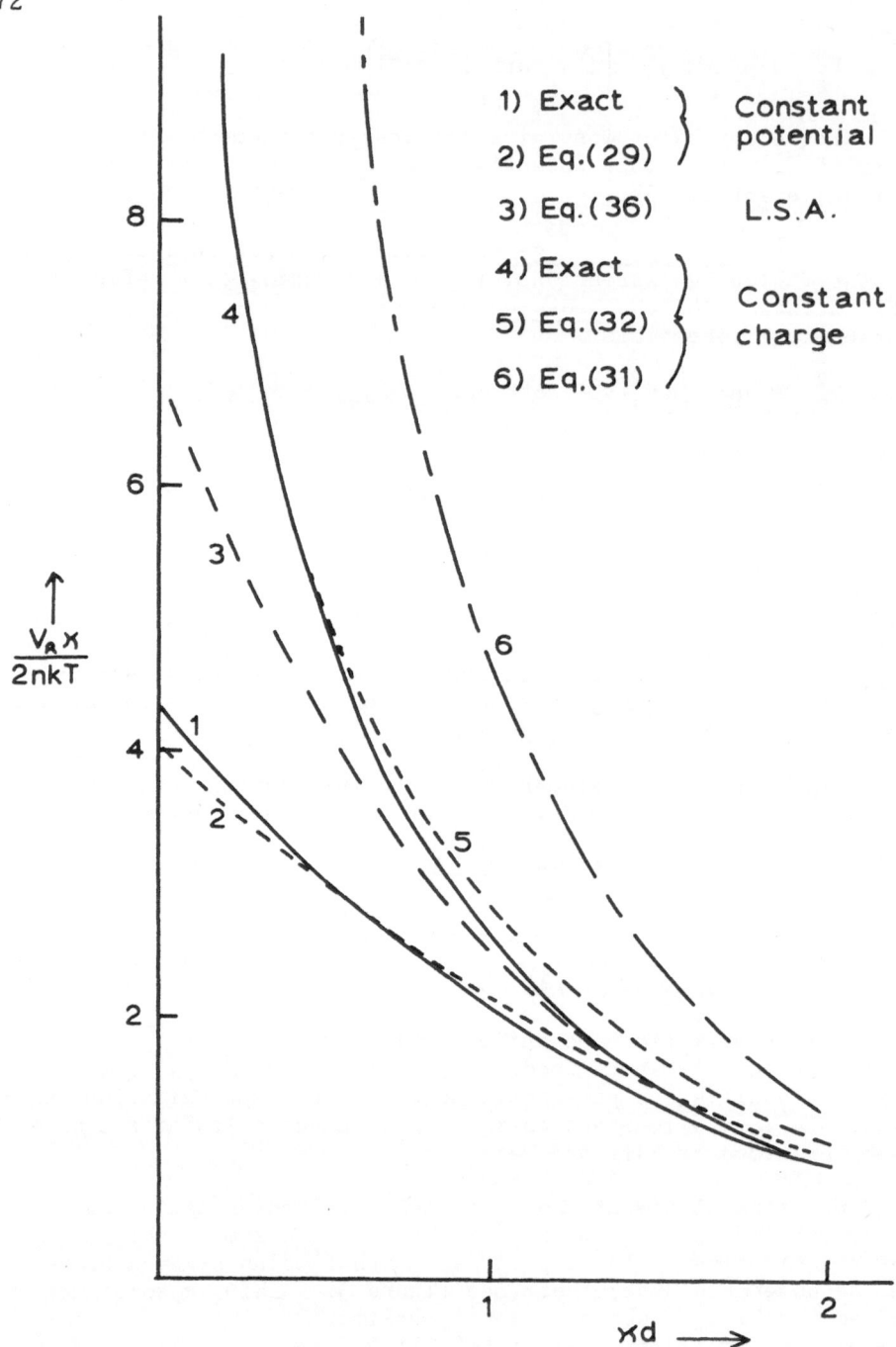

FIGURE 4 Interaction of flat plates. Comparison of
approximate expressions with exact results.

The minimum potential can, again, be found by differentiation:

$$y_m = 8(Y_1 Y_2)^{\frac{1}{2}} \exp (- \varkappa d/2) \qquad (34)$$

The force then follows from eq(26) (if $y_m < 1$) and the energy of interaction from eq(27):

$$V_R = \frac{64nkT}{\varkappa} Y_1 Y_2 \exp (- \varkappa d) \qquad (35)$$

This expression can only be used if the plate potentials are not greatly different, i.e. when the minium potential does not occur very close to either plate.

For equal potentials eq(35) reduces to the frequently-used expression:

$$V_R = \frac{64nkT}{\varkappa} Y_1^2 \exp (- \varkappa d) \qquad (36)$$

Since the LSA only applies at large plate separations ($\varkappa d \gg 1$), where the constant potential and constant charge assumptions lead to similar results, V_R in eqs(35) and (36) is not given a superscript ψ or σ. The main virtue of these expressions is that they can be used for high plate potentials provided that the separation is large.

The various approximations are compared with exact values for the constant potential and constant charge flat plate interaction energies in figure 4. Only the case of equal plate potentials is considered with $y_1 = 2$. Lower potentials give better agreement in all cases. The energy is plotted in dimensionless form, as $V_R \varkappa /2nkT$, against the dimensionless separation $\varkappa d$. Note that eq(31) is a very poor approximation in the constant charge case.

The interaction of spherical charged particles could, in principle, be treated by considering the overlap of spherical double layers. This is only possible by numerical techniques and, again, tabulated results are available[3,27].

A simpler, through approximate, method is that of Derjaguin[35], who considered the interaction of spheres as a summation of interactions between concentric, parallel rings on the sphere surfaces, as in figure 5. The method is only valid when the minimum separation of the spheres, d , is much smaller than the radii, so that only rings close to the line of centres play any part in the interaction. The results based on several approximate flat plate expressions are given below:

74

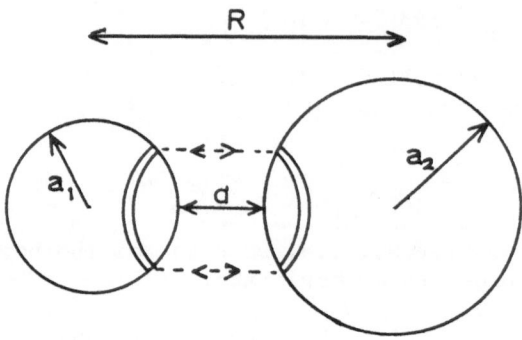

FIGURE 5   Interaction of two spheres.

For constant potential, from eq(28)

$$V_R^{\psi} = \frac{a_1 a_2}{(a_1 + a_2)} \frac{2\pi nkT}{\varkappa^2} (y_1^2 + y_2^2) \left[\frac{2 y_1 y_2}{y_1^2 + y_2^2} \ln\left(\frac{1+\exp(-\varkappa d)}{1-\exp(-\varkappa d)}\right) \right.$$

$$\left. + \ln\left[1 - \exp(-2\varkappa d)\right] \right] \qquad (37)$$

Eq(37) was first derived by Hogg et al[32] and, for equal spheres reduces to:

$$V_R^{\psi} = \frac{4\pi a_1 nkT\, y_1^2}{\varkappa^2} \ln\left[1 + \exp(-\varkappa d)\right] \qquad (38)$$

$$\left(= 2\pi \varepsilon a_1 \psi_1^2 \ln\left[1 + \exp(-\varkappa d)\right] \right)$$

At constant charge the expression for unequal spheres, based on eq(30) is just like eq(37) except for a negative sign between the two terms in square brackets. For equal spheres:

$$V_R^{\sigma} = \frac{-4\pi a_1 nkT y_1^2}{\varkappa^2} \ln\left[1 - \exp(-\varkappa d)\right] \qquad (39)$$

These approximate constant charge expressions are based on the assumption of low potentials and are hence unreliable at close approach (see figure 6). At present there is no convenient expression for spheres based on eq(32).

The LSA expressions, eqs(35) and (36) lead to:

$$V_R = \frac{a_1 a_2}{(a_1+a_2)} \frac{128\,\pi\,nkT}{\varkappa^2} Y_1 Y_2 \exp(-\varkappa d) \tag{40}$$

and for equal spheres:

$$V_R = \frac{64\,\pi\,a_1\,nkT}{\varkappa^2} Y_1^2 \exp(-\varkappa d) \tag{41}$$

All of the sphere expressions so far have been based on the assumption that the separation, d , is much smaller than the sphere radii. For most practical problems this leads to little error since the range of electrical interaction is usually quite small. However, when this condition is not fulfilled, an expression due to McCartney and Levine[36] may be used. This is for the case of equal spheres of small, constant potential, and where the double layer is fairly thin in comparison with the sphere radius ($\varkappa a_1 > 5$):

$$V_R^{\psi} = \frac{8\pi a_1 nkT\, y_1^2}{\varkappa^2} \frac{(R-a_1)}{R} \ln\left[1 + \frac{a_1}{R-a_1} \exp\left[-\varkappa(R-2a_1)\right]\right] \tag{42}$$

where R is the distance between the centres of the spheres. Eq(42) reduces to eq(38) when d(= $r-2a_1$) $\ll a_1$.

For equal spheres, with $y_1=2$, results from eqs(38), (39) and (41) are compared with exact computations[27] of V_R^{ψ} and V_R^{σ} in figure 6. As before, the interaction energy is plotted in dimensionless form, this time as $V_R \varkappa^2/4\pi a_1 nkT$. For the interaction of spheres, the constant potential and constant charge assumptions do not lead to such widely different results as in the case of flat plate interaction (although the difference becomes more significant at lower potentials). Eq(39) is clearly a very poor approximation and should not be used.

In practice, neither constant potential nor constant charge is likely to be a correct assumption and a situation somewhere between these two extremes would be expected. For this reason, the simple LSA expressions might be the most appropriate, since both eq(36), for plates, and eq(41), for spheres, give results which are intermediate between those from exact calculations in the constant potential and constant charge cases.

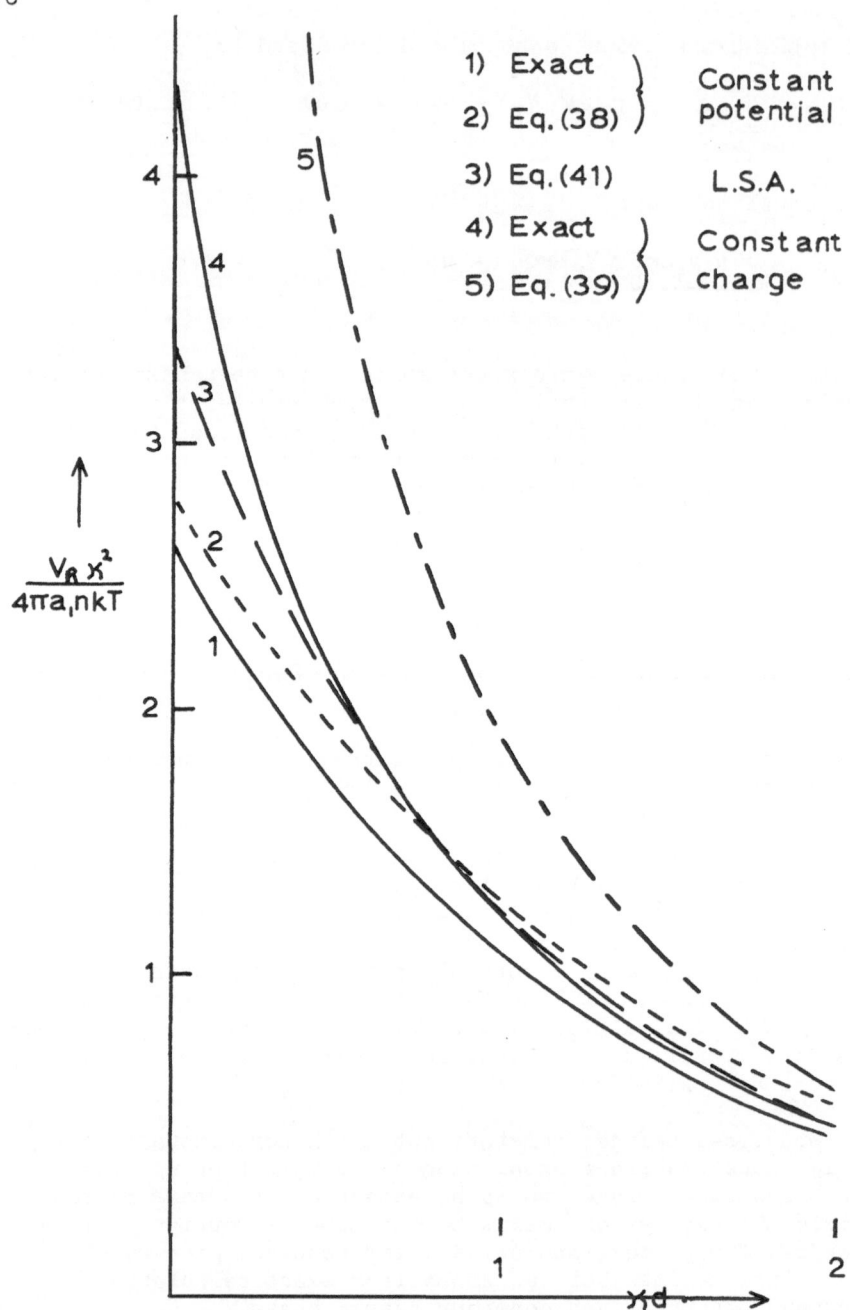

FIGURE 6 Interaction of equal spheres. Comparison of
approximate expressions with exact results.

In a granular filter, the interaction between a suspension particle and a filter grain can often be treated as that between a sphere and an infinite flat surface[37]. Application of Derjaguin's method to this situation leads to an interaction energy exactly twice that for two spheres of equal radii, a, at the same separation, d, and with potentials, y_1 and y_2, equal to those of the sphere and flat surface respectively. This can be easily verified by putting $a_2 = \infty$ in eq(37) or (40)

3b van der Waals interaction

Attractive forces between atoms and molecules were postulated over a century ago by van der Waals to explain the non-ideality of real gases and vapours[38]. These forces can be of three types:

1) If both molecules have permanent dipole moments, μ_1 and μ_2 they will tend to orient themselves so as to give the most favourable electrical attraction. The thermal energy of the molecules opposes this orientation to some extent and the interaction energy of the two molecules at a distance, R, is found to be:

$$U = -\frac{2}{3kT}\frac{\mu_1\mu_2}{R^6}$$

(43)

This is the ORIENTATION or KEESOM effect.

ii) If one of the molecules has a permanent dipole moment, a further dipole can be induced in a neighbouring molecule, depending on its polarizability, α, again giving an attraction:

$$U = -\frac{1}{R^6}\left(\alpha_1\mu_2^2 + \alpha_2\mu_1^2\right)$$

(44)

This is the INDUCTION effect (Debye, 1920).

iii) Many molecules have no permanent dipole but they still attract each other. this is a result of fluctuating dipoles which arise from the random motion of electrons within a molecule. Although only transient, such dipoles can induce dipoles in nearby molecules to give an attraction. This is now called the DISPERSION effect and was first treated by London in 1930. The energy between two similar molecules is given by:

$$U = -\frac{3}{4}\frac{h\nu_1\alpha_1^2}{R^6}$$

(45)

where h is Planck's constant and ν_1 is a characteristic frequency of the molecule.

Between particles, similar interactions can occur, but only the dispersion contribution is (approximately) additive, and effective over a significant distance. For this reason, van der Waals interaction between particles is often referred to as Dispersion interaction.

Hamaker[39] showed how such interactions between molecules could be integrated to give the energy of interaction between particles of various shapes. The results for parallel flat plates and for spheres are of special interest and are given below. These expressions are based on the assumption of complete additivity of intermolecular interactions and contain a constant, A_{12}, which depends only on physical properties of the interacting materials 1 and 2. This is known as the HAMAKER CONSTANT.

For <u>flat plates</u> the attraction energy is:

$$V_A = - A_{12}/12 \pi \ d^2 \tag{46}$$

and for <u>two spheres</u>

$$V_A = - \frac{A_{12}}{6d} \frac{a_1 a_2}{(a_1 + a_2)} \tag{47}$$

Eq(47) only applies when the separation between the spheres, d is much smaller than their radii. This is rarely a severe limitiation since the interaction is almost always too weak to be significant at larger separations.

For equal spheres, eq(47) becomes:

$$V_A = -A_{12} \ a_1/12d \tag{48}$$

and for a sphere and flat surface (i.e. $a_2 = \infty$):

$$V_A = - A_{12} \ a_1/6d \tag{49}$$

The result for the sphere-flat plate case is twice that for the interaction of two equal spheres, just as was found for double layer interaction by Derjaguin's method.

It is important to note that the distance dependence of the attraction between particles is very different to that between atoms and molecules, the latter showing an inverse sixth power relationship. The fact that the attraction falls much less steeply with distance for particles means that van der Waals interaction is more significant in colloidal systems than was originally thought.

In order to calculate interactions in practical cases, a value of the appropriate Hamaker constant must be found. Hamaker[39] originally wrote the constant as:

$$A_{12} = \pi^2 N_1 N_2 B_{12} \qquad (50)$$

where N_1 and N_2 are the numbers of molecules per unit volume of the materials and B_{12} is the coefficient of r^{-6} in the expression for dispersion energy between the molecules (cf. eq(45)).

If the dominant contribution to the interaction comes from a fairly narrow band of frequencies in the ultra-violet region ($> 10^{15}$Hz), it is possible to express the Hamaker constant in terms of optical properties of the materials. In such a case Gregory[40] has shown that:

$$A_{12} = \frac{27}{32} \frac{h v_1 v_2}{v_1 + v_2} \left(\frac{n_1^2 - 1}{n_1^2 + 2}\right)\left(\frac{n_2^2 - 1}{n_2^2 + 2}\right) \qquad (51)$$

where v_1 and v_2 are the characteristic "dispersion frequencies" of the media and n_1 and n_2 are the limiting refractive indices. For similar materials:

$$A_{11} = \frac{27}{64} h v_1 \left(\frac{n_1^2 - 1}{n_1^2 + 2}\right)^2 \qquad (52)$$

This expression has also been derived by Tabor and Winterton[41].

The required frequency and limiting refractive index can be obtained for each material if reasonably accurate values of refractive index at various frequencies in the optical range are available[40]. It is then unnecessary to have any information on the molecular nature of the materials, which would be required for the calculation of Hamaker constants from eq(50).

Although involving only bulk properties of the materials, eqs(51) and (52) are based on the assumption of complete additivity of intermolecular interactions. This assumption is avoided in the treatment of Lifshitz[42], which is based entirely on macroscopic properties of the materials. No details can be given here, but if the assumption of a dominant contribution from the ultra-violet region is again made, it can be shown[40], that the Lifshitz approach leads to an effective Hamaker constant, given, for similar materials, by:

$$A_{11} = \frac{0.230 \, h v_1 \left(n_1^2 - 1\right)^2}{\left(n_1^2 + 1\right)^{\frac{3}{2}} \left(n_1^2 + 2\right)^{\frac{1}{2}}} \qquad (53)$$

In the limit $n_1 \rightarrow 1$ (i.e. for rarefied media) eqs(52) and (53) give identical results. For materials with refractive index in the range 1.5 to 2, eq(52) gives results about 20% higher than eq(53), which may be due to the assumption of additivity, inherent in eq(52). For a few materials, direct measurements of van der Waals forces have been made (e.g. Tabor and Winterton[49]), which give results of the same order as those from eqs(52) and (53), but it is not yet possible to be certain of the validity of either.

Some doubt has recently been cast on this simple approach to dispersion interaction by the assertation[43] that, for polar materials (especially water), frequencies other than those in the ultra-violet region may make important contributions to the total interaction. Furthermore, the whole concept of a Hamaker "constant" in colloid interaction has been called into question[44]. Some of these conclusions have been disputed[45] and for the present, it can be assumed that eqs(46)-(49) are applicable in most cases although the actual value of A_{12} may be uncertain and may have to be treated as an adjustable parameter.

When the interacting particles are separated by a third medium, such as water, eqs(46)-(49) can still be used, but the Hamaker constant has to be modified as follows:

$$A_{132} = A_{12} + A_{33} - A_{13} - A_{23} \qquad (54)$$

where the subscript 3 refers to the intervening medium. It is usual to multiply the right-hand side of eq(54) by a factor to account for the transmission of the dispersion force through the separating medium, and Schenkel and Kitchener[56] used $1/n_3$, i.e. about 0.56 for water. Visser[46] found this factor to be incorrect and suggested that it should be replaced by a factor of 1.6 for interactions in water. In view of the uncertainties it might be better to ignore this factor until more information is available and use the uncorrected expression, eq(54).

For the interaction of similar particles, making the assumption that $A_{13} \sim \sqrt{A_{11}A_{33}}$, eq(54) becomes

$$A_{131} = (\sqrt{A_{11}} - \sqrt{A_{33}})^2 \qquad (55)$$

It is apparent from eq(55) that, if A_{11} and A_{33} are fairly close, the "composite" Hamaker constant A_{131} may be very small. Furthermore, because of the uncertainties already mentioned, especially for interactions in water, values of A_{131} must be still more uncertain. Visser[46] has pointed out that a 20% error in A_{11} can lead to a 500% error in A_{131}. It is thus not surprising that the majority of experimental values of Hamaker constants in aqueous systems (mainly from colloid stability studies) are not in accord with theoretical predictions[40,46].

As a rough guide, Hamaker constants for materials in water are generally found to lie in the range 0.1 to 10 x 10^{-20} J. At the lower end of this range are materials such as polymers whose optical properties do not differ greatly from those of water and at the upper end are metals. Intermediate values ($\sim 10^{-20}$ J) are found for crystalline solids such as silver iodide and various oxides. Compilations of Hamaker constants are available in works already cited[7,40,48], but it must be emphasized again that current theoretical uncertainties cast considerable doubt on many previous calculations.

3c Combined interaction

If, in a colloidal dispersion, the only forces between particles are van der Waals attraction and electrical repulsion, the total interaction can be calculated simply by adding the two contributions. Assuming the particles to be equal spheres, radius 1 μm, surface potential 25mV (i.e. y $\sim$ 1 for 1-1 electrolytes at 25°C) in a solution with $1/\varkappa$ = 10 nm (e.g. 10^{-3} M NaCl), the electrical energy V_R has been calculated from eq(41) and plotted in figure 7. In both cases the energy is given in units of kT (~ 4.2x 10^{-21} J) to facilitate comparison with the thermal energy of the particles. The total interaction V_t is shown as a broken line.

These curves would also apply to the interaction of a sphere, radius 0.5 μm, with an infinite flat surface, provided that the surface potentials were both 25 mV and the same value of Hamaker constant could be assumed.

A number of points emerge from figure 7 which are important in understanding colloid stability. The most important feature is that the V_t curve passes through a maximum, because of the different dependence of V_A and V_R on separation. At very close approach the attraction becomes very great and ultimately overcomes the repulsion. This means that, for the particles to come into contact, they must be able to overcome an ENERGY BARRIER. The height of this barrier, V_{max}, determines the stability of the dispersion. In the case shown in figure 8 the height is about 200kT and, since the chance of a particle surmounting such a barrier is virtually zero, the suspension would be stable indefinitely. If the barrier were reduced to zero there would be nothing to prevent contact of the particles, and flocculation would occur at a rate governed by the collision frequency of the particles. With a barrier of moderate height (say, 10-20 kT) only a small fraction of collisions would result in attachment and the suspension would flocculate slowly. For a slowly

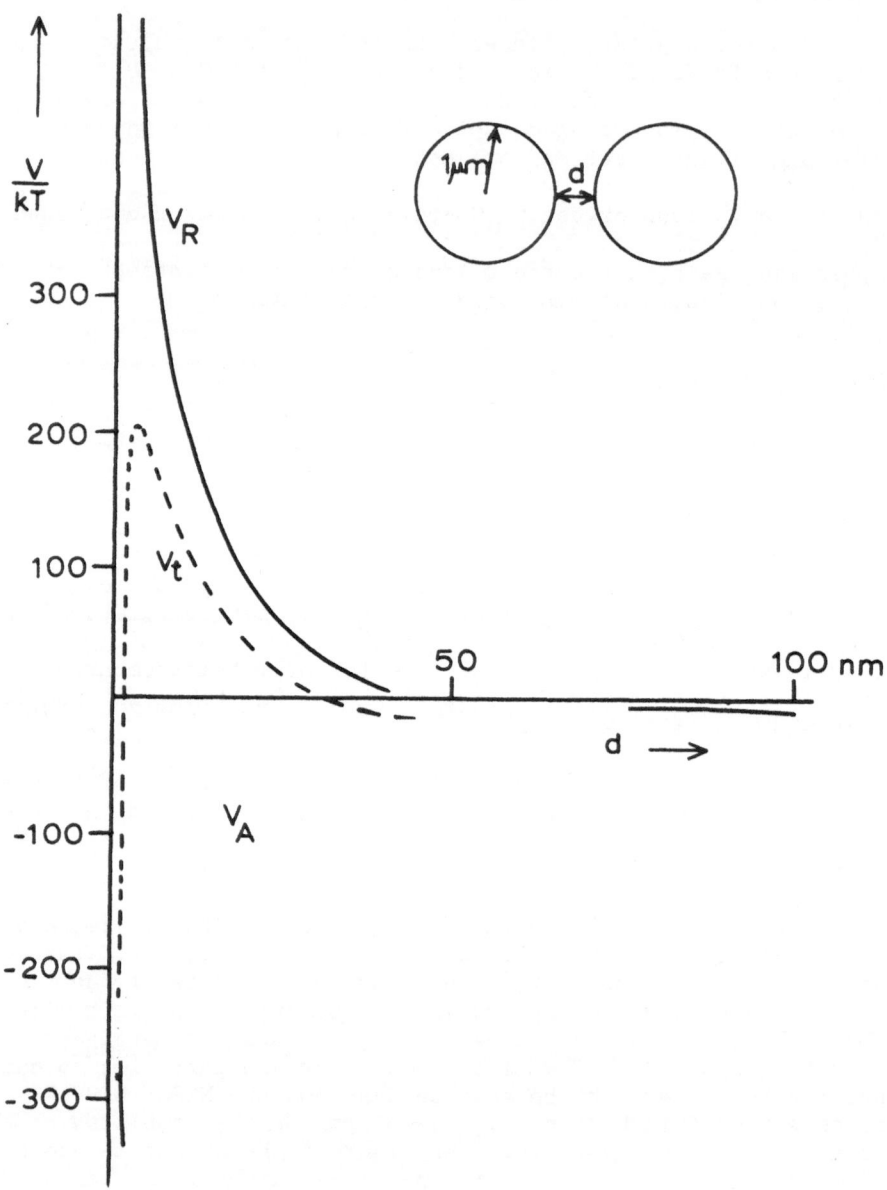

FIGURE 7   Combination of electrical repulsion (V_R) and
van der Waals attraction (V_A) to give a total
interaction curve (V_t).

flocculating suspension, the rate is slower by a factor W than the most rapid flocculation rate (i.e. that when V_{max} = 0). W is the STABILITY RATIO and 1/W can be regarded as the proportion of collisions resulting in attachment. Fuchs[47] showed how the stability ratio can be calculated if the complete form of the interaction curve is known. A much simpler, approximate method of Reerink and Overbeek[48] leads to:

$$W \sim \frac{1}{2 \varkappa a} \quad \exp(V_{max}/kT) \tag{56}$$

In the case discussed above, this expression leads to the result $W \sim 4 \times 10^{86}$ - and hence indicates a very stable suspension!

Both V_A and V_R are directly proportional to particle radius and so the shape of the curves in figure 7 would be similar for smaller particles but the value of V_{max} would be less. Since the average thermal energy of a particle is independent of particle size this means that small particles should be less stable than large ones if all other conditions are equal. (Note - this conclusion is only valid for BROWNIAN or PERIKINETIC flocculation, and does not apply to ORTHOKINETIC flocculation induced by velocity gradients).

In order to reduce colloid stability and hence to promote flocculation, or attachment to filter grains, V_{max} must be reduced, preferably to zero. One method of achieving this is to increase the ionic strength of the solution and hence to increase $\varkappa$. From eq(41) it is clear that this would influence V_R in two ways - the pre-exponential factor would become less and the range of the repulsion would be reduced because of the $\exp(-\varkappa d)$ term. The latter effect is a result of the reduced "thickness" of the diffuse layer around the particles and would cause the V_R curve in figure 7 to be compressed. Adding salts with multivalent ions is most effective because of the effect of Z on $\varkappa$ (see eq(5)). ·For unsymmetrical electrolytes (e.g. $CaCl_2$) the most effective flocculants are those with highly charged counterions, so that, for negative particles, $CaCl_2$ would be better than Na_2SO_4.

The most common method of testing colloid stability is to add increasing amounts of a salt until the concentration required for rapid flocculation, c_f, is found. If it is assumed that, under these conditions, V_{max} = 0, then it is possible to show[3], from eqs(41) and (48), that:

$$c_f = \text{constant} (\ \gamma_i^4/z^6) \tag{57}$$

For high values of surface potential, $Y_1 \sim 1$ and eq(57) predicts that the flocculation concentration should be inversely proportional to the sixth power of the valency of the ions as predicted by the empirical SCHULZE-HARDY rule. While not obeyed quantitatively in many cases, this is a useful rough guide and the flocculation concentrations of, for instance, (Na^+, Ca^{2+} and La^{3+}) with negative colloids are of the order of: 100, 2 and 0.1 mM l^{-1} respectively.

Of course, the addition of salts usually causes the Stern potential ψ_δ to fall, as shown in figure 2a, and so the repulsion is further reduced. When an ion is specifically adsorbed on the particles the value of ψ_δ can be drastically reduced (see figure 2b) and flocculation can be brought about without any great increase in ionic strength. For instance, minute traces of certain hydrolyzed metal ions[49] or cationic polymers[50] can cause flocculation of negative particles. In such cases, charge reversal often occurs and the particles can be restabilized if an excess of the flocculant is added.

The V_t curve in figure 7 implies that, once the energy barrier is overcome, the attraction between particles becomes infinitely strong and hence that flocculation should be irreversible. However, this conclusion is often invalidated by the fact that real particles are not perfectly smooth and by the presence of short-range repulsive forces.

A final feature of figure 7 worth mentioning is the presence of a rather shallow minimum in the V_t curve at fairly large separations. This is the SECONDARY MINIMUM and is due to the fact that the electrical repulsion falls with distance more rapidly than the van der Waals attraction. The depth of the minimum in figure 7 is about 15kT and this should be sufficient to cause the particles to flocculate, forming rather weak aggregates. The effect would only be significant for fairly large particles, as in the present example. This phenomenon has been invoked to account for anomalous flocculation results[51], but has yet to receive convincing experimental verification.

It should be remarked that, in spite of many attempts, a satisfactory demonstration of the validity of the DLVO picture of colloid interactions has not been achieved. Of the many examples, flocculation of latex particles by salts[52] and deposition of particles on a rotating disc[53] give results at variance with theoretical predictions. However, as mentioned in the Introduction, this is more likely due to the choice of various parameters, rather than to any fundamental misconceptions of the theory.

The choice of spheres to represent colloidal particles is open to question in some cases, especially for plate-like particles (e.g. some clays). However, use of equations for flat plates rather than spheres leads to the difficulty that particle size cannot easily be incorporated in the calculations. Generally, the qualitative conclusions are similar, whichever model is adopted.

The presence of adsorbed layers around particles may considerably influence colloidal stability and, since this is not treated in the DLVO theory, may possibly explain some of the anomalies mentioned previously. The effect of adsorption layers will be considered very briefly in the next section.

3d Adsorbed layers

Consider the simple model in figure 8, of two spheres of substance 1, each surrounded by a layer of substance 2 and immersed in medium 3. There are several ways in which the adsorbed layers could influence the interaction:
i) The structure of the electrical double layer may be modified, for instance, by the restricted access of counterions to the particle surfaces. The effect might be to displace the Stern plane outwards and hence to give a greater value of potential just outside the adsorbed layer than would be expected if the latter were absent. The electrical repulsion between the particles and hence the colloidal stability would be increased. This effect has been discussed by Vincent et al[74]. Of course, these considerations apply to <u>uncharged</u> adsorbed layers; with charged layers the electrical repulsion would be more profoundly affected.

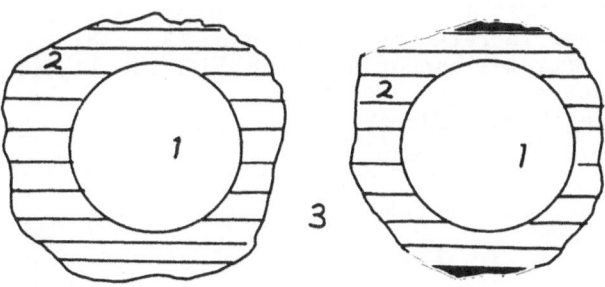

ii) It is likely that substance 2 would have a different Hamaker constant than substance 1 and so the van der Waals interaction between the particles would be affected by the adsorbed layers. Vold[55] has treated this case in some detail. In most cases, the adsorbed layer would be more like the dispersion medium than the particles, so the attraction would be reduced and the colloidal stability increased.

iii) When the adsorbed material consists of fairly long, flexible molecules, close approach of the particles could only occur if some interpenetration of the adsorbed layers were possible. Such interpenetration often results in a strong repulsion between the particles, frequently called "steric repulsion", although this term covers a number of possible types of interaction.

In most practical cases, the "steric" effect is more important than i) or ii), especially in non-aqueous dispersions such as paints, where adsorbed polymers are widely used to maintain stability. In aqueous systems "steric stabilization" is often associated with highly hydrated adsorbed molecules, as mentioned in the Introduction. Difficulties associated with the dewatering of sludges from biological effluent treatment are due to the presence of hydrophilic substances which are synthesised by certain organisms. In such cases, quite drastic treatment, such as freezing, may be necessary to cause flocculation.

Some attempts have been made to produce a quantitative theory of steric stabilization (see Lyklema[4] for a good review), but these are not yet sufficiently far advanced to give a detailed understanding of the process. However, for most purposes, it can be assumed that the interaction only becomes significant when the particle separation is about twice the thickness of the adsorbed layer (i.e. the adsorbed layers of two particles are in contact) and that the repulsion rises very steeply at smaller separations. The most important parameter is the thickness of the adsorbed layer in relation to the particle radius. If this is large, the particles cannot approach sufficiently close for van der Waals attraction to hold them together and the suspension is stable. With thinner adsorbed layers, permanent aggregates can be formed.

REFERENCES

1) Ives, K.J. and Gregory, J., Basic concepts of filtration. Proc. Soc. Water Treat. Exam., 16, 147, 1967.

2) Derjaguin, B.V. and Landau, L.D., Theory of the stability of strongly charged lyophobic sols and of the adhesion of strongly charged particles in solutions of electrolytes. Acta Physicochim URSS 14, 633. 1941.

3) Verwey, E.J.W. and Overbeek, J.Th.C., Theory of the stability of lyophobic colloids, Elsevier, Amsterdam, 1948.

4) Lyklema, J., Principles of the stability of lyophobic colloidal dispersions in non-aqueous media, Advan. Colloid Interface Sci. 2, 65, 1968.

5) La Mer, V.K., Introduction to coagulation symposium, J. Colloid Sci. 19, 291, 1964.

6) e.g. Mirnik, M., Discussion remarks, Disc. Faraday Soc. 42, 101, 1966.

7) Guggenheim, E.A., The conceptions of electrical potential difference between two phases and the individual activities of ions, J. Phys. Chem. 33, 842, 1929.

8) Andelman, J.B., Ion selective electrodes - theory and application in water analysis, J. Water Poll. Contr. Fed. 40, 1844, 1968.

9) Levine, S. and Smith, A.L., Theory of the differential capacity of the oxide/aqueous electrolyte interface, Disc. Faraday Soc. 52, 290, 1971.

10) Parks, G.A., The isoelectric points of solid oxides, solid hydroxides and aqueous hydroxo complex systems. Chem. Rev. 65, 177, 1965.

11) Parks, G.A., Aqueous surface chemistry of oxides and complex oxide minerals, Adv. Chem. Ser. 67, 121, 1967.

12) van Olphen, H., An Introduction to Clay Colloid Chemistry, Interscience, New York, 1963.

13) Gouy, G., Sur la constitution de la charge electrique a la surface d'un electrolyte, J. Phys. 9, 457, 1910.

14) Chapman, D.L., A contribution to the theory of electro-capillarity, Phil. Mag. 25, 475, 1913.

15) Loeb, A.L., Overbeek, J.Th.G. and Wiersema, P.H.
The Electrical Double Layer around a Spherical Colloid
Particle. M.I.T. Press, Cambridge, Mass., 1960.

16) Stern, O., Zur Theorie der Elektrolytischen Doppelschicht,
Z. Elektrochem. 30, 508, 1924.

17) Grahame, D.C., The electrical double layer and the theory
of electrocapillarity, Chem. Rev. 41, 441, 1947.

18) Levine, S., Mingins, J. and Bell, G.M., The discrete-ion
effect in ionic double layer theory, J. Electroanal. Chem.
13, 280, 1967.

19) Sparnaay, M.J., The Electrical Double Layer, Pergamon Press,
Oxford, 1972.

20) Smith, A.L., Electrical phenomena associated with the solid-
liquid interface, in Dispersion of Powders in Liquids,
Parfitt, G.D., Ed., Elsevier, London, 1969, 39.

21) de Groot, S.R., Mazur, P. and Overbeek, J.Th.G.,
Nonequilibrium thermodynamics of the sedimentation potential
and electrophoresis, J. Chem. Phys., 20, 1825, 1952.

22) Shaw, D.J., Electrophoresis, Academic Press, London, 1969.

23) Jacazio, G., Probstein, R.F., Sonin, A.A. and Yung, D.
Electrokinetic salt rejection in hyperfiltration through
porous materials.Theory and experiment, J. Phys. Chem.
76, 4015, 1972.

24) Wiersema, P.H., Loeb, A.L. and Overbeek, J.Th.G.,
Calculation of the electrophoretic mobility of a spherical
colloidal particle. J. Colloid Sci. 22, 78, 1966.

25) Cooper, F.C., Mees, Q.M. and Bier, M., Water purification by
forced-flow electrophoresis. J. Sanit. Eng. Div. ASCE.
91, SA6, 13, 1965.

26) Devereux, O.F. and de Bruyn, P.L., Interactions of Plane-
Parallel Double Layers, M.I.T. Press, Cambridge, Mass., 1963.

27) Honig, E.P. and Mul, P.M., Tables and equations of the diffuse
double layer repulsion at constant potential and at constant
charge, J. Colloid Interface Sci. 36, 258, 1971.

28) Derjaguin, B.V., A theory of the heterocoagulation, interaction and adhesion of dissimilar particles in solutions of
electrolytes, Disc. Faraday Soc., 18, 85, 1954.

29) Bierman, A., Electrostatic forces between nonidentical colloidal particles, J. Colloid Sci. 10, 231, 1955.

30) Frens, G., The Reversibility of Irreversible Colloids, Thesis, Utrecht, 1968.

31) Frens, G. and Overbeek, J.Th.G., Repeptization and the theory of electrocratic colloids, J. Colloid Interface Sci., 38, 376, 1972.

32) Hogg, R., Healy, T.W. and Fuerstenau, D.W., Mutual coagulation of colloidal dispersions, Trans. Faraday Soc. 62, 1638, 1966.

33) Gregory, J., An approximate expression for the interaction of diffuse electrical double layers at constant charge, JCS Faraday I submitted February 1973

34) Bell, G.M. and Peterson, G.C., Calculation of the electric double layer force between unlike spheres, J. Colloid Interface Sci. 41, 275, 1972.

35) Derjaguin, B.V., Theorie des Anhaftens kleine Teilchen, Kolloid Z. 69, 155, 1934.

36) McCartney, L.N. and Levine, S., An improvement on Derjaguin's expression at small potentials for the double layer interaction energy of two spherical colloidal particles, J. Colloid Interface Sci. 30, 345, 1969.

37) Ives, K.J. and Gregory, J., Surface forces in filtration, Proc. Soc. Water Treat. Exam. 15, 93, 1966.

38) Chu, B., Intermolecular Forces, Interscience, New York, 1966.

39) Hamaker, H.C., The London-van der Waals attraction between spherical particles, Physica, 4, 1058, 1937.

40) Gregory, J., The calculation of Hamaker constants, Adv. Colloid Interface Sci. 2, 397, 1969.

41) Tabor, D. and Winterton, R.H.S., The direct measurement of normal and retarded van der Waals forces, Proc. Roy. Soc. A312, 435, 1969.

42) Lifshitz, E.M., Theory of molecular attractive forces between solids, Soviet Physics JETP, 2, 73, 1956.

43) Parsegian, V.A. and Ninham, B.W., Applications of the Liftshitz theory to the calculation of van der Waals forces across thin lipid films, Nature, 224, 1197, 1969.

44) Parsegian, V.A. and Ninham, B.W., Toward the correct calculation of van der Waals interactions between hypophobic colloids in an aqueous medium, J. Colloid Sci. 37, 332, 1971.

45) Nir, S., Rein, R. and Weiss, L., On the applicability of certain approximations of the Lifshitz theory to thin films, J. Theor. Biol. 34, 135, 1972.

46) Visser, J., On Hamaker constants: a comparison between Hamaker constants and Lifshitz-van der Waals constants, Adv. Colloid Interface Sci. 3, 331, 1972.

47) Fuchs, N., Uber die Stabilitat und Aufladung der Aerosole, Z. Physik, 89, 736, 1934.

48) Reerink, H. and Overbeek, J.Th.G., The rate of coagulation as a measure of the stability of silver iodide sols, Disc. Faraday Soc. 18, 74, 1954.

49) Matijevic, E., Janauer, G.E. and Kerker, M., Reversal of charge of hyophobic colloids by hydrolyzed metal ions J. Colloid Sci. 19, 333, 1964.

50) Gregory, J., Flocculation of polystyrene particles with cationic polyelectrolytes, Trans Faraday Soc. 65, 2260, 1969.

51) Wiese, G.R. and Healy, T.W., Effect of particle size on colloid stability, Trans. Faraday Soc. 66, 490, 1970.

52) Ottewill, R.H. and Shaw, J.N., Stability of monodisperse polystyrene latex dispersions of various sizes, Disc. Faraday Soc. 42, 154, 1966.

53) Hull, M. and Kitchener, J.A., Interaction of spherical colloid particles with planar surfaces, Trans. Faraday Soc. 65, 3093, 1969.

54) Vincent, B., Bijsterbosch, B.H. and Lyklema, Competitive adsorption of ions and neutral molecules in the Stern layer on silver iodide and its effect on colloid stability. J. Colloid Interface Sci. 37, 171, 1971.

55) Vold, M.J., The effect of adsorption on the van der Waals interaction of spherical colloidal particles, J. Colloid Sci. 16, 1, 1961.

56) Schenkel, J.H. and Kitchener, J.A., A test of the Derjaguin-Verwey-Overbeek theory with a colloid suspension, Trans. Faraday Soc. 56, 161, 1960.

FILTRATION PRETREATMENT

Richard J. Akers

Loughborough University of Technology

INTRODUCTION

The throughput of a filtration process may be considerably enhanced by submitting the material to some form of treatment prior to the actual filtration.   This pretreatment may make an otherwise difficult separation possible.   The types of pretreatment available are those which modify the properties of the filter cake to increase permeability and/or prevent blinding of the medium and those which perform a preliminary concentration or "thickening" which reduces the load on the filter itself. Among the former techniques are those of coagulation with salts, flocculation with polyelectrolytes and the use of body aids and filter precoats.   The aim of the coagulation and flocculation reactions is to increase the state of aggregation of the particles comprising the filter cake.

The words flocculation and coagulation have long been used indiscriminately to describe the coagulating effect of ions on hydrophobic colloids.   La Mer[1] has suggested that the word coagulation be reserved for the effect brought about by reducing the zeta potential of a particle by changes in electrolyte concentration and that flocculation be reserved for the effect induced by long chain organic polymers which act by forming "bridges" between the solid particles.   This usage is not universally accepted but will be used here for the convenience of differentiating between two distinct processes, although they may both be induced by the same reagent.

A very full bibliography up to 1963 is given in "Solid-liquid

Separation"[1] and a more specialised review of flocculation by polymers by Audsley[2].

The flow of fluid through a cake of aggregated particles will be concentrated around the envelopes of the aggregates which will have the effect of reducing the magnitude of the specific surface term in the Kozeny-Carman equation and hence the resistance to fluid flow. Because the aggregates may be very open, i.e. porosities may be in excess of 90%, an increase in cake compressibility may well accompany this type of pretreatment. In the case of depth filtration, the increase in particle will enhance inertial capture and may lead to both a reduction in the rate of filter blockage and an increase in filter efficiency. The use of salts to coagulate suspensions is outlined in this chapter as it has been dealt with in detail by Dr. Gregory. The action of a body aid is to increase the permeability of the cake by altering the particle size distribution to one which has an increased porosity and hence permeability. This has been dealt with by Prof. Heertjees.

Particles may be concentrated by allowing them to sediment in either a gravity field, or if this is not sufficient, a centrifugal field. Other types of thickeners are met with such as the Shriver thickening filter press, but these will not be dealt with here.

An alternative, and increasingly more important, way of increasing the state of aggregation of suspended particles is to flocculate them with a solution of high molecular weight polymer. This technique is discussed in the chapter, as are those employed in thickening.

CHEMICAL PRETREATMENT

The Coagulation of Suspensions

The stability of a suspension is determined by the balance between the attractive "short range" or van der Waals forces and the repulsive electrical forces between the particles.[3] Surrounding a particle in a fluid is a region where an imbalance of ionic concentrations leading to an electrical potential gradient exists. The electrical potential at the hypothetical plane of shear between the part of this double layer attached to the particle and the bulk fluid is know as the Zeta potential and may be determined by studying the electrokinetic behaviour of the particle; e.g. electrophoretic mobility, streaming potential, etc. As a rough rule of thumb a zeta potential of about 40 mV is just sufficient to cause stabilisation by mutual repulsion of

the particles. It is well known that increasing the ionic concentration of the suspending fluid will bring about a decrease in
zeta potential and hence stability and that this effect increases
markedly with an increase in the charge carried by the ions used
(Schulze-Hardy Rule).

This effect has been made use of in the choice of practical
coagulating agents, the most common containing Fe^{+2}, Al^{+3}, Fe^{+3}
or Ca^{+2} and applied as aluminium sulphate sodium aluminate,
ferrous sulphate, chlorinated ferrous sulphate, lime, aluminium
chloride and aluminium chlorohydrate. The choice of a particular reagent is governed by the nature of the solid to be separated,
its previous history, the pH, concentration of other components,
ionic and non ionic, temperature, etc. Hence the actual choice
is largely governed by experience and semi-empirical testing and
is in large part dictated by economic considerations and the
acceptability of the coagulant in the processes to which it is
applied. The problem of acceptability is particularly important
in the effluent treatment field and potable water treatment.

As coagulation is a process brought about by binary collisions of particles, it is a kinetically second order process, i.e.
the rate depends on the square of the concentration of the solid[4]
and may therefore be very slow for dilute suspensions as encountered in potable water treatment. One way of surmounting this
difficulty is to use a coagulant such as aluminium sulphate which
slowly hydrolyses and precipitates aluminium hydroxide which is
itself coagulated and aids in coagulation of the unwanted solids
by both increasing the solids concentration and by entrainment.
With dilute suspensions the flocs produced may be very small and
may penetrate filter media such as deep bed sand filters. The
combination of coagulants and flocculants is being used for these
systems and shows promising results in maintaining filtration
rates whilst increasing the life of the filter bed.

The flocculation of Suspensions by Polymers

Natural polyelectrolytes have long been used as flocculants
although it is only recently that the term polyelectrolyte has
come into common use. The crushed nuts of an Indian tree[5] have
been used for cleaning water for centuries and isinglass has a
long history in the brewing industry. The action of these materials depends on the presence of water soluble polymers. Other
polymers used as flocculants are activated silica, soluble starches,
gelatine, guar gum and sodium alginate and more recently synthetic
polymers, mainly based on derivatives of polyacylamide, have been
developed. The loose open floc structures obtained with these
reagents are to be compared with the deposits of small sediment

volume and poor filtrability which may be produced by coagulants.

Effects related to the nature of the flocculating agent, the particles being flocculated, and operating conditions on the flocculation reaction are best thought of in terms of the overall flocculation reaction mechanism. Ruehrwein[6] proposed that the principal effect of polyelectrolyte flocculants was to form 'bridges' from one particle to the next as shown in Fig. 1.

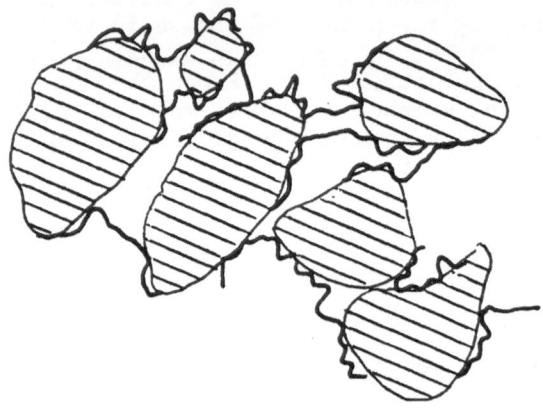

Fig.1   Particles Flocculated by Polymer Bridging.

This mechanism requires that the polymer chain be adsorbed from solution on to one particle and that, when another particle comes within close enough range for the extended polymer chain to be adsorbed on to it, a physical bridge will be formed between the two particles. This elementary floc then grows by bridging with other particles until an optimum floc size is reached. As discussed later, the flocculation reaction, unlike the coagulation reaction, is irreversible, hence it is misleading to think of the flocs as reaching an equilibrium size. The results of much subsequent work have confirmed this bridging hypothesis, and, indeed, Audsley[7] has published electron micrographs which it is possible to interpret directly as particles with polymer bridges between them. In general the coagulation reaction brought about by reduction of zeta potential is not important in polyelectrolyte flocculation. This is shown both by the efficiency of neutral polyelectrolytes as flocculants and also by the high efficiencies exhibited in some systems where the flocculant carries the incorrect electrical charge with respect to that of the material being flocculated. The main effect of highly charged polyelectrolytes on electrical double layer properties is to increase, not decrease, the zeta potential. For example, low molecular weight, i.e. 10,000 to 50,000, polyacrylates are used as dispersants, not flocculants. Because of their considerable amount of non-charged organic skeleton they are able to adsorb non-specifically, giving rise to an increase in zeta potential or even to charge reversal. On ground chalk whiting in water for example, sodium polyacrylates will give zeta potentials of about -60mV and calculation of the potential energy curve for the system shows that this is sufficient to account for the stabilization observed[8]. Charge effects may be more important with some lower molecular weight flocculating polyelectrolytes such as the polyethylene imines. In these materials the polymer chain length is insufficient for bridging to occur to any considerable extent, and alternative mechanisms such as zeta potential reduction must be invoked. Investigation has shown this to be true.

Flocculation is best represented in terms of a sequence of elementary reactions. It will be considered that a small quantity of a relatively concentrated solution of polyelectrolyte flocculant is to be added to a much larger volume of the suspension of the solid to be flocculated. The reaction sequence proposed is dispersion of the flocculant in the liquid phase of the suspension, diffusion of flocculant to the solid-liquid interface, adsorption of flocculant on to the solid surface, collision of particle bearing adsorbed polymer with another particle, adsorption of free polymer chain on to second particle forming bridge, and subsequent collision and adsorption reactions leading to build up of floc.

a. <u>Dispersion in the Suspension</u>. Polyelectrolytes in solution exhibit high viscosities and low diffusion rates because of their high molecular weights. This first step in the reaction sequence is analogous to a diffusion rate controlling reaction in solid-fluid catalysis. As the adsorption reaction is usually very much faster than the diffusion reaction, it is necessary to disperse the phases one into another mechanically. In practice, this is usually accomplished by having a short, vigorous premixing stage during the process to enable dispersion to be carried out in the shortest practical time.

b. <u>Diffusion to the Solid-liquid Interface</u>. The factors influencing this stage of the reaction are identical to those in Section a, above.

c. <u>Adsorption at the Solid Surface</u>. The bridging hypothesis requires that the polyelectrolyte chain be adsorbed on the solid surface at only a few points of attachment, with the bulk of the chain projecting into the liquid phase away from the solid-liquid interface. This model of adsorption of a flexible chain carrying many possible points of attachment has been treated in detail by Silberberg[9]. Among other things, Silberberg shows that this model of adsorption gives rise to an adsorption isotherm of the Langmuir type although not for the same reasons as in the Langmuir case; that the amount of adsorbed material increases with the chain length, i.e. the molecular weight of the adsorbed molecules, and that the number of points of attachment of the chain to the solid surface increases until an equilibrium compressed layer is achieved. This is shown diagrammatically in Fig. 2

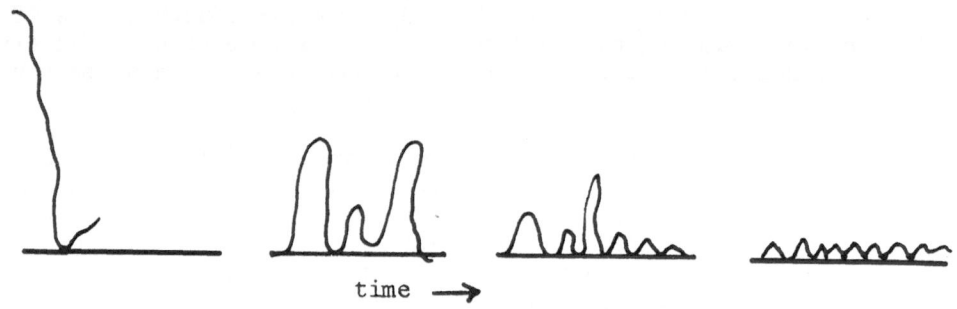

time →

Fig. 2. Compression of Adsorbed Polyelectrolyte layer.

It also follows from this theory that the nature of the adsorption process is dependent on the configuration of the polyelectrolyte in solution which in its turn depends upon the structure of the polyelectrolyte, on the solvent power of the solvent for the poly-electrolyte, on the solution pH, and on the concentrations of other ions present.

Because of the difficulties of obtaining materials of narrow molecular weight range and the lack of a really good analytical technique, little precise data is available concerning the adsorption of polyacrylamide derivatives. However, what is known is generally consistent with the Silberberg model. The initial adsorption is rapid and irreversible as predicted. Although the energy of adsorption of each link may be quite low and the adsorption at that particular point reversible, the probability of a large number of links leaving the surface at any one time becomes so low as to be discounted. It has been claimed that the high molecular weight components are selectively adsorbed from a

solution of wide molecular weight range. Silberberg claims that
this is more a question of the time taken to reach equilibrium
and that when equilibrium is reached there is no molecular weight
selectivity.

Bonding forces between the polymer and the solid surface
depend on the nature of both materials and the conditions within
the solution, but the most commonly invoked adsorption mechanism
is hydrogen bonding with evidence of ionic charge effects and
non-specific van der Waal's adsorption in some systems. Although
the energies involved in the hydrogen and non-specific bonding
mechanisms are very low for each active site, and this is confirmed by the very low temperature dependence of the adsorption
isotherms, the overall energy over many sites is sufficient to
account for the phenomena observed.

d.   Collision of Polymer-coated Particle with Another. It would
be expected that the rate of flocculation of particles having
loops of polymer at their surface would follow the Smoluchowski
type of rate equations for coagulation reactions.   These equations
were all derived for coagulating systems, but as they apply to
irreversible processes, they might be expected to be obeyed more
strictly by flocculating systems having made allowances for the
increased collision diameter of the particles due to the presence
of the polymer loops extending from their surfaces.

e.   Subsequent Collisions Leading to Flocs.   The points discussed in Section d, above would also apply to the final stages
of the reaction.   Sutherland[10] has described a computer simulation of floc structure on the basis of the Smuluchowski model.
The irregularly shaped flocs of very high porosity obtained are
consistent with those observed in actual flocculation reactions.

Factors Effecting Polyelectrolyte Flocculation

A detailed study of the flocculation of silica by polyacrylamides has been made by Linke and Booth[11], and whilst systems
differ widely in their behaviour, their conclusions are generally
valid and illustrate the properties of flocculating systems.

(a) Polymer concentration

(i)   At optimum dosage practically all the polymer is
adsorbed.

(ii) The optimum dosage is proportional to the solid
surface area over a wide range.

(iii) The stability of the flocs increases with solids
concentration.

98

La Mer and Healey[1] have shown that optimum flocculation occurs when half the available solids are covered. This requires a grossly excessive amount of flocculant assessed by economic considerations and the actual dosage required must be determined by experiment.

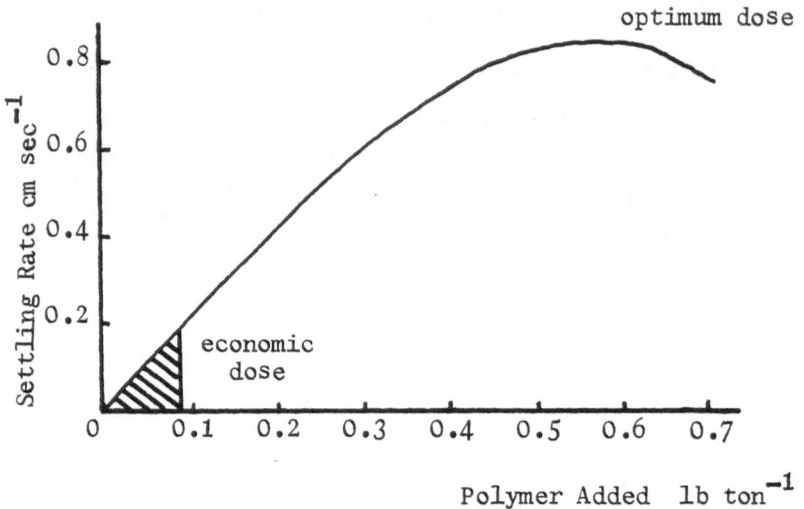

Fig. 3.   Effect of Flocculant Dose.

(b) Agitation.   The effects of agitation are complex. Prolonged agitation may lead to folding back of polymer on to the surface, also poor flocculation and high shear rates would lead to mechanical disruption of flocs.   However, a degree of agitation is necessary to bring the particles close enough for bridging to occur.   Practical systems are usually designed to ensure a rapid dispersal of the flocculant in the pulp followed by slow agitation to encourage flocculation.

(c) pH.   pH will affect the amount and type of charged site present on a solid surface and the amount of free charge and hence the configuration in solution of the polymer.   In general, fully flexible isolated macromolecules will exist in solution as randomly coiled chains.   If, however, charged sites are introduced, mutual electrostatic interaction will cause the chain to become extended and acquire a degree of rigidity.   These changes will, in general, lead to enhanced flocculation.

Although, as Linke and Booth show, a cationic polymer tends to be more efficiently adsorbed on to an anionic surface than a neutral, than an anionic polymer, the adsorption step is largely

due to non specific van der Waals mechanisms and may not be
strongly influenced by considerations of charge. Manufacturers
claim that the choice of reagent is not governed solely by considerations of pH and that many reagents are effective over a
wide range of conditions.

(d) Molecular Weight of Flocculant. Increase in molecular
weight leads to an increase in the optimum flocculant/solid ratio.
This leads to stronger flocs but is offset by the economic disadvantages of higher dose rates and the increased cost of higher
molecular weight polymers. This increase in efficacy with
molecular weight has also been correlated with a decrease in
solubility of the polymer, the optimum flocculation being found
when the polymer is about to be precipitated from solution. This
has been demonstrated by correlating the effectiveness of gelatine
and glue as flocculants with the degree of cross-linking
introduced.

(e) Other factors. Among other factors affecting flocculation are ionic strength of the liquid phase and the number and
distribution of active groups on the polymer.

Specific Polyelectrolyte Flocculants

(a) Polymers based on polyacrylamide. These materials may
be prepared by the free radical polymerisation of acrylamide

$$-\left[\begin{array}{l} CH_2 - CH - \\ \quad\quad | \\ \quad\quad C = O \\ \quad\quad | \\ \quad\quad NH_2 \end{array}\right]_n \quad\quad \text{polyacrylamide}$$

Polyacrylamide itself is neutral and it may be given anionic
character either by partial hydrolysis or copolymerisation with
acrylic acid

$$-\quad CH_2 - CH - CH_2 - CH - CH_2 - CH -$$
$$\quad\quad\quad | \quad\quad\quad\quad | \quad\quad\quad\quad |$$
$$\quad\quad\quad C=O \quad\quad\quad C=O \quad\quad\quad C=O \quad\quad \text{an anionic copolymer}$$
$$\quad\quad\quad | \quad\quad\quad\quad | \quad\quad\quad\quad |$$
$$\quad\quad\quad OH \quad\quad\quad NH_2 \quad\quad\quad NH_2$$

and cationic character by quarternisation of some of the amino groups in the chain.

$$- CH_2 - CH - CH_2 - CH - CH_2 - CH -$$

$$\begin{array}{ccc} C{=}0 & C{=}0 & C{=}0 \\ NH_2 & {}^{+}NH_2R & NH_2 \end{array}$$

Polyacrylamides are available in either neutral cationic or anionic forms with molecular weights up to several million and with up to 35% of the acrylamide residues replaced by charged sites. They are very effective flocculants, rather expensive but used at very low dose rates. Some manufacturers are marketing special grades for potable water treatment with very low residual monomer concentration. Other polymers are used as flocculants but on a smaller scale.

(b) Guar Gums[2]. These are neutral water soluble polysaccharides useful as flocculants. They are non ionic in character and are not markedly influenced by variations in pH and ionic strength. They have been used in uranium ore processing. Care must be taken to prevent enzymic degradation of their solutions on storage. This can be prevented by adding reagents such as citric or oxalic acid.

(c) Glue and Gelatine.[2 12] These are widely used for mineral pulps since they are cheap and easily available. On a weight basis they are much less effective than the polyacrylamides but may be more economical. Gelatine is unique as a flocculant in that it will flocculate almost any suspension. This could in part be due to its multiplicity of ionic groupings, their zwitterion character and the many hydrogen bonding sites. Cross linking has been shown to profoundly influence its action.

(d) Starches[2]. These very cheap reagents have long been used in treating colliery wash waters. Starches from different sources differ widely in their properties. This difference is almost certainly not due to differences in phosphate content as was once claimed but to the molecular weight, the relative amounts of amyloses and amylopectins and the amount of cross linking.

(e) Alginates[2]. Sodium alginate, extracted from seaweed has been used as a flocculant, particularly in potable water treatment where its lack of toxicity is a considerable advantage. In hard water it has the disadvantage of precipitating as calcium alginate.

(f) Tannins.[2] Tannin derivatives have been used as flocculants in raw water, sewage and general effluent treatment situations. They are subject to degradation reactions on storage and are most effective under conditions of acid pH.

Miscellaneous Techniques

(a) Freezing.[13] Freezing and thawing has been successfully applied to the difficult organic sludges found in sewage and water treatment. Although technically successful the technique is too expensive for sewage but has been given a trial at a water works[8]. In this plant, the cost of treating a 4% solids sludge was only 0.09d. per 1000 gallons more than conventional lagooning. The operating conditions are critical. The sludge is converted to fine hard grains which are easily separated.

(b) Heating.[13] Sewage sludges have been successfully treated by "pressure cooking" at about $400^{\circ}F$ and 150p.s.i. However, providing the necessary amount of heat is an expensive process.

(c) Aeration. Injection of air has been shown to bring about flocculation in many cases but in, for example, some digested sludges, inhibits flocculation.

(d) Mechanical and Ultrasonic Vibration.[14] Ultrasonic vibration has been successfully applied in ore beneficiation techniques, sewage processes and colliery wash plants. However, there is much dispute as to the efficiency of the process, some workers claiming no benefit at all. Coagulation is a reversible process and at any particular shear rate there must be an equilibrium between shear induced coagulation and dispersion. This would appear to be very dependent on the geometry of the system, the frequency of the vibration and the properties of the slurry. The same conditions would apply to mechanical vibration.

(e) Electric and Magnetic Treatment. Electrolysis with aluminium or ion electrodes has been shown to have a coagulating effect on slurries. This is due to release of Al^{+3} and Fe^{+3}. Magnetic powders added to slurries which are then exposed to a magnetic field have also been shown to encourage coagulation.

(f) Radiation Techniques. In the period 1960-64 the United States Atomic Energy Commission published a series of reports[15] on the effect of ionising radiations on sewage sludges and montmorillonite clays. From the observed increase in sedimentation rates it was concluded that the technique would be valuable if the isotope costs could be reduced to about $1/6$ of the current figure.

Mention is made in the literature of the use of U.V. radiation, but no details are given.[13]

(g) <u>Conclusions</u>. Apart from the freezing technique, the economics of which are open to debate, the other techniques are at present at the speculative stage. Mechanical and ultrasonic vibration, aeration and probably heating are means of performing mechanical work on the system whilst heating, electrical and radiation techniques affect the surface properties of the solid materials.

Assessing the Performance of Flocculants

Settling rates, filtration rates and sediment volumes have all been used to assess flocculants. Although these phenomena are all interrelated, they do not always measure the same physical property of the system. It is therefore necessary to settle on one reproducible procedure for comparative testing.

The easiest and one of the most common methods is the measurement of settling rate. When a flocculated material sediments, a curve of the type shown in Fig. 2 is obtained, from

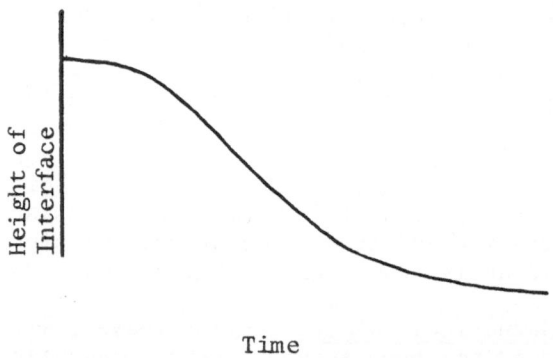

Fig. 4.   A sedimentation curve.

which may be obtained the settling rate and sediment volume. Place 500 ml of suspension in a glass measuring cylinder, add the flocculant as a 0.05% solution made up freshly from a stock 1 solution, stopper the tube or close with the hand and gently invert and restore to the vertical position three times. Start a stop watch.   Record the height of the mudline as a function of time every 2 cm. or so until compaction occurs.   Plot the results to give the settlement curve.   To select the correct dosage, use differing quantities of flocculant solution until the optimum quantity is found.

La Mer[1] has advocated the use of refiltration rates in the study of flocculation and has claimed that a relationship of the form

$$Q - Q_o = K \theta^4 (1 - \theta^4) \qquad\qquad (1)$$

where Q = filtration rate

Q_o = filtration rate for untreated material

θ = fraction of surface sites covered by polymer

is uniquely characteristic of a flocculated system.   However, later work by Kitchener[16] does not support this conclusion.

Most workers on flocculation have been concerned with the study of sedimentation processes and have not studied in detail the effect on filtration of the various parameters involved in flocculation reactions.

If filtration tests are required, it has been claimed that a vacuum filter leaf will give more reliable results than a Buchner funnel, although both are used.   Under standard conditions the amount of filtrate produced is measured at regular time intervals and it may also be instructive to measure the residual water content of the cake.

A very convenient test for flocculant selection is the Capillary Suction Time (C.S.T.) test of Baskeville and Gale.[17] The C.S.T. is the time taken for a certain amount of liquid from the suspension being tested to diffuse into a sheet of filter paper through the suspension/filter interface.   A very convenient apparatus is commercially available in which this time is measured electrically.   Packham[18] has found that this test correlates well with other assessments of filtration performance.

THICKENING

Gravity thickening

The performance of a solid–liquid separation device may be

considerably improved if the suspension to be separated is concentrated or 'thickened' prior to being fed to the separator.
This thickening may be carried out by means of continuous centrifuges, hydrocyclones and gravity thickeners. The choice of
technique will depend on the particle size, feed and discharge
concentrations, slurry properties. This contribution will
describe the sizing of continuous thickeners.

    An idealised model of a settling suspension may bve con-
sidered as having two interfaces, the upper one between clear
liquor and the sedimenting particles and the lower one between
the sedimenting particles and a layer of compacted sediment.

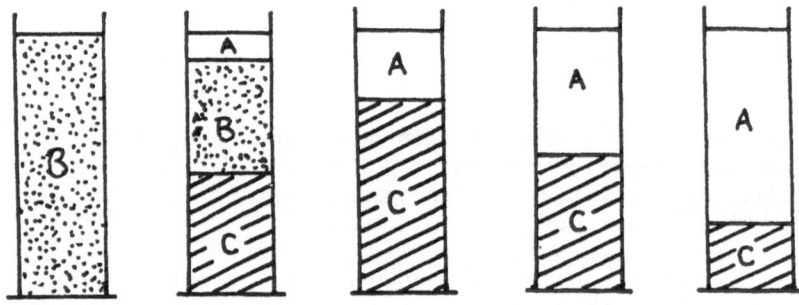

Fig. 5.   Batch Sedimentation.

In the case of a suspension of solids with a wide size distribution, a gradual concentration gradient is found. If, however,
the suspension is coagulated of flocculated the sedimentation is
much closer to the ideal model. The height of the upper boundary
as a function of time may be plotted as a sedimentation curve
(Fig. 6) in which it can be seen that during free settling the
upper interface moves downwards at a constant rate until it meets
the lower interface. This interface will then move downwards
increasingly slowly until the sediment achieves its equilibrium
compactness. With some highly compressible sediments this
equilibrium may take a considerable time to attain.

    The velocity of free sedimentation is a function of concen-
tration of the suspension. An isolated particle will fall at a
rate described by Stokes' law provided that the sedimentation

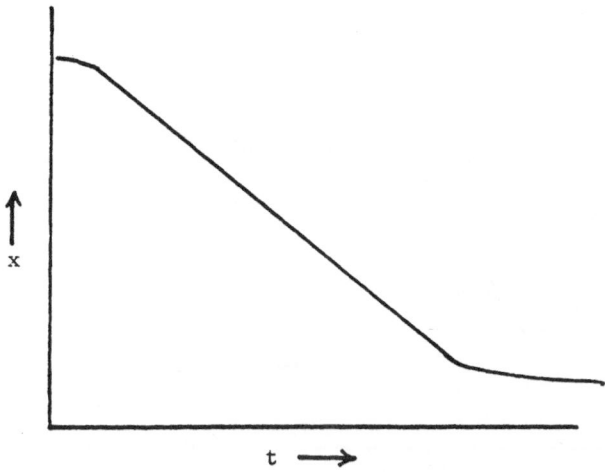

Fig. 6.   A Settling Curve.

velocity corresponds to a sufficiently low Reynolds number, say
< 2.   Even ignoring complications like wall effects, increasing
the particle concentration leads to complex changes in sedimentation behaviour.   At low concentrations hydrodynamic interaction
between particles, e.g. 'pair' formation will lead to an enhance-
ment of the sedimentation rate, whilst at higher concentrations
the return displacement flow of liquor will lead to an apparent
reduction in sedimentation velocity because the particles are
moving downwards in a fluid with a net upward velocity.   For
systems of wide particle size range there is the additional com-
plication of large particles catching up with and 'interacting'
with smaller ones and in flocculated systems there is the limited
flow of fluid within and through the floc so that the floc may be
considered as a sedimenting envelope.   It is because flocs tend
to reach an equilibrium size that their sedimentation behaviour
is closer to that of the idealised model shown in Fig. 5.

A suspension may be characterised by measuring the sedimentation velocity as a function of concentration, see Fig. 7.

Design from Sedimentation Data — Coe and Clevenger

This method was used by Coe and Clevenger[19] in the first
published method of thickener design.   Consider a layer of pulp
in the sedimentation test where the particle concentration is c
so that rate of fall of solids in the sedimenting layer is given
by

$$W = A\ uc \qquad\qquad (2)$$

106

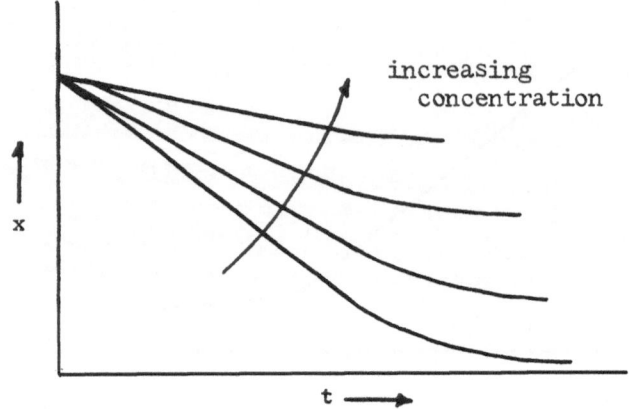

increasing
concentration

x

t →

Fig. 7.   Settling Rate as a Function of Concentration.

where W is the weight of solids sedimenting at velocity u over a
cross sectional area A.

Hence: $\quad \dfrac{A}{W} = \dfrac{(\frac{1}{c})}{u}$ $\hspace{6cm}$ (3)

For a continuous thickener the mass balance requires that the
rate of removal of the underflow must be taken into account,

hence: $\quad \dfrac{A}{W} = \dfrac{(\frac{1}{c})}{u + v}$ $\hspace{6cm}$ (4)

where v is the downward velocity due to removal of the underflow.
As the settling rate is a function of concentration only and the
concentration in the settling zone is considered constant

$$\frac{A}{W} = \frac{\frac{1}{c_u}}{(u_u + v)}$$ $\hspace{6cm}$ (5)

where u_u corresponds to concentration c_u

Hence by elimination we obtain

$$\frac{A}{W} = \frac{(\frac{1}{c} - \frac{1}{c_u})}{u - u_u} \tag{6}$$

and if the underflow is considered as exhibiting no sedimentation, i.e. $u_u = 0$

$$\frac{A}{W} = \frac{\frac{1}{c} - \frac{1}{c_u}}{u} \tag{7}$$

Hence from this procedure the area of thickener A that is required for a given weight of material W may be calculated from a determination of u, the sedimentation velocity.

Obtaining the Batch Flux Curve. In order to save the need to measure the sedimentation velocity at all concentrations, Kynch[20] proposed the construction of a batch flux curve. This analysis is based on the assumption that the sedimentation velocity is a function solely of the local solids concentration. Consider a mass balance on a layer of unit cross sectional area in the sedimenting system

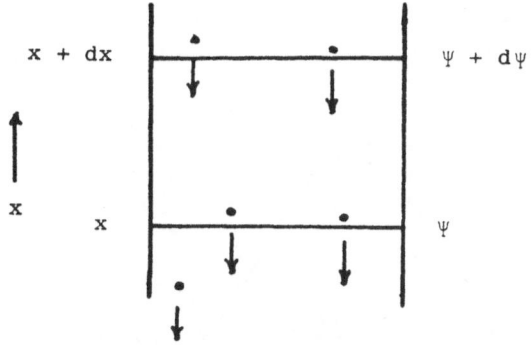

where dx/dt is the velocity of propagation of a zone of constant concentration within the mud zone and ψ is the solids flux at level x. By mass balance the rate of accumulation of material is the rate difference between the rate of mass in and out of the zone, hence:

$$\psi + (\frac{\partial \psi}{\partial x}) \, dx - \psi = \frac{\partial (c \, dx)}{\partial t} \tag{8}$$

i.e.
$$\frac{\partial \psi}{\partial x} = \frac{\partial c}{\partial t} \tag{9}$$

108

but $\quad c \;=\; f(x,t)$

$$\therefore \quad dc \;=\; \left(\frac{\partial c}{\partial x}\right) dx \;+\; \left(\frac{\partial c}{\partial t}\right) \; dt \tag{10}$$

but in a zone of constant composition, $\quad dc = 0$

$$\therefore \quad \left(\frac{\partial c}{\partial x}\right) \frac{dx}{dt} \;+\; \frac{\partial c}{\partial t} \;=\; 0 \tag{11}$$

hence by substitution

$$\left(\frac{\partial c}{\partial x}\right) \frac{dx}{dt} \;+\; \frac{\partial \psi}{\partial x} \;=\; 0 \tag{12}$$

i.e. $\quad \dfrac{dx}{dt} \;=\; \dfrac{-\partial \psi}{\partial c} \tag{13}$

consequently the velocity of propagation of the zone of constant concentration is given by

$$v \;=\; - \; \partial \psi / \partial c \tag{14}$$

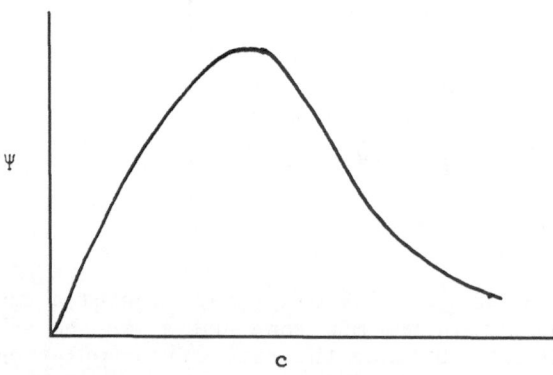

Fig. 8.   The Batch Flux Curve

since $\left.\begin{array}{l}\psi = 0 \\ c = 0\end{array}\right\}$ and $\left.\begin{array}{l}\psi = 0 \\ c = \text{max} \\ u = 0\end{array}\right\}$ the curve will show a maximum.

Determination of Concentration at the Interface of mud zone in a batch sedimentation. Consider a batch sedimentation system of initial concentration and height c_o and x_o. After a time t when the following rate period has been reached, a zone of constant concentration c has reached the interface after moving upwards at a rate v.

Since the zone has reached the interface, all the solid material must have passed through it with a relative velocity $(u + v)$. The simple mass balance follows:

$$c_o \, x_o \, A \; = \; c(u + v) \, t \, A \tag{15}$$

i.e.
$$c \; = \; \frac{c_o \, x_o}{ut + vt} \tag{16}$$

Since u is the slope of the tangent to the sedimentation curve $(ut + vt)$ can be seen from the following construction to be x_t, the intercept of the tangent with the x axis.

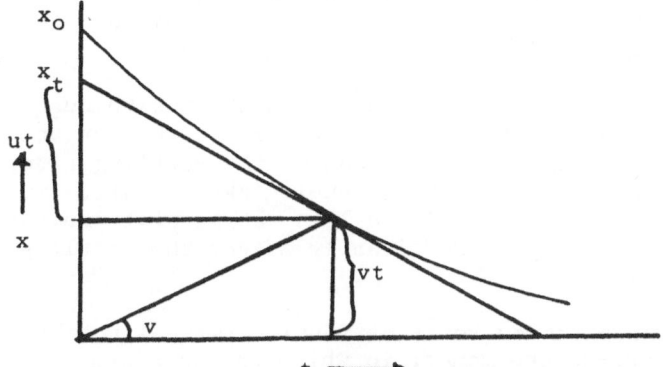

Fig. 9.   Construction to Determine Interlacial Concentration.

Hence the concentration, d, of the slurry is given by:

$$c \; = \; \frac{c_o \, x_o}{x_t} \tag{17}$$

and the velocity of sedimentation, u, is the negative of the slope of the tangent.

$$u \; = \; - \frac{dx}{dt} \tag{18}$$

Deviations from Ideal Behaviour.   The Kynch analysis will apply if the sedimentation rate is a function of the local concentration only.   Frequently this is not the case and the consequent deviations from the ideal may be for the following reasons:

(i)   At very low concentration (less than about 5% ....) in dispersed non-flocculated systems, size segregation occurs. The settling rate, which is dependent on both particle size and concentration, will then become a function of position and time as well as of concentration.

(ii)   In some flocculated systems, age may affect the degree of flocculation and, hence, the sedimentation rate.

(iii) In flocculated systems, neighbouring flocs will come into contact with each other if the concentration is high enough. When this occurs, a complete floc network (or sludge) may be formed.   The presence of a boundary (like the wall of the sedimentation vessel) will tend to support the slurry and so reduce the sedimentation rate.   This can strongly influence the behaviour in the sedimentation test vessel of such materials as sewage sludge.   Since there is mechanical contact between the floc particles, the depth of the sediment can strongly influence the settling velocity.

(iv)   Channeling (the formation of rivulets in the mud through which return-flow liquor can return easily to the surface) can occur.   This, of course, speeds up the settling rate in the mud zone.   Because the channels take a finite time to develop, any form of agitation like raking will prevent their formation and may significantly affect the settling rate.

The shape of the batch curve is obviously critical to the design of a thickener, so the manner in which the data are obtained is important.   Three methods of obtaining data from batch tests have been considered by Scott:

The consequences of Kynch's analysis are

i   That a zone of constant concentration is propagated upwards at a constant rate.   This permits scale up of data.

ii   That this rate is defined by the equation

$$v = \frac{dx}{dt} = -\frac{\partial \psi}{\partial c} \tag{19}$$

iii   That a batch flux curve can be obtained from one sedimentation experiment by the following simple construction

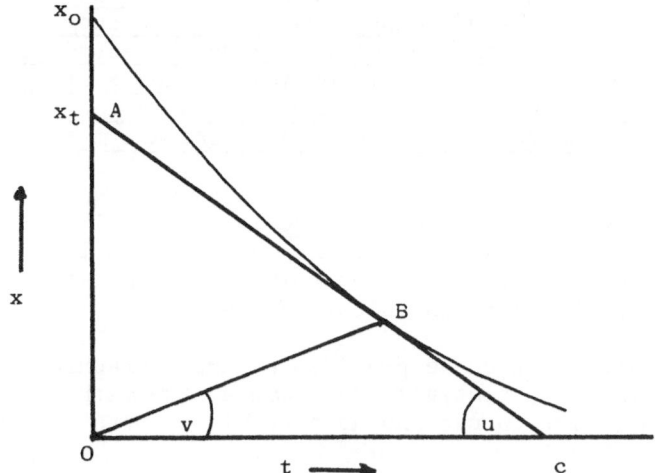

Fig. 10.   Construction to Obtain Batch Flux Curve
from Sedimentation Curve.

AC represents the sedimentation velocity u, drawn as a tangent to
the batch settling curve, A the slurry concentration $\frac{c_o \, x_o}{x_t}$
where c_o and x_o refer to the initial conditions.

$$u(c) \;=\; \frac{dx_t}{dt} \quad \text{and} \quad c \;=\; \frac{c_o \, x_o}{x_t} \quad \text{may be obtained and}$$

hence $\Psi = cu$.

Hence a batch flux curve (Fig. 9) may be drawn

(i)   The observation of the settling rate at different concen-
       trations (the Coe and Clevenger method).   A dilute slurry
       is tested, some clear liquid is removed by decanting to
       increase the concentration of the slurry and a second test
       is performed.   The procedure is repeated until a complete
       flux curve has been plotted.   Data obtained by this tech-
       nique lead to a thickener which is too small.

(ii)   Yoshioka[22] recommends that the slurry is stirred during
       sedimentation.   This reduces the rate of sedimentation and
       stimulates more closely the conditions in a real continuous
       thickener.   This leads to a more realistic design.

Scott[23] [24] reported: For non-ideal pulps in which the settling rate is not a unique function of solids concentration, the estimated size of thickener required to perform a specific duty is likely, in general, to be a function of the method selected to measure settling rates and, because of the various modes of settling observed for a flocculated pulp, the most reliable method will be the one in which the actual thickener conditions are most closely simulated.

Use of the Flux Curve - Yoshioka Method.   Yoshioka[22] has described a method for calculating thickener areas which is similar to that of Coe and Clevenger but being based on batch flux curves is easier and more convenient to use.

The use of a batch flux curve provides a simple graphical procedure which, although essentially the same as the method described above, is easier to use and as a result is more popular.

As a result of discharge, there will be a bulk movement of thickener contents downwards with a velocity v (where $v = D/A$). Hence there is an additional flux cv through a section of the thickener where the concentration is c.   If the batch flux corresponding to this concentration is ψ_b, then a solids continuity equation can be written as:-

$$ F\, c_f \;=\; A\,(\psi_b + vc) \;=\; D\, c_D \qquad\qquad (20) $$

For unit area

$$ \frac{(F)}{(A)}\, c_f \;=\; \psi_b \;+\; vc \;=\; v\, c_D \qquad\qquad (21) $$

Equation (20) shows that the discharge concentration is inversely proportional to the discharge rate.   Also it depends only on c and not on the depth. c must be constant in the zone immediately below the feed inlet.   This shows that the mechanism is independent of the height of the vessel.

The flux for a continuous thickener is seen from equation (21) to be the sum of the batch flux and the flux for discharge, vc:

i.e. ψ total $= \psi$ batch $+$ vc $\qquad\qquad (22)$

The concentration at the minimum flux will be the 'capacity limiting concentration' and it is, therefore, important to examine the conditions as ψ min (see Figure 11).

It is worth noting here that at this condition $v = 0$, and from the Kynch theory

$$- \frac{d\Psi}{dc} = v = 0 \qquad (23)$$

Feed flux at capacity $= \Psi \min = v\, c_D$

$$\therefore \quad c_D = \Psi \min / v \qquad (24)$$

Since the slope of the discharge line is v, the discharge concentration is c_{D1} on the graph. The maximum discharge concentration can thus be found simply from the above diagram for a given thickener area and discharge rate. However, for a given value of c_D and discharge rate it is difficult from this construction to find the area required, since v is found by dividing the discharge rate by A.

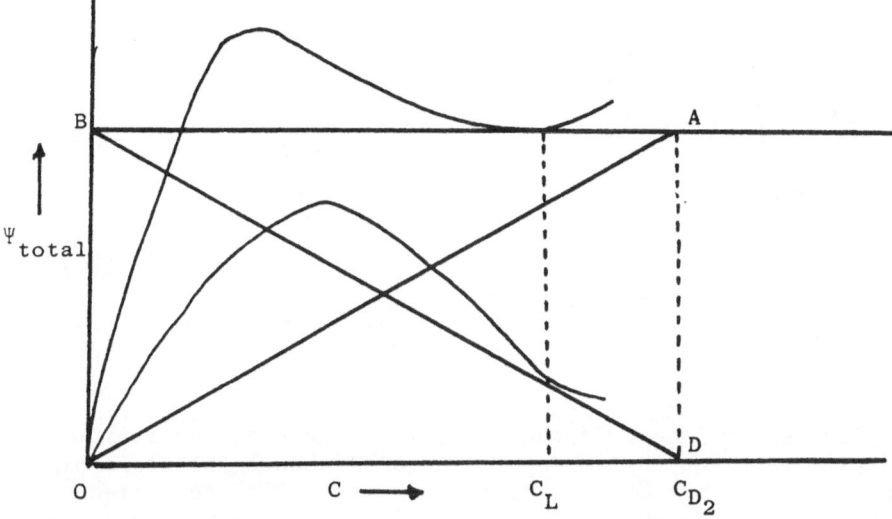

Fig. 11.   Flux Curve for a Continuous Thickener.

Yoshioka overcame this difficulty by using the ingeniously simple construction shown in Figure 12.

If c_{D1} is the discharge concentration, draw a straight line through c_{D1} which is a tangent to the batch flux curve. The line should meet the axis at Ψ min.

Then since $\Psi \min = \frac{F}{A}\, c_f = \frac{D}{A}\, c_D$ the area can easily

114

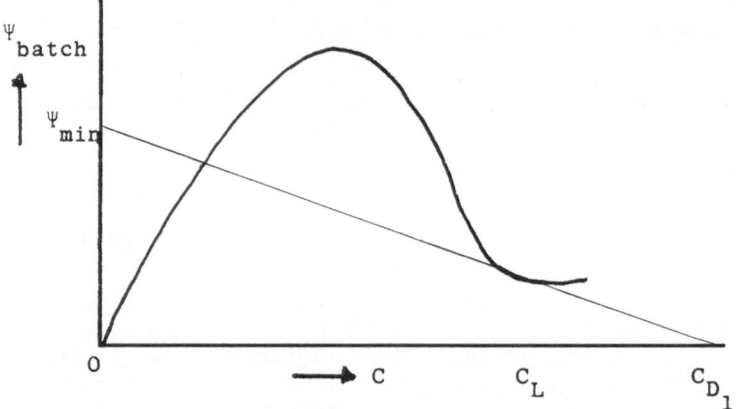

Fig. 12.   Yoshioka's Construction.

be determined.   The theory behind the Yoshioka construction can
be seen by considering Figure 12

At the limiting conditions of flux

$$\frac{d\Psi}{dc} \text{ total } = 0 \tag{25}$$

From equation (22)

$$\frac{d\Psi}{dc} \text{ batch } + v = 0 \tag{26}$$

$$\therefore \quad \frac{d\Psi}{dc} \text{ batch } = -v. \tag{27}$$

Thus the gradient of the batch flux curve at the limiting
condition is minus the slope of line OA.   Hence the slope of
line OA is the same as that of BD but with a different sign.
Since it is also known that the limiting conditions Ψ min $= vc_{D1}$,
then we can see that the position of the line is determined
as that which will coincide with BD.

It can be seen from the above that the Yoshioka construction
is the easiest to use in determining the area of a thickener, and
it is certainly the recommended technique.

It should be remembered that the depth of a system which is
flocculated and therefore likely to be under compression will
influence the settling rate and thus the position of the flux
curve.   Such an effect would influence the area required for
thickening.

The thickeners considered by Scott and Alderton were not, of course, controlled by thickening rate, but rather by the rate of clarification.

Depth of Thickener.   Robins[25], Hassett[26] and Yoshioka[22] say that the depth of a thickener does not affect its operation. Their ideas are based, of course, on the assumption that absolutely steady state conditions prevail; and in practice a finite depth is necessary to allow for fluctuations in feed conditions. It is not universally accepted that depth has no effect; both Moncrieff[27] and Fitch[28] recently proposed methods of determining this parameter.

It was accepted for some time that whilst the thickener throughput was determined by its area, the concentration of the underflow was determined by the 'detention time'.   This concept of detention time is meaningless.   In continuous thickening the underflow concentration is affected by the draw-off rate which in turn affects the detention time.   Low underflow rates are often associated with greater thickener depths and hence greater weight of solids per unit area.

There are, of course, a number of different regions in a thickener as shown in the diagram below:-

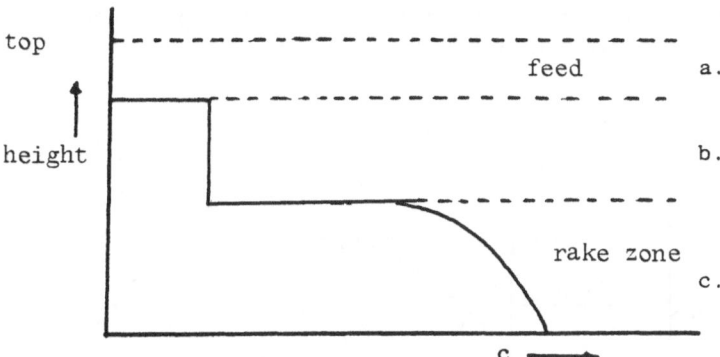

Fig. 13.   Zones in a Thickener.

The depth of a region (a), the clarification region, will constitute a design problem only if the feed dilution is so great that a cohesive solids structure is not formed.   A zone settling region (b) will be observed.

If there is compression in the thickener then the problem of predicting the depth (c) on the basis of batch tests remains unsolved.   Fitch has proposed a procedure, but it is untested and so will not be discussed here.   The only reliable methods, in fact, are to conduct a pilot plant test or use a semi-continuous procedure.

In practice, the height of a thickener is usually between 4 and 10 feet depending on the scale of the operation. It is unlikely that the height of a compaction zone will exceed 3 feet.

THE HYDROCYCLONE

If the particles in a suspension are too fine to sediment at an appropriate rate for gravity thickening or if space is at a premium, the body force on the sedimenting particles may be increasing by subjecting them to a centrifugal field. An additional advantage of centrifugal devices is that they do not involve a very high liquid hold-up, for example, a thickener may contain in excess of 1,000,000 gallons. The centrifugal field may be generated by causing the fluid to take a rotating flow path, as in the hydrocyclone or to be contained in a rotating drum as in the centrifuge. All separation devices have a finite efficiency for any given particle size and operating condition, and this efficiency must be taken into account in specifying a particular machine for a given application. In general the efficiency of such a device decreases with decreasing particle size. The efficiency of such a device is conveniently expressed in the form of a grade efficiency curve where the fraction separation at a particular size is plotted against particle size.

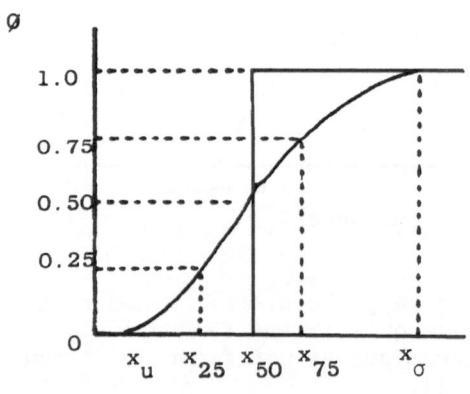

Fig. 14. A Grade Efficiency Curve.

In Figure 14 the curve $\phi(x)$ represents the coarse grade efficiency of a separating device, i.e. the fraction ϕ of feed material of particle size x being retained in the coarse fraction, or in the case of a thickener, in the underflow. The vertical line at x_{50} represents the efficiency of an ideal separator, all particles $> x_{50}$ being retained and all $< x_{50}$ passed. The aim of any separator designer is to obtain a grade efficiency curve as close to this ideal as possible. From the curve it can be seen that there is no separation of material finer than x_u, and complete retention of that greater than x_σ. A simpler and widely used criterion of efficiency is the ratio ϕ_{75}/ϕ_{25}, and an even more simple one frequently used in the discussion of centrifugal separators is ϕ_{50}, the cut point.

The essential features of the hydrocyclone are shown in Figure 15.

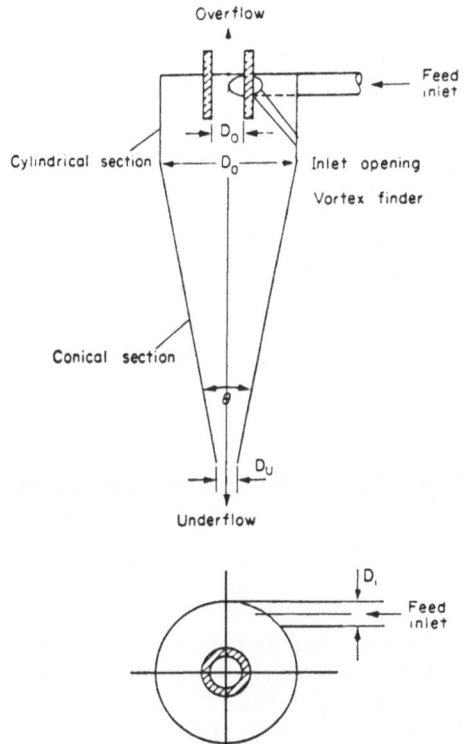

Fig. 15.   The Hydrocyclone.

The main vessel is coni-cylindrical with a tangential inlet to
the cylindrical portion.   At the base of the cone is an outlet
for the underflow fraction whilst a cylindrical pipe, the vortex
finder projects downwards into the cylindrical upper portion.
This pipe removes the overflow.   In operating the cyclone means
will be provided to throttle the underflow as the flow rate and

hence fluid pressure energy which produces the rotary fluid motion
enables separation to occur.   Figure 16 shows an idealised version
of the flow path within a cyclone.

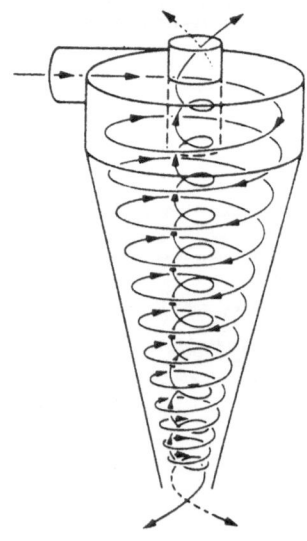

Fig. 16.   Idealised Flow Pattern Within a Hydrocyclone.

The fluid entering by the tangential inlet flows downwards in a
spiral in the outer part of the device.   Within this is a return
upward flow rotating in the same direction, the two flows being
separated by the locus of zero velocity (LZVV).   A small fraction
of the downward flow adjacent to the walls is removed through the
bottom exit and will carry with it all those particles that have
migrated outwards, due to the centrifugal force they experience.
This ideal flow is complicated by eddy flow and short circuit
flow.   It can be seen that this flow pattern causes considerable

shear and this is of practical significance in considering the hydrocyclone in relation to flocculated suspensions. Because of mass conservation considerations, the tangential velocity of the fluid within a cyclone increases with decreasing radius, which also creates a shear gradient. This is an important consideration in calculating the efficiency of cyclones and in a significant point of difference from centrifuge theory.

Bradley[29] has discussed cyclone theory in detail and has put forward the following relationships

$$x_{50} \propto D^x \quad \text{where} \quad 1.36 \leqslant x \leqslant 1.52 \quad \text{and D is the cyclone diameter}$$

$$\Delta P \propto D^y \quad \text{where} \quad P \text{ is the pressure drop and} -3.6 \leqslant y \leqslant -4.1$$

i.e. smaller x_{50}, the smaller the cyclone and the greater the pressure drop for a given flow.

Other relations given by Bradley are

Volumetric flow rate $Q = D^z \quad 1.8 \leqslant z \leqslant 2.0$

Diameter of feed aperture D_i

small d_{50}, small D_i, but high ΔP

large D_i, low efficiency, high throughput

suggests $D_i \backsim D/6$ to $D/7$

Diameter of vortex finder D_o

$D/8 \leqslant D \leqslant D/2.3$

Length of vortex finder

long vortex finder — reduces short circuit flow but causes entrainment of unseparated large particles in outer vortex.

Bradley suggests about $D/3$.

Two general approaches to efficiency have been considered. The 'equilibrium' orbit theory considers the orbit at which a particular size of particle experiences equilibrium between the

centrifugal force and Stokes' drag. The critical size of
particle is that which has an orbit of the same radius as the
locus of zero velocity in the vessel (LZVV).

The 'residence time' approach, initiated by Rosin, Rammler
and Intelmann[20] and developed by Rietma[31] considers the time taken
by a d_{50} particle from the inlet to the cyclone wall. The particle velocity is the terminal velocity in the centrifugal field
such that the residence time t is given by

$$t = \frac{L}{V_t} \qquad (28)$$

where L is the horizontal distance traversed across the fluid
stream. For a spiral having constant velocity at the inlet

$$t = \frac{2 \pi R_A N}{V_i} \qquad (29)$$

where R_A is the average radius of fluid stream, N – number of
turns in spiral, and V_i inlet velocity

$$\therefore \quad \frac{t}{V_t} = \frac{2 \pi R_A N}{V_i} \qquad (30)$$

hence assuming Stokes drag

$$V_t = \frac{d_{50} \Delta P \ V_i^2}{18 \mu R_A} \qquad (31)$$

Rietma defined a dimensionless 'characteristic cyclone
number' Cy_{50} which is independent of pressure drop and Stokes
number.

$$C_{y_{50}} = \frac{d^2_{50} \ \Delta \rho \ L \ \Delta P}{\mu \ \rho \ Q} \qquad (32)$$

where ρ is fluid density. For cyclones of certain proportions
this equation is capable of predicting d_{50} fairly accurately.

Many other equations have been published and are discussed by
Bradley.[29]

To summarise, the hydrocyclone is an inexpensive separating device for the 200-5 um particle size region.   For very fine particles very small cyclones are required to give the required value of L. Cyclones as small as 1 cm. diameter are used and to obtain the necessary throughput large batteries of parallel cyclones are used.   A significant disadvantage of the hydrocyclone is its considerable power consumption, its poor grade efficiency characteristics and tendency to wear by abrasive materials.

The centrifuge

The type of centrifuge most commonly thought of for continuous thickening is the scroll discharge centrifuge.   As shown in Fig. 17, this consists of a rotating coni-cylindrical body within which the suspension is centrifuged.   The overflow discharges from

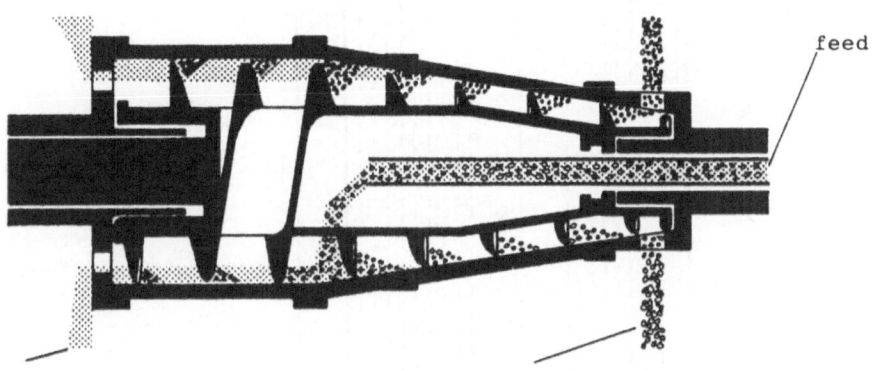

feed

liquid discharge   solids discharge

Fig. 17.   The Scroll Discharge Centrifuge.

the cylindrical end of the bowl over a weir which is usually adjustable and the underflow is withdrawn by pulling it up the sloping "beach" with a screw which rotates at a slightly different speed to the bowl.   This type of centrifuge is a complex and relatively expensive item of equipment but is capable of a high throughput in a small space for moderate power consumption.   It is also much more capable of accepting variations in flow rate than the hydrocyclone.

A particle within a centrifugal field experiences a force of

$$F = \frac{m}{g} \frac{r}{w^2} \tag{33}$$

where m – mass of particle
       r – distance from axis of rotation
       g – acceleration due to gravity
       w – angular velocity = revolutions per second x π

However, the flow in a continuous centrifuge is so complex that it is not practical to attempt to calculate a priori the performance of a centrifuge. Ambler[32] suggested a parameter, the Σ (sigma) value which may be used to scale up the performance of test centrifuges.

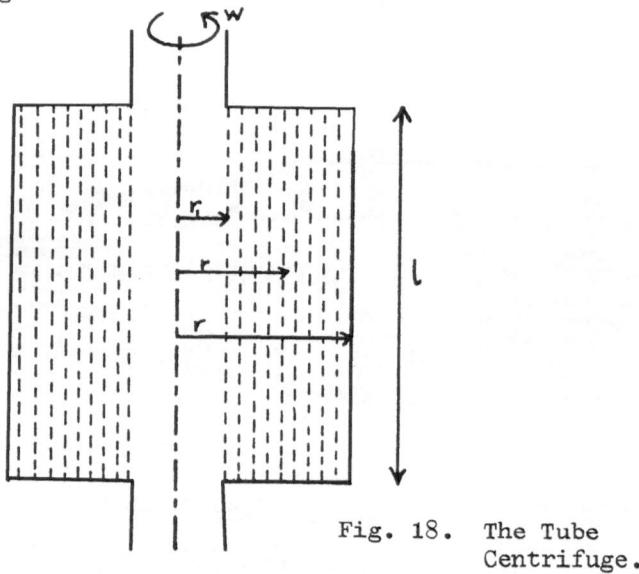

Fig. 18. The Tube Centrifuge.

Consider a solid bowl centrifuge, length L containing a cylindrical body of liquid, inner radius r_1 and outer radius r_2, the radius of the bowl. Consider also that the fluid is moving through the bowl at a volumetric flow rate of   , such that a bowl of volume V gives a residence time of $t = V/Q$. The settling velocity of the particle is given by balancing the centrifugal force it experiences and the Stokes drag it experiences, i.e.

$$\frac{dr}{dt} = \frac{\Delta\rho \, d_{50} \, w^2 \, r}{18\mu} \tag{34}$$

where d_{50} is the particle size having a grade efficiency of 0.50, $\Delta\rho$ the density difference between particles and fluid and μ the fluid viscosity

integrating between r and r_2 gives

$$t = \frac{18\,\mu}{\Delta\rho\; d_{50}\; w^2}\; \ln\; \frac{r_2}{r} \tag{35}$$

but the time available is the residence time of the fluid in the bowl

$$\text{i.e.}\quad Q = \frac{V\,\Delta\rho\;\; d_{50}\; w^2}{18\;\mu\; \ln\; \frac{r_2}{r}} \tag{36}$$

for 50% cut off, all particles outside radius

$$r = \left(\frac{r_2^2 + r_1^2}{2}\right)^{\frac{1}{2}} \qquad \text{will be removed,}$$

hence

$$Q = \frac{2\,V\,\Delta\rho\;\; d_{50}\; w^2}{18\;\mu\; \ln\; \left(\dfrac{2\,r_2^2}{r_2^2 + r_1^2}\right)} \tag{37}$$

In a gravity field the same particles would have a terminal velocity of

$$V_t = \frac{d_{50}\;\Delta\rho\; g}{18\;\mu} \tag{38}$$

hence

$$Q = \frac{2\,V_t\; w^2\; V}{g\ln\left(\dfrac{2\,r_2^2}{r_2^2 + r_1^2}\right)} \tag{39}$$

$$\text{i.e.}\quad Q = 2\,V_t\,\Sigma \tag{40}$$

124

$$\Sigma = \frac{w^2\,V}{g\,\ln\left(\dfrac{2\,r_2^2}{r_2^2 + r_1^2}\right)} \tag{41}$$

where

is the area of an equivalent gravity settling tank to achieve the same throughput. The treatment makes the following simpli-fying assumptions:—

(a) that the fluid passes through the bowl in plug flow.

(b) that the particles experience Stokes drag.

(c) that there is no slip in the liquid.

In a real centrifuge there will be considerable turbulence, for example, caused by the incoming fluid and the rotation of the discharge screw, hence its efficiency may be considerably less than the calculated Σ value. For such a centrifuge an experimental efficiency may be evaluated by comparison of the measured efficiency and calculated Σ but the effect of scale up on this efficiency coefficient is not understood.

REFERENCES

1. La Mer, V.K. and Healy, T.W. in Solid Liquid Separation, ed. Poole, J.B. and Doyle, D. H.M.S.O. London, 1966.

2. Audsley, A. The Flocculation of Suspensions with Organic Polymers. D.S.I.R. Report NCL/Dep 5, London, 1965.

3. Overbeek, J. Th., in Colloid Science, Chapter 6, Ed. Kruyt, H.R., Elsevier, Amsterdam, 1952.

4. Overbeek, J. Th. C., ibid. chap. 7.

5. Packham, R.F., Proc. Soc. Water Treat. Exam., 16, 18, 1967.

6. Ruehrwein, R.A., and Ward, D.W., Soil Science, 73, 485, 1952.

7. Audsley, A. and Fursey, A., Nature, 208, 753, 1967.

8. Akers, R.J. and Riley, P.W., unpublished data.

9. Silberberg, A. J. Phys. Chem., 66, 1884, 1962, J. Chem. Phys. 46, 1005, 1967; 48, 2535, 1968.

10. Sutherland, D.N.   J. Colloid. Interface Sci.,   25, 373, 1967.

11. Linke, W.F. and Booth, R.D.   Trans. Amer. Inst. Min. Met. Eng.,   217, 364, 1959.

12. Kragh, A.M. and Langston, W.B.   Report C21,   British Gelatine and Glue Research Association, 1960.

13. Doe, P.W., Benn D. and Bays, L.R.   J. Inst. Water Engr. 19, 251, 1965.

14. Lyon, W.A.   Sewage and Industrial Waste 23, 1084, 1951. Husman, W., Gesund heitzung.   73, 127, 1952. Imaczumi, T. and Inone, T.   Nippon Koggo Kaishi, 76, 43, 1960.

15. Nuclear Science Abstracts 15, Nos. 7656; 4002, 1961; 16, No. 15047, 1967; 17, Nos. 2007; 16057, 1963; 18, No.5362, 1964.

16. Slater, R.W. and Kitchener, J.A.,   Disc Farad. Soc., 42, 267, 1966.

17. Baskeville, R.C. and Gale, R.S.   Wat. Pollut. Control, 67, 233, 1968.

18. Packham, R.F., Br. Polym. J., 4, 305, 1972.

19. Coe, H.S., Clevenger, G.H., Trans. Am. Inst. Min. Met. Eng., 55, 356, 1916.

20. Kynch, G.J., Trans. Farad. Soc., 48, 166, 1951.

21. Talmage, W.P. and Fitch, E.B., Ind. Eng. Chem., 47, 38, 1955.

22. Yoshioka, N. et al., Kagaku Kogaku, 21, 66, 1957.

23. Scott, K.J.   Trans. Inst. Min. Metall., 77, C85–97, 1968.

24. Scott, K.J., Ind. Eng. Chem. (Fundamentals) 7(4), 582, 1968.

25. Robins, W.H.M., Trans. Inst. Chem. Eng., 42, 158, 1964.

26. Hassett, N.J., Ind. Chem. Manuf. 73, 729, 1965.

27. Moncrieff, A.G.   Bull. Inst. Min. Metall. 73, 729, 1965.

28. Fitch, E.B.   Ind. Eng. Chem., 58 (10), 18, 1966.

29. Bradley, D., The Hydrocyclone, Pergamon, 1965.

30.   Rosin, P., Rammler, E., and Intelmann, W., _Zeit. ver_
      Deutsch. Eng., 26, 433, 1932.

31.   Rietma, K.   _Cyclones in Industry_, Chap. 4, Elsevier, 1961.

32.   Ambler, C.M., _Chem. Eng. Progr._   48, 150, 1952.

SYMBOLS

A   Mass rate of solids flow

C   Mass/volume concentration of solids

C_v Volume/volume concentration of solids

D   discharge rate

D   diameter of a cyclone

D_i diameter of cyclone inlet

D_o diameter of cyclone vortex finder

F   volumetric feed rate

F   centrifugal force

g   gravitational constant

K   a constant

L   radial distance traversed by particle in a cyclone

m   mass of a particle

N   number of revolutions described by an element of fluid
        in a cyclone

Δ P   pressure drop

Q   volumetric flow rate

Q   filtration rate

Q_o filtration rate for bed without flocculant

Q(x)   fraction of particles greater than size x

r   distance of particle from radius of rotation

R_A	average fluid radius in a cyclone
t	time or residence time
u	settling velocity of particles in a batch test
v	velocity imposed on suspension due to underflow in a thickener
v	velocity of upward propagation of constant concentration zone in a batch thickener
V_t	terminal particle velocity
V_i	inlet velocity in cyclone
W	mass rate of solids
n	fluid viscosity
ρ	fluid density
Σ	Ambler's factor
θ	fraction of surface covered by polymer
ψ	solids flux
$ψ_{min}$	limiting solids flux
ω	angular acceleration

Subscripts

D	discharge
F	feed
i	interface
L	limiting
o	initial

ACKNOWLEDGEMENT

These notes have been prepared with the assistance of material prepared by my colleagues Dr. J. I. T. Stenhouse and Mr. A.S. Ward.

FLOW THROUGH POROUS MEDIA AND FLUID-PARTICLE HYDRODYNAMICS

Lloyd A. Spielman

Division of Engineering and Applied Physics,
Harvard University, Cambridge, Mass. 02138

SYMBOLS

a	sphere radius (m)
a_p	particle radius (m)
a_s	spherical grain radius (m)
A_s	porosity function for Happel's model (dimensionless)
b	cell radius (m)
$F_1(H)$, $F_2(H)$, $F_3(H)$	functions characterizing hydrodynamic interactions (dimensionless)
h	minimum separation between particle and collector (m)
$H = h/a_p$	dimensionless separation
$\vec{i}$	unit vector
k_o	Kozeny constant (dimensionless)
L	apparent pore length (m)
L_e	tortuous pore length (m)
p	pressure (kg m^{-1} s^{-2})
ΔP	dynamic pressure change (kg m^{-1} s^{-2})
r	radial coordinate (m)
R	circular tube radius (m)

R_h	hydraulic radius (m)
Re	Reynolds number (dimensionless)
s	internal surface per unit bed volume (m^{-1})
$\vec{u}$	fluid velocity (m s^{-1})
$\vec{u}_{sh}$	shear velocity field tangential to collector (m s^{-1})
$\vec{u}_{st}$	axisymmetric stagnation velocity field normal to collector (m s^{-1})
U	superficial velocity (unless otherwise stated) (m s^{-1})
U'	interstitial velocity (m s^{-1})
U_∞	velocity far from single sphere (m s^{-1})
Δx	thickness of section of porous bed (m)
α	bed solidity (dimensionless)
γ	ratio of cell radius to sphere radius (dimensionless)
ε	bed porosity (dimensionless)
θ	angle from forward incidence of flow (radians)
θ_p	angular position of particle (radians)
κ	permeability (m^2)
μ	viscosity (kg m^{-1} s^{-1})
τ	stress (kg m^{-1} s^{-2})
ψ	stream function (m^3 s^{-1})

INTRODUCTION

Two important factors affecting the economics of water
filtration are effectiveness of particle removal and hydrodynamic resistance to flow. These factors are in turn closely
associated with the mechanics of pore flow. Particle removal is
in large part governed by the interaction of various physical
and chemical forces, such as Brownian motion, gravity settling
and colloid chemical attraction and repulsion, with hydrodynamic
forces in the neighborhood of the collecting surfaces. The
hydrodynamic resistance of deep filter beds results from the
integrated action of hydrodynamic stresses on the stationary bed
matrix. In fact, much about both aspects of deep bed filtration
can be understood using theoretical models which focus on the
detailed flow of water-borne particles around a single characteristic bed grain.

It is important to recognize that because most of the
filtered particles are in the range of submicrons to tens of

microns (1 micron = 10^{-4} cm), while the bed grains are usually
no smaller than a few tenths of a millimeter, that the particles
are much smaller than the pores or grains and so can penetrate
to considerable bed depths before encountering surface and being
captured. To deal in an effective way with both bed resistance
and the movement of particles in porous media, requires consideration of the fundamental equations governing particle-fluid
hydrodynamics, especially those describing small scale flows.

STOKES EQUATIONS

General

The equations generally governing isothermal incompressible
fluid flow are the full Navier-Stokes equations,

momentum: $\rho\dfrac{D\vec{u}}{Dt} = \rho\left(\dfrac{\partial\vec{u}}{\partial t} + \vec{u} \cdot \nabla\vec{u}\right) = -\nabla p + \mu\nabla^2\vec{u}$ (1)

continuity: $\nabla \cdot \vec{u} = 0$ (2)

Here $\vec{u}$ is the fluid velocity field, p the dynamic pressure field,
ρ the constant fluid density and μ, the constant fluid viscosity.

Because Eq. (1) is nonlinear, its solution can be obtained
only under suitable approximations or ideal circumstances. If
one considers the flow within a small scale geometry, the magnitudes of the inertial acceleration terms can be small compared
with the viscous terms in Eq. (1). Because the ratio of inertial
to viscous terms is estimated to order of magnitude by the
Reynolds number, one obtains

$$Re = \frac{\ell V\rho}{\mu} \sim \frac{\rho\,|D\vec{u}/Dt|}{\mu\,|\nabla^2\vec{u}|} \ll 1$$ (3)

Here ℓ is a characteristic length and V a characteristic
velocity. For pore flow ℓ is on the order of the grain or pore
dimension and V the pore velocity. When the inequality (3) is
satisfied, the left hand terms of Eq. (1) may be neglected, in
which case Eqs. (1) and (2) simplify to Stokes equations of
creeping flow:

$$\nabla p = \mu\nabla^2\vec{u}$$ (4)

$$\nabla \cdot \vec{u} = 0$$ (2)

It should be stressed that for porous media flow the smallness of

Re in (3) results mainly from the smallness of the characteristic
length ℓ, in contrast with those of larger scale flows which
occur in geophysical, atmospheric or most hydraulic situations.
Furthermore, whereas the inequality (3) is on occasion only
marginally satisfied or even violated for pore flow, it is
almost always satisfied for the disturbance flows of the water-
borne particles, whose micron dimensions assure correspondingly
smaller Reynolds numbers, with fluid inertia playing an even
more minor role.

Linearity and Superposition

The applicability of the Stokes Eqs. (4) and (2), rather
than the full Navier-Stokes Eqs. (1) and (2), implies major
simplifications in both the mathematics of equation solving as
well as the developing of scaling criteria by dimensional analysis. Because Eqs. (4) and (2) are linear, they are often re-
sponsive to the variety of existing methods suited to linear
partial differential equations, such as separation of variables.
One of the most powerful tools used to solve problems described
by Eqs. (4) and (2) is the method of linear superposition of
solutions. Clearly, if each of the velocity and pressure fields
$\vec{u}_1$, p_1 and $\vec{u}_2$, p_2 satisfy Eqs. (4) and (2), respectively, then
their linear combinations $\vec{u} = \vec{u}_1 + \vec{u}_2$, $p = p_1 + p_2$ also give a
solution. Using this property one can construct complicated
flows describing the hydrodynamics of particles near one another
or near collectors, from component solutions governing relatively
simpler flows. An example of this is construction of the flow
field describing simultaneous translation and rotation of a
particle, by linear superposition of the flow fields governing
its translation alone and its rotation alone. Another example
is describing the translation of a particle toward or away from a
solid planar surface at some oblique angle, by superposition of
the flows governing its movements respectively perpendicular
and tangential to the surface; if, in addition, the particle is
rotating, then its isolated rotational flow field can be superimposed too, and so forth. Of course, great care must be taken
to make certain that all the boundary conditions add up properly
so the resultant flow field is precisely the one desired.

Forces and Flow Reversal Symmetry

To compute the hydrodynamic forces acting on an object
immersed in a given flow field, one usually integrates the local
hydrodynamic stresses acting over its surface. The local hydrodynamic stresses are straightforwardly related to the velocity
and pressure fields given by the solution of the governing
equations under appropriate boundary conditions. In Cartesian

coordinates the local stresses for incompressible viscous flow are given by

$$\tau_{xx} = p - 2\mu \frac{\partial u_x}{\partial x} \qquad \text{(i)}$$

$$\tau_{yy} = p - 2\mu \frac{\partial u_y}{\partial y} \qquad \text{(ii)}$$

$$\tau_{zz} = p - 2\mu \frac{\partial u_z}{\partial z} \qquad \text{(iii)}$$

$$\tau_{xy} = \tau_{yx} = -\mu \left(\frac{\partial u_x}{\partial y} + \frac{\partial u_y}{\partial x} \right) \qquad \text{(iv)}$$

$$\tau_{yz} = \tau_{zy} = -\mu \left(\frac{\partial u_y}{\partial z} + \frac{\partial u_z}{\partial y} \right) \qquad \text{(v)}$$

$$\tau_{xz} = \tau_{zx} = -\mu \left(\frac{\partial u_z}{\partial x} + \frac{\partial u_x}{\partial z} \right) \qquad \text{(vi)}$$

in which p is the thermodynamic pressure. The stresses given by Eqs. (i) through (vi) are linearly related to the viscosity and velocity gradients. This follows for the normal stresses given by Eqs. (i) through (iii) because the pressure also is seen to be proportional to μ and velocity gradients via Eq. (4). If a given flow field satisfies Eqs. (4) and (2), then so must that corresponding to reversal of the velocities and pressure gradient. Eqs. (i) through (vi) then imply that all drag forces exerted by the flow field are simply reversed also.

Quasistatic Property

Another convenient property of Eqs. (4) and (2), which is not in general possessed by Eqs. (1) and (2), is that the former are quasistatic. That is, their time varying flows may be viewed as a smooth sequence of instantaneous steady state flows. To see how this property can be used, let us consider the well-known Stokes resistance formula for the drag under steady movement of a sphere with radius a, moving with constant velocity U through an unbounded stationary fluid having viscosity μ, which says the particle experiences a steady drag force of magnitude

$$F = 6\pi\mu aU \qquad (5)$$

If the particle is permitted to accelerate such that its velocity is an arbitrary specified function of time, U(t), then Eqs. (4) and (2) imply immediate extension of Eq. (5) to describe the time varying drag force as

$$F(t) = 6\pi\mu a U(t) \tag{6}$$

This extension results because the condition (3) assures us that fluid inertia is effectively absent so the fluid responds and adjusts to time variations instantaneously. Of course, if the particle is accelerated too suddenly, or brought to such large velocities that the inequality (3) is violated, then Eqs. (4) and (2) won't apply and one must resort to the nonlinear Eqs. (1) and (2) for a realistic description.

Dimensional Analysis and Scaling

We notice that because the inertia terms are absent, the fluid density does not appear explicitly in Eqs. (4) and (2). This greatly simplifies the development of scaling criteria through dimensional analysis because the fluid density does not usually have to be included in our list of parameters. For instance, let us assume we did not know the Stokes formula, Eq. (5), but wished to obtain as much information as possible about the relationship of drag force to the other parameters, without undertaking solving Eqs. (4) and (2), as Stokes did. Careful consideration of the relevant parameters appearing in Eqs. (4) and (2) and their boundary conditions, tells us our list should include F, U, a, and μ, but not ρ. From these parameters, only one dimensionless group can be formed, thus we obtain

$$\frac{F}{\mu a U} = \text{constant} \tag{7}$$

Dimensional analysis gives nearly the entire formula (5) and the detailed solution of Eqs. (4) and (2) merely gives the value of the dimensionless constant = 6π. Had we been less perceptive in inspecting Eqs. (4) and (2), and conservatively included the fluid density ρ in our list, we would have obtained instead

$$F = \mu a U \, \mathcal{F}(\frac{aU\rho}{\mu}) \tag{8}$$

in which $\mathcal{F}$ is an unknown function of the Reynolds number and which conveys far greater ambiguity concerning the desired relationship among the parameters than Eq. (7). It is a characteristic of Stokes flows, that drag forces and stresses are directly proportional to viscosity and velocity, with the coefficient of proportionality depending on geometry. For a fuller discussion of

Stokes equations see Happel and Brenner.[1]

PACKED BED HYDRODYNAMICS

Relationship of Stokes Flow to Darcy's Law

By using the simple ideas concerning dimensional analysis outlined above with some further plausible arguments, we can derive Darcy's law for flow through porous media. In its simplest form Darcy's law for one dimensional flow may be stated as

$$U = -\frac{\kappa}{\mu}\frac{\Delta P}{\Delta x} \tag{9}$$

In Eq. (9), U is the superficial velocity = volume flow rate/cross sectional area of bed, μ the viscosity, ΔP is the dynamic pressure difference across thickness Δx of porous medium, and κ is the hydraulic permeability, which is experimentally found to be a property of the porous solid. In what follows, the porous solid is taken to be macroscopically uniform in the x-direction, but not having any special microscopic geometry. That is, the solid is not in particular assumed to be a bundle of tubes or a regular array of spheres, but can be of any degree of microscopic complexity so long as its overall bulk character is uniform.

Straightforward reasoning shows that the overall pressure difference ΔP should be directly proportional to the bed depth Δx. This is because ΔP times the bed cross sectional area measures the net force on the opposite faces of the bed and must be equal to the total drag force exerted over all the microscopic surface inside the bed, since it is equal and opposite to the total force necessary to hold the bed fixed. It therefore follows that doubling the bed depth, Δx, will double the pressure drop, because it doubles the amount of porous solid over which the internal drag force is exerted, hence ΔP must be directly proportional to Δx.

ΔP and Δx should therefore enter the final expression only as their ratio, $(\Delta P/\Delta x)$. If we now require that the small pore Reynolds number condition (3) be satisfied so the microscopic flow is in the Stokes regime and Eqs. (4) and (2) govern, then in accord with the previous discussion the fluid density ρ should not appear explicitly in the end result. The quantities which do appear should therefore include only $(\Delta P/\Delta x)$, U and μ, as well as a potentially long list of independent geometric parameters. These geometric parameters would formally appear in a very complicated expression describing the internal solid surface at which the no-slip boundary condition for Eqs. (4) and (2) would be applied (we don't really have to be able to write down all these geometric parameters or the equation of the surface, but only recognize that such an expression applies). In the list of geometric parameters, there must be at least one characteristic length, say ℓ, since even the simplest imaginable pore shapes require one parameter to describe them (e.g. circular). The list of quantities appearing in the final expression would then look like:

$$\frac{\Delta P}{\Delta x}, \ U, \ \mu, \ \ell, \ \ell', \ \ell'', \ \ldots, \ \text{other lengths, angles, etc.}$$

Dimensional analysis then gives

$$-\frac{\mu U}{\ell^2 (\Delta P / \Delta x)} = f(\frac{\ell'}{\ell}, \ \frac{\ell''}{\ell}, \ \ldots, \ \text{other geometric ratios}) \qquad (10)$$

in which the left hand side of Eq. (10) is a dimensionless group and the right hand side a dimensionless function of geometric ratios.

But Eq. (10) can be rewritten as Eq. (9) if we interpret the permeability in the latter as

$$\kappa = \ell^2 \ f(\frac{\ell'}{\ell}, \ \frac{\ell''}{\ell}, \ \ldots) \qquad (11)$$

Eq. (11) shows the permeability has dimensions of length squared and depends only on the geometry of the porous solid. It also follows from the flow reversal principle that reversing the direction of flow through a porous solid cannot alter its permeability.

The foregoing derivation of Darcy's law clearly shows the key assumption which underlies it. Namely, that because the pores are small, the pore Reynolds number is usually small so fluid inertia effectively plays no role in the dynamics of flow. If not, Darcy's law doesn't apply. A common misconception about porous media says that as flow rate increases Darcy's law first breaks down upon the onset of turbulence in the pores. In fact, pore Reynolds numbers rarely become large enough for turbulent flow in the pores. Breakdown of Darcy's law really marks the onset of inertial forces in laminar flow, which occurs at pore Reynolds numbers on the order of 1-10.[2,3] For very complicated solid geometries, especially consolidated porous solids, the best route to determining the permeability is by direct measurement, using Eq. (9). Usually U is plotted against ΔP to get a straight line and κ calculated from the measured slope.

Kozeny-Carman Theory[1,2,3]

Over the years there have been many attempts to relate permeabilities to the geometry of the porous solid by using special models. One of the most widely used of these theories is the Kozeny-Carman development which is outlined below.

The average velocity for laminar flow through a straight circular tube is given by the well-known formula:

$$U_{avg} = \frac{Q}{A} = -\frac{R^2 \Delta P}{8\mu L} \tag{12}$$

where Q is the volume flow rate, A the tube cross sectional area, R the tube radius and L the tube length. For noncircular tubes, Eq. (12) has been generalized to

$$U_{avg} = -\frac{R_h^2 \Delta P}{k_o \mu L} \tag{13}$$

In Eq. (13), R_h is the hydraulic radius defined as (flow cross sectional area/wetted perimeter). For a circular tube $k_o = 2$ and $R_h = R/2$. For straight tubes with noncircular cross sections, such as rectangles, ellipses, etc., whose aspect ratios are not very different from unity, the coefficient k_o is Eq. (13) varies from about 2.0 to 2.5 and so may be considered as roughly independent of shape. Assuming the porous solid to be a bundle of irregularly shaped, straight channels, the above definition of hydraulic radius gives

$$R_h = \epsilon/s \tag{14}$$

where ϵ is the voids fraction and s is the internal surface area per unit volume of bed. It is then argued that the effective pore length is really somewhat greater than the bed depth because the fluid travels a tortuous path, thus the apparent length L in Eq. (13) should be replaced by L_e where the tortuosity factor is $L_e/L > 1$. The interstitial velocity U' is related to the superficial velocity U by

$$U' = U/\epsilon \tag{15}$$

However, it is then argued that because the tortuous fluid path is longer than L by the factor L_e/L, the velocity along the tortuous path must be correspondingly greater than that for travel straight through which is given by Eq. (15), hence the proper velocity to use in Eq. (13) is

$$U'' = U'(\frac{L_e}{L}) = \frac{U}{\epsilon}(\frac{L_e}{L}) \tag{16}$$

Setting U_{avg} = U'' and L = L_e in Eq. (13) gives,

$$U'' = - \frac{R_h^2 \Delta P}{k_o \mu L_e} \qquad (17)$$

Now substituting Eqs. (14) and (16) into Eq. (17) gives,

$$U = -(\frac{L}{L_e})^2 \frac{\varepsilon^3}{k_o \mu s^2} \frac{\Delta P}{\Delta x} \qquad (18)$$

Now taking k_o = 2.5 and using Carman's assumption,

$$\frac{L_e}{L} = \sqrt{2}, \qquad (19)$$

gives

$$U = - \frac{\varepsilon^3}{5.0 \mu s^2} \frac{\Delta P}{\Delta x} \qquad (20)$$

The numerical factor in Eq. (20) is in fair accord with experiment for unconsolidated granular beds in which the pores do not vary too greatly in size. One of the most important uses of Eq. (20) is in determining the internal surface areas of porous materials from permeability data. It also indicates how changes in packing density should affect permeability.

The Kozeny-Carman theory has received much criticism, largely undeserved since it correlates bed resistance data for a wider class of porous media than any other permeability theory. On the other hand, to develop theories of particle removal by granular beds requires a more detailed picture of the flow field near the collecting surfaces within the bed than the Kozeny model affords. Progress to overcome this has been made adopting Happel's cell model to say more about the microscopic flow field. Instead of viewing a packed bed as a bundle of tortuous channels as the Kozeny theory does, the cell models view the bed grains as an assemblage of interacting, but essentially individual spheres, with the flow field about an average sphere being described in detail. Moreover, by summing up the drag forces acting on the individual bed grains, the cell model also permits self-consistent prediction of bed permeabilities, which agree with data for unclogged media at least as well as the Kozeny equation. It also yields predictions of bed expansion in backflow as well as hindered settling of suspensions. To analyze the cell model, however, requires a closer look at solutions of Eqs. (4) and (2) which describe the fluid mechanics of particles in general.

Flow Through Assemblages of Spheres

Lamb's general solution. Lamb[4] gave a general solution to
Eqs. (4) and (2), suited to treating boundary value problems in
which velocities are prescribed on spherical surfaces. Lamb's
solution takes the form

$$p = \sum_{n=-\infty}^{\infty} p_n \quad ,$$

$$\vec{u} = \sum_{n=-\infty}^{\infty} [\nabla \times (\vec{r}\chi_n) + \nabla\phi_n + \frac{(n+3)}{2\mu(n+1)(2n+3)} r^2 \nabla p_n$$

$$- \frac{n}{\mu(n+1)(2n+3)} \vec{r}p_n]$$

in which p_n, χ_n and ϕ_n are each solid spherical harmonics which
are determined from the specified boundary conditions. Happel
and Brenner[1] illustrate the adaptation of Lamb's general solution
to treat a variety of boundary value problems. Although Lamb's
solution provides a general approach to such problems, many
axisymmetric flow problems involving spheres can be solved using
the simpler, though more restricted method which follows.

Stokes solution for a single sphere. A number of important
boundary value problems in axisymmetric Stokes flow may be treated
using a simple general solution of Eqs. (4) and (2), first obtained by Stokes. Among these are uniform flow past an isolated
rigid sphere and the cell models mentioned previously, as well as
circulating droplets and particles moving by electrophoresis.[1,5]

Eqs. (4) and (2) may be simplified by introducing a stream
function ψ such that

$$u_r = - \frac{1}{r^2 \sin\theta} \frac{\partial \psi}{\partial \theta} \quad , \qquad u_\theta = \frac{1}{r \sin\theta} \frac{\partial \psi}{\partial r} \qquad (21)$$

In Eqs. (21), r and θ are spherical coordinates. Eq. (2) is then
automatically satisfied and eliminating the pressure between the
r- and θ-components of Eq. (4) gives

$$[\frac{\partial^2}{\partial r^2} + \frac{\sin\theta}{r^2} \frac{\partial}{\partial \theta} (\frac{1}{\sin\theta} \frac{\partial}{\partial \theta})]^2 \psi = 0 \qquad (22)$$

For solutions of the form,

$$\psi = f(r) \sin^2\theta \quad , \qquad (23)$$

substitution into Eq. (22) gives

$$f(r) = \frac{A}{r} + Br + Cr^2 + Dr^4 \qquad (24)$$

in which A, B, C, D are integration constants to be determined from boundary conditions.

For an isolated sphere with no-slip at its surface and uniform flow at infinity, the boundary conditions are

$$u_r = 0 , \quad u_\theta = 0 \text{ at } r = a$$

or, equally, from Eq. (21),

$$\frac{\partial \psi}{\partial \theta} = \frac{\partial \psi}{\partial r} = 0 \text{ at } r = a \qquad (25)$$

and $u_r \rightarrow -U_\infty \cos \theta$, $u_\theta \rightarrow U_\infty \sin\theta$ as $r \rightarrow \infty$

or, using Eq. (21) we have equivalently,

$$\psi \rightarrow \frac{1}{2} U_\infty r^2 \sin^2\theta \text{ as } r \rightarrow \infty \qquad (26)$$

The constants in Eq. (24) are then determined as

$$A = \frac{1}{4} U_\infty a^3 , \quad B = -\frac{3}{4} U_\infty a ,$$

$$C = \frac{1}{2} U_\infty , \quad D = 0$$

This gives the velocity field,

$$\frac{u_r}{U_\infty} = -[1 - \frac{3}{2}(\frac{a}{r}) + \frac{1}{2}(\frac{a}{r})^3]\cos\theta ,$$

$$\frac{u_\theta}{U_\infty} = [1 - \frac{3}{4}(\frac{a}{r}) - \frac{1}{4}(\frac{a}{r})^3]\sin\theta \qquad (27)$$

and the pressure distribution,

$$p = p_0 + \frac{3}{2} \frac{\mu U_\infty}{R}(\frac{a}{r})^2 \cos\theta \qquad (28)$$

The total normal stress is given in spherical coordinates by

$$\tau_{rr} = p - 2\mu \frac{\partial u_r}{\partial r} \qquad (29)$$

and the tangential stress,

$$\tau_{r\theta} = \tau_{\theta r} = -\mu [r\frac{\partial}{\partial r}(\frac{u_\theta}{r}) + \frac{1}{r}\frac{\partial u_r}{\partial \theta}] \tag{30}$$

Integration of Eqs. (29) and (30) over the entire sphere surface gives the drag force, Eq. (5). Equating the drag force to the weight minus buoyancy gives the well-known Stokes law for the terminal settling velocity of an isolated particle,

$$U_s = \frac{2}{9}\frac{a^2 \Delta\rho g}{\mu} \tag{31}$$

where $\Delta\rho$ is the density difference between particle and fluid and g is the acceleration due to gravity.

Happel's cell model. Happel[1,6] treated the problem of flow through an assemblage of spheres by assuming a typical sphere to be enclosed within a spherical envelope of radius b, whose volume corresponds to the voids ratio in the overall assemblage, i.e.,

$$\frac{a}{b} = \gamma = \alpha^{1/3} = (1 - \varepsilon)^{1/3} \tag{32}$$

where α is the solidity (volume fraction spheres) and ε the porosity. He then used the general form given by Eqs. (23) and (24), retaining the no-slip surface conditions (25), but instead of the isolated flow condition (26), used boundary conditions at the envelope, r = b, to fix all the constants in Eq. (24). Thus he takes

$$u_r = -U \cos\theta , \quad \tau_{r\theta} = 0 \quad \text{at} \quad r = b \tag{33}$$

with $\tau_{r\theta}$ given by Eq. (30). The first of conditions (33) sets the radial component of velocity equal to that corresponding to the superficial velocity U (or, equivalently, to the velocity U of the assemblage as it moves through the fluid). The second of conditions (33) assumes the envelope at r = b to be a free surface, which can be justified in some sense by arguing that a free surface of a different shape, but equivalent volume, must exist for regular arrays of equal spheres. This determines all the constants and gives the flow field near the sphere. The force on the sphere can then be evaluated as previously for the isolated sphere.

The apparent arbitrariness of the free surface assumption is made clear by considering an alternate condition used by Kuwabara[7] in his cell model. Instead of the vanishing shear condition (33) Kuwabara assumed vanishing vorticity, i.e.,

$$\frac{\partial u_\theta}{\partial r} - \frac{1}{r}\frac{\partial u_r}{\partial \theta} + \frac{u_\theta}{r} = 0 \quad \text{at} \quad r = b \tag{34}$$

Whether the vanishing shear or the vanishing vorticity assumption is more correct cannot convincingly be answered on theoretical grounds, but is better judged by comparison with experiment.

Happel's model gives for the drag on each sphere,

$$F = \frac{4\pi\mu a U (3+2\gamma^5)}{(2-3\gamma+3\gamma^5-2\gamma^6)} \tag{35}$$

where γ depends on voids fraction through Eq. (32). In the limit that the voids fraction ϵ tends to unity, $\gamma \to 0$, and Eq. (35) appropriately reduces to Eq. (5) for the isolated sphere.

One can now use Eq. (35) to predict pressure drop through a packed bed of equal spheres. Equating the force difference due to pressure on the opposite faces of a thickness of bed Δx, to the sum of the drag forces on all the spheres in the thickness, gives

$$\Delta P = -\frac{\alpha \Delta x}{\frac{4}{3}\pi a^3} F \tag{36}$$

Substituting from Eqs. (35) and (32) and rearranging, gives

$$U = -[\frac{2}{9}\frac{a^2}{\gamma 3}(\frac{3-9\gamma/2+9\gamma^5/2-3\gamma^6}{3+2\gamma^5})]\frac{\Delta P}{\mu \Delta x} \tag{37}$$

The bracketed term in Eq. (37) is a function of a and $\gamma = (1 - \epsilon)^{1/3}$ only and corresponds to the Darcy permeability, κ, defined by Eq. (9). Eq. (37) is found to be in good agreement with experiment, closely agreeing with the Kozeny equation (20) in the porosity range, $0.4 < \epsilon < 0.7$. At higher porosities Happel's model is superior to Eq. (20) because the former reduces to an assemblage of isolated spheres whereas Eq. (20) does not. On the other hand Kuwabara's model leads to a stronger dependence on porosity, giving somewhat higher pressure drops than observed.[1] This supports Happel's model as giving the more realistic flow field near a typical grain.

Happel's model may also be applied to assemblage settling, giving

$$\frac{U}{U_o} = (\frac{3-9\gamma/2+9\gamma^5/2-3\gamma^6}{3+2\gamma^5}) \tag{38}$$

in which U is the hindered settling velocity and U_o that given
by Eq. (31). Here too agreement with experiment appears to be
good.

Happel and Brenner[1] discuss the use of the cell model to
describe fluidized bed behavior during the expanded bed phase.

PARTICLE-COLLECTOR INTERACTIONS

Here we consider theoretical aspects of particle motion near
a much larger collecting grain of radius a_s. This discussion is
not intended to deal in a complete manner with collection
mechanisms, but to outline how fluid mechanical effects enter the
particle capture process. Early treatments of particle capture
assume the particles move with the undisturbed fluid velocity
except for the action of external forces such as van der Waals
attraction or gravity. However, recent treatments[8,9] consider
the exact Stokes disturbance flow field created by the particle
in proximity to the collector. The particle is taken to be pro-
pelled by the undisturbed flow near the collector rather than
artificially superimposed upon it. The entrained particle freely
translates and rotates as it should according to its equations of
motion under the hydrodynamic and external forces which act upon
it. The particle thus creates a locally confined hydrodynamic
disturbance which is governed by Stokes Eqs. (4) and (2). The
boundary conditions are taken to be the undisturbed flow field
far from the particle with no slip at both spherical particle
and collector surfaces and no net force or torque acting on the
particle (all inertia is neglected). Because the curvature of
the collector is so much smaller than that of the particle, the
former is approximated as a plane surface in the neighborhood of
the particle. Also, external field forces such as van der Waals
attraction, double layer repulsion, and gravity, can be included
in the overall force balance. In this way, both external and
hydrodynamic interactions are simultaneously taken into account
in a rigorous manner. Neglect of inertia is justified by the
smallness of the particle and its Reynolds number.

The flow field very near the spherical collector can be
obtained by expanding Eq. (23) in Taylor series about the surface.
This gives, to lowest order,

$$\psi = \frac{3}{4} A_s U(r - a_s)^2 \sin^2\theta \tag{39}$$

144

which is restricted to small $(r - a_s)/a_s$, where a_s is the grain
radius. In Eq. (39), A_s is a dimensionless parameter characterizing the flow model. For an isolated sphere in Stokes flow with
a uniform velocity U at infinity, $A_s = 1$. For a spherical grain
within a packed bed, A_s is a known function of bed porosity.
Happel's model for flow around a characteristic grain gives

$$A_s = \frac{2(1-\gamma^5)}{2-3\gamma+3\gamma^5-2\gamma^6} \tag{40}$$

Let us now define a system of local cylindrical coordinates
$\tilde{\omega}$ and z whose origin is on the collector surface at $r = a_s$,
$\theta = \theta_p$; θ_p is the angle corresponding to the center position of an
entrained particle. The origin of the coordinate system thus
changes position as the entrained particle moves around the collector (Fig. 1). By straightforward transformations, the undisturbed flow field can be expressed as

$$\vec{u}_o = \vec{u}_{st} + \vec{u}_{sh} \tag{41}$$

where

$$\vec{u}_{st} = \frac{3}{2} \frac{A_s U \cos\theta_p}{a_s^2} (\tilde{\omega}z\vec{i}_{\tilde{\omega}} - z^2\vec{i}_z)$$

and

$$\vec{u}_{sh} = \frac{3}{2} \frac{A_s U \sin\theta_p}{a_s} (z\vec{i}_y)$$

Here $\vec{i}_y$, $\vec{i}_{\tilde{\omega}}$, and $\vec{i}_z$, are unit vectors in the y, $\tilde{\omega}$, and z direc-
tions respectively. The above expression for the undisturbed
field is correct through terms of lowest order in $\tilde{\omega}$ and z. In
Eq. (41) the undisturbed flow field near the collector has been
decomposed into two additive fields, each of which satisfies

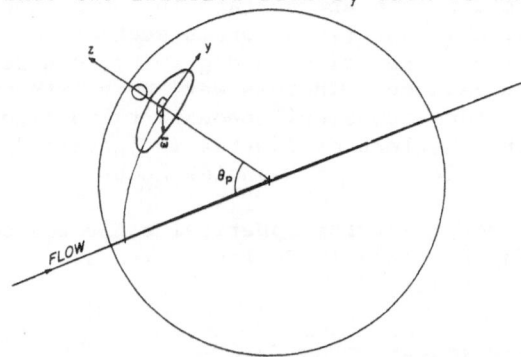

Fig. 1. Local coordinates as particle moves around collector.

Eqs. (4) and (2) separately. The field u_{st} is axisymmetric about the z-axis and has a stagnation point at $\varpi = 0$, z = 0; u_{sh} is a uniform shear field directed parallel to the collector surface. Letting h be the minimum separation between the entrained particle and the locally flat collector surface, the particle center is located at $\varpi = 0$, z = z_p = a_p + h, $\theta = \theta_p$. For Eq. (41) to give the boundary condition on the disturbance field, the particle must be so small compared with the collector, that within separations where the particle deviates appreciably from an undisturbed streamline, the collector can be approximated as a planar wall (except inasmuch as its geometry determines the undisturbed flow). Thus $a_p << a_s$ and Eq. (41) applies only near the moving origin and outside the region of the disturbance.

The movement of the entrained particle and its corresponding disturbance flow field are now decomposed into the fields corresponding to its normal and tangential motions separately. This is permitted because the governing Eqs. (4) and (2) are linear and all the velocity boundary conditions (at the particle surface, obstacle surface and far from the particle) are arranged to be additive. The method of superposition of solutions discussed previously is used to construct the solution for the particle freely moving near the collector. Also, because the creeping flow equations (4) and (2) are quasistatic, they apply at any instant as the entrained particle proceeds along its trajectory.

The disturbance flow corresponding to the z-directed particle motion may further be decomposed into two additive flows. These are summarized in Table I. In one such flow, the particle moves in the z-direction under the influence of an instantaneously applied normal force F_n, which, for the present may be viewed as unspecified, with the velocity field taken to vanish far from the particle and no-slip at both the particle and the effectively planar collector surfaces. The particle motion in this Stokes flow is given by

$$\frac{dh}{dt} = \frac{F_n F_1(h/a_p)}{6\pi\mu a_p} \tag{42}$$

The dimensionless function $F_1(h/a_p) = F_1(H)$ is known for all H from the exact solution of Stokes equations given by Brenner.[10] The function $F_1(H)$ is shown graphically in Figure 2. In a second flow contributing to the z-directed motion of the particle, the particle is taken to be held fixed in a field which becomes the axisymmetric velocity field $\vec{u}_{st}$, given by Eq. (41), far from the particle, again with no-slip at both the particle and 'planar' collector surfaces. Because of the axisymmetry of this flow, the particle experiences a purely z-directed force,

TABLE I. Summary of Superimposed Flow Fields Giving Resultant z-Directed Motion

	Particle Stationary in Axisymmetric Flow	Motion under Applied Force in Otherwise Stationary Fluid	Resultant Flow
Velocity field outside disturbance	$\dfrac{3}{2}\dfrac{A_s\,U\,\cos\theta}{a_s^2}P(\tilde{\omega}z\vec{i}_{\tilde{\omega}} - z^2\vec{i}_z)$	0	$\dfrac{3}{2}\dfrac{A_s\,U\,\cos\theta}{a_s^2}P(\tilde{\omega}z\vec{i}_{\tilde{\omega}} - z^2\vec{i}_z)$
Velocity at particle surface	0	$\dfrac{dh}{dt}\vec{i}_z$	$\dfrac{dh}{dt}\vec{i}_z$
Velocity at collector surface	0	0	0
Force exerted on particle by fluid	$F_{st}\vec{i}_z$	$-F_n\vec{i}_z = -(F_{ext} + F_{st})\vec{i}_z$	$-F_{ext}\vec{i}_z$

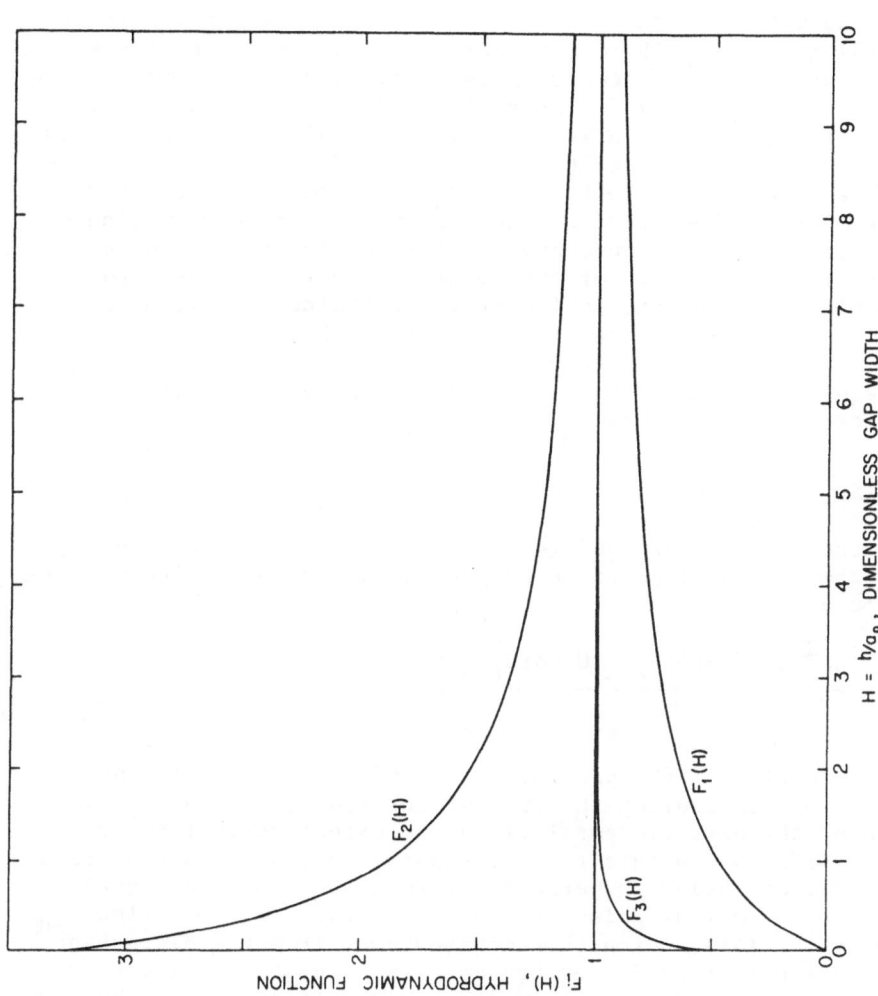

Fig. 2. Functions characterizing hydrodynamic interactions between particle and collector.

$$F_{st} = \frac{-6\pi\mu a_p^3 A_s U \cos\theta_p}{a_s^2} \frac{3}{2}(H + 1)^2 F_2(H) \tag{43}$$

The dimensionless function $F_2(H)$ is known for all $H = h/a_p$ from the exact solution of Stokes equations given by Goren[11] and Goren and O'Neill[12] and also is shown graphically in Figure 2. Eq. (43) giving F_{st} is subsequently made use of in the force balance with F_n which occurs in reconstituting the original flow field. Let $F_{ext}(H)$ be an external field force, which for simplicity, will be taken to have a z-component only. That is, it is directed perpendicular to the collector and depends only on the distance of the particle from the surface, for instance, surface forces of colloidal origin. The resultant z-directed motion of the particle and its accompanying fluid motion can now be obtained by linear superposition of the separate flows described above. Neglecting inertia, the motion of the entrained particle is obtained by combining the flows such that the net force on the particle is zero. This requires that the applied force of Eq. (42) equals the sum of the hydrodynamic force of Eq. (43) and the external force, i.e.,

$$F_n = F_{st} + F_{ext} \tag{44}$$

Substituting Eqs. (42) and (43) into Eq. (44) and rearranging, gives the motion of particle perpendicular to the collector surface at any instant:

$$\frac{6\pi\mu a_p^2}{F_1(H)} \frac{dH}{dt} = \frac{-6\pi\mu a_p^3 A_s U \cos\theta_p}{a_s^2} \frac{3}{2}(H + 1)^2 F_2(H) + F_{ext}(H) \tag{45}$$

The motion of the particle tangential to the collector is obtained by considering the part of the flow field which describes motion of the particle parallel to the effectively planar collector surface. In this flow, the particle undergoes free rotation and translation (experiences zero net torque and force) as it creates a confined disturbance in the uniform shear flow $\vec{u}_{sh}$ given by Eq. (41), which is recovered away from the disturbance. The solution to this Stokes boundary value problem was obtained by Goldman, Cox and Brenner,[13] and the resulting expression for the induced θ-directed particle velocity is

$$a_s \frac{d\theta_p}{dt} \cong u_{p_\theta} = \frac{a_p}{a_s} A_s U \sin\theta_p \frac{3}{2}(H + 1) F_3(H) \tag{46}$$

This induced velocity is purely θ-directed (Fig. 1). The dimensionless function $F_3(H)$ is shown in Figure 2. The deviations of F_1, F_2, and F_3 from unity in Fig. 2, reflect the strengths of the particle-collector hydrodynamic interactions.

Eqs. (45) and (46) are the differential equations which describe the normal and tangential translation of the entrained particle in the vicinity of the collector surface. The equation describing the particle trajectories is obtained by eliminating the time t between Eqs. (45) and (46). The resulting equation is

$$\frac{3}{2}(H + 1) \frac{F_3(H)}{F_1(H)} \sin\theta_p \frac{dH}{d\theta_p} = -\frac{3}{2}(H + 1)^2 F_2(H) \cos\theta_p$$
$$+ \frac{a_s^2 F_{ext}(H)}{6\pi\mu a_p^3 A_s U} \tag{47}$$

The numerical solution of Eq. (47) has been reported to predict collection by London-van der Waals attraction and gravitational external forces[8,9] and compared with experiment.[14] Its solution for capture by combined London attraction and electrical double layer repulsion has been recently reported.[15]

REFERENCES

1. Happel, J. and Brenner, H., Low Reynolds Number Hydrodynamics, Prentice-Hall, Englewood Cliffs, N.J., 1965.

2. Bird, R.B., Stewart, W.E., Lightfoot, E.N., Transport Phenomena, John Wiley and Sons Inc., New York, 1960, chap.6.

3. Scheidegger, A.E., The Physics of Flow through Porous Media, University of Toronto Press, 1960.

4. Lamb, H., Hydrodynamics, Cambridge University Press, 1932; reprint Dover Publications, New York, 1945, 594.

5. Levich, V.G., Physicochemical Hydrodynamics, Prentice-Hall, Englewood Cliffs, N.J., 1962.

6. Happel, J., AIChEJ, 4, 197, 1958.

7. Kuwabara, S., J. Phys. Soc. Jap., 14, 527, 1959.

8. Spielman, L.A. and Goren, S.L., Environ. Sci. Technol., 4, 134, 1970; 5, 254, 1971.

9. Spielman, L.A. and FitzPatrick, J.A., J. Colloid Interface, 42, 607, 1973.

10. Brenner, H., Chem. Eng. Sci., 16, 242, 1961.

11. Goren, S.L., J. Fluid Mech., 41, 619, 1970.

150

12. Goren, S.L. and O'Neill, M.E., <u>Chem. Eng. Sci.</u>, 26, 325, 1971.

13. Goldman, A.J., Cox, R.G., and Brenner, H., <u>Chem. Eng. Sci.</u>, 22, 637, 653, 1967.

14. FitzPatrick, J.A. and Spielman, L.A., <u>J. Colloid Interface Sci.</u>, May 1973.

15. Spielman, L.A. and Cukor, P.M., <u>J. Colloid Interface Sci.</u>, 43, 51, 1973.

CHARACTERISTICS OF UNCONSOLIDATED FILTER MEDIA

D. Leclerc

Département Génie Chimique, Ecole Nationale Supérieure
des Industries Chimiques, 1 rue Grandville
54000 Nancy France

NOTATION LIST

a	cylinder length	L
a_c	bed specific surface	L^{-1}
a_p	particule specific surface	L^{-1}
A	total pores area	L^2
b	cylinder diameter	L
b'	constant	$M^{-1}L^1T^2$
c	constant	$-$
C	co-ordination number	$-$
C_c	co-ordination number	$-$
d, d_1, d_2	sphere diameters	L
N_1, N_2	number of spheres of diameter d_1 or d_2	$-$
p	mercury pressure	$ML^{-1}T^{-2}$
P	equilibrium pressure	$ML^{-1}T^{-2}$
P_o	vapour pressure	$ML^{-1}T^{-2}$
r	pore radius	L
R	universal constant of gas	$ML^2T^{-2}\,mole^{-1}$
R_s	sphere radius	L
T	temperature	$^{\circ}K$
v	molal volume	$L^3\,mole^{-1}$
V	adsorbed volume	L^3
V_a	apparent volume of filter media	L^3
V_m	first strata volume of adsorbed molecules	L^3
V_s	solid volume of filter media	L^3
V_1^s, V_2	diameter d_1 or d_2 sphere volume	L^3

V'_1, V'_2	increase volume brought by sphere of diameter d_1 or d_2	L^3
y_E	volumetric fraction of small particules at the entectic state	–
y_1	volumetric fraction of small particules	–
$\alpha = \dfrac{a}{d}$	reduced length of cylinder	–
$\beta = \dfrac{b}{d}$	reduced diameter of cylinder	–
ε	filter media porosity	–
ε'	porosity of disturbed zone	–
ε_o	porosity of random clore packing	–
ε_1, ε_2	porosity of pure packing of beads of diameters d_1 or d_2	–
σ	liquid surface tension	MT^{-2}
θ	angle of contact	rad

Unconsolidated filter media can be classified in two categories. Firstly thick beds of grains in which particules are retained. These are generally called deep bed filters. Secondly beds
formed by gradual accumulation of particules on a filter cloth,
these are called filter cakes. Whatever the filtration mecanism,
determination of its performance requires a knowledge of the filter media and its properties. These can be obtained in two ways
which will be described. Either by direct measurement of porous
media characteristics such as porosity, specific area, pore diameters, co-ordination or by estimation of these characteristics
from grain size distributions. It is not possible to predict entirely the charatceristics of packing from a full grain size distribution and here we will only consider binary mixtures.

MEASURED PARAMETERS OF POROUS MEDIA

Porosity

Porosity is the volume fraction ε occupied by the fluid phase. We will see later that for regular packings of identical spheres, porosity may be calculated directly. It is important to notice that porosity is independent of absolute bead size and only
depends on the shape and geometric disposition of these beads. For
filter media, the determination of these characteristics is too
complicated and it is only possible to determine a mean porosity.
Generally a measurement of the apparent volume V_a of the filter
media and the solid volume or fluid volume V_s is sufficient

$$\varepsilon = \frac{V_a - V_s}{V_a} \tag{1}$$

For an isotropic, homogeneous and random medium in any cross
section the area void fraction exposed is equal to the volume
fraction occupied by the pores. Using this theorem, known as Du-
puit's law, porosity can be measured on photograph of a cross section of a porous medium from the pore area exposed. However the
use of this method for fibrous or ordered media which do not satisfy the previous conditions can lead to erroneous results.

An other technique is the measurement of mercury penetration
under pressure into pores using a mercury porosimeter. Generally
used for consolidated media, this method gives other information
as will be described further on.

Most filter cakes are compressible and their porosities vary
with the local pressure of the fluid. In such cases a study with

a permeability-compressibility cell is useful. Its description has been made by Tiller.[1]

Porosity, easy to define, more delicate to measure must be used with precaution on media with large pore size distribution. It is necessary to adapt the measurement method to the porosity controlling the studied phenomena. For example micro-porosity can have little effect in fluid flow characteristics and be a very important parameter for washing performance.

Specific area of porous media

Specific area of porous media, is the ratio of total area of pores to the apparent volume of the porous media

$$a_c = \frac{A}{V_a} \tag{2}$$

For unconsolidated granular media of mean specific area a_p

$$a_c = a_p (1 - \varepsilon) \tag{3}$$

If beads have no micro-porosity, this definition does not present any ambiguity. On the contrary, operations with mass transfert phenomena such as cake washing will be in addition affected by the internal area of grains.

The most frequently used methods are adsorption and permeability measurements.

Permeability will be the subject of an other lecture.

By physical adsorption of gas on a solid and determination of the volume V_m occupied by the first stata of adsorbed molecules, specific area of solids can be calculated. If the adsorption isotherm is of the langmuir type, the relation beetween volume of gas adsorbed V and the pressure P is :

$$\frac{P}{V} = \frac{1}{V_m b'} + \frac{P}{V_m} \tag{4}$$

b' is a constant.

More frequently, the isotherm is of the Brunauer, Emmet and Teller type :

$$\frac{P}{V(P_o - P)} = \frac{1}{V_m c} + \frac{c - 1}{c} \frac{1}{V_m} \frac{P_o}{P} \tag{5}$$

where P_o is vapour pressure and c a constant.

Other gas adsorption methods exist which have been described by Allen.[2]

Liquid phase sorption can be also used for surface area measurement. This method, described by Orr and Delavalle,[3] is based on the fact that in a solution of two miscible elements one is more sorbed than the other.

Pore size distribution

The pores space in a random packing of particules has a very complicated geometry which is impossible to describe rigorously. However the mean size of "pores" and the distribution around this mean are important parameters for describing the medium. The methods for measurement of specific area can be used to determine pore size parameters providing an hypothesis is made on the shape of the pores. This hypothesis can be to consider the porous volume as constituted of circular section capilaries. Two methods are currently used to obtain the pore size distribution curve : gas desorption and mercury porosimetry.

Described by Allen,[2] the gas desorption method consist in measuring the quantity of desorbed gas when pressure decreases under isothermal conditions. In the pores of the medium, evaporation of molecules is not as easy as on the free surface of a liquid because of the lowering of vapour pressure above the meniscus formed by the condensed liquid in the pores. This pressure is given by the Kelvin equation

$$\text{Log} \frac{P}{P_o} = - \frac{2 \sigma v \cos \theta}{r \, RT} \tag{6}$$

where P is the mesured pressure, P_o the saturated vapour pressure, σ the surface tension, v the molal volume, r the radius of pores, θ the angle of contact.

Pore sized distribution may be found by determining the volume penetration of mercury into pores under increasing pressure p.

Mercury will enter all pores larger than those of radius r where

$$r = \frac{2 \, \sigma \, \cos \, \theta}{p} \tag{7}$$

σ is the surface tension, θ the angle of contact. With relation (7) experimental curve of volume penetration vs p gives the volume fraction of pores with a radius greater than a given value r. For pores larger than a micron, the apparatus is relatively simple as it does not request high pressure. Relation (7) applies well to other liquids which can be substituted to mercury.

Co-ordination

The co-ordination number is the number contacts a given particule has with others in the packing. Co-ordination number is of practical interest in numerous phenomena where contacts between beads take place. Such is the case for crevice site of retention in deep bed filtration. The difficulty in measuring such a parameter led to the definition of a neighbourhood co-ordination which extends notion of contact to a zone around the particule. Co-ordination number has only been determined in the case of packs of spherical beads. Measurement is made by impregnation of the beds with a liquid, either paint or acid (depending on the nature of the particules) which will mark the contact zones. Other methods consist in determining the precise geometrical position of each particule and its neighbours, or in studying polyhedra obtained by isotropic compression of deformable particules.

GRANULAR PACKED TEXTURE

Ordered packing of monosized spherical beads

Ordered packing presents analogies with crystals : they are constituted by repetition of an elementary pattern presenting symetrical characteristics. The systematic study of regular packings can be made from two dimensional assemblage of spheres of radius R_s of equal co-ordination in a plane where their centres are the apexes of polygons with a $2 R_s$ side. By packing identical planes, an ordered three-dimensional packing can be obtained of which the co-ordination number C and porosity are given in table 1. Regular packing can be considered too as a packing of identical polyhedra : the faces of the polyhedra of contact points on a given sphere are the planes tangent to that sphere at all the points of contact between that sphere and its neighbours.

Crystalline system.	C	Porosity
Rhombohedral different orientation	12	$1 - \dfrac{\pi}{3\sqrt{2}} = 0,2595$
Tetragonal spheroidal	10	$1 - \dfrac{\pi}{3\times1,5} = 0,3020$
Orthorhombic	8	$1 - \dfrac{\pi}{3\sqrt{3}} = 0,3955$
Cubic centered	8	$1 - \dfrac{\pi\sqrt{3}}{8} = 0,3198$
Cubic	6	$1 - \dfrac{\pi}{6} = 0,4764$

Table 1   Relation between porosity and co-ordination number in regular packings.

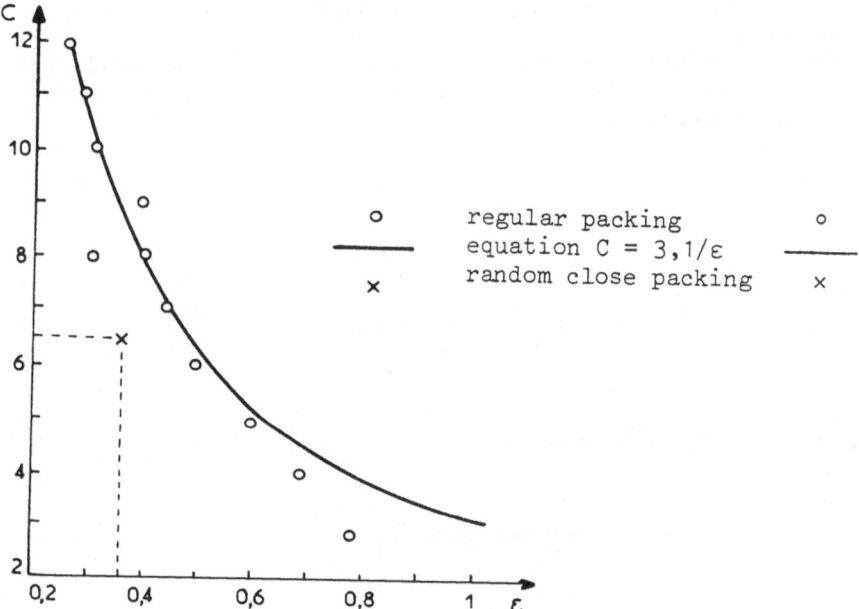

Fig. 3   Co-ordination number vs porosity

Random packing of monosized spherical beads

   <u>Porosity</u>. Random packing of monosized spherical beads cannot
be defined with as much precision as can regular packings. However
it is possible to measure reproducible parameters characterizing
random packing. By pouring spheres into a container, the packing
state obtained is not very reproducible. Porosity depends to a
large extent on the method of packing. However it is never greater
than 0.42, the value obtained in a fixed bed deposited when the
flow of fluid to a fluidised bed is stopped. This packing called
loose packing is not very stable. With small quantities of energy
supplied by shocks or vibrations the porosity decreases continou-
ly till it reaches a certain level which is called close packing.
The stability of this state is relative since it can be destroyed
by increasing energy to produce more or less compact packings which
depend on the nature, amplitude and frequency of the vibrations or
the shocks.

   The most precise measurement of porosity of a random close
packing of monosize spheres far from any wall is $\varepsilon_o = 0.36$.

   <u>Wall effect</u>. A wall is a surface which disturbs a certain
thickness of packing near which porosity is 1. For example considering a plane wall and the first d/2 thick state in contact with
the wall, even in the case where the arrangement along this wall
is the most compact regular packing, porosity will not be less than
$\varepsilon = 0.3956$, therefore superior to ε_o. In all cases, the wall creates a zone of high porosity which spreads into the packing as a
damped sinusoid (Fig. 1).

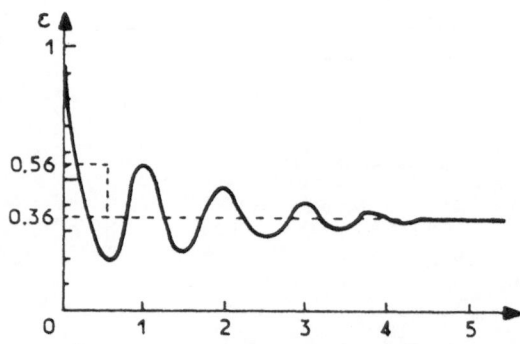

Fig. 1 Variation of local porosity.

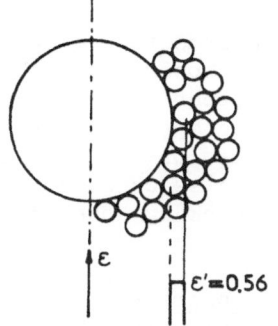

Fig. 2 Model for variation of
local porosity around a big sphere.

In theoretical and experimental studies, using numerous litterature results Ben Aim and Le Goff[4] showed that the wall effect can globaly be taken into account by supposing a disturbed zone of thickness d/2 and of porosity $\varepsilon' = 0.56$ (Fig. 2).

However, it is necessary to remember that for transfer phenomena occuring on a granular scale, the wall effect is felt to 4 or 5 beads diameters.

Co-ordination. The co-ordination of a random close packing is relatively well defined though of course the co-ordination number is a mean value as co-ordination varies from one sphere to the other in the packing. Ben Aim and Le Goff[5] collected litterature results which can be resumed in the following way :

contact co-ordination (distance between centers = d) : 6 to 7
neighbourhood co-ordination (distance between centers < 1,1 D) :
8 to 9
number of geometric neighbours (distance between centers < $\sqrt{2}$ D) :
13 to 14.

It can be seen that the co-ordination for random close packing is below the curve corresponding to regular packings (Fig. 3).

Tesselation is the way of filling space with ordered packings of identical cells. Only cell presenting symetry of 2nd, 3rd, 4th or 6th order allow regular packing. Desordered tesselations of space are frequently observed in nature : animal or vegetal tissues, bubble texture in foam, grain joint in alloys. In all these polyhedra a high frequency of pentagonal faces can be observed. In particular Bernal[6] studied polyhedra obtained by compression of random packing of spheres. He determined that the mean number of faces of these polyhedra are 13.6 and that 60 % of all these faces are pentagons. Therefore it appears that the random characteristics of tesselation is connected with the important number of pentagonal faces.

Random packing of binary mixtures

Binary mixtures of beads present an intermediate complexity between the ideal case of a bed of identical spheres and the real case of a bed constituted of more or less dispersed sized grains. They may help the understanding of filtration operations such as blocking of a filter media by small particles or the action of a filter aid in filter cake structure. All studies on random packings of binary mixtures deal with compact packing, the only one which is suffcently reproducible.

Mixture of two monosized fractions of spherical beads. A binary mixture of spherical beads has a lower porosity than a monosized packing. There is a minimum porosity for a given mixture composition which varies with the diameter ratio d_1/d_2 ($d_1 < d_2$). This minimum is more pronounced as the ratio d_1/d_2 is smaller. By analogy with binary liquid mixtures, we call the mixture corresponding to the maximal compacity the "Entectic state". This Entectic state is the limit between two composition domains particularly clear when the diameter ratio is lower than 0.2 (Fig. 4).
- The first domain corresponds to a small volumetric fraction of small diameter particles. If particules are of smaller diameter than the dimensions of the caverns formed between the big beads, they will not affect the apparent volume of the bed as they will be confined in the caverns. $d_1 = 0.22 \ d_2$ corresponds to the maximal diameter of a small bead which can be contained in a tetrahedral cavern. These caverns corresponding to the volume limited by four big spheres in contact represents 75 % of the porous volume in a close random packing(Delachambre)[7]. When the Entectic state is obtained, small particules form a random close packing in each cavern formed by the big particules and the porosity of the mixture is near ε_o^2.

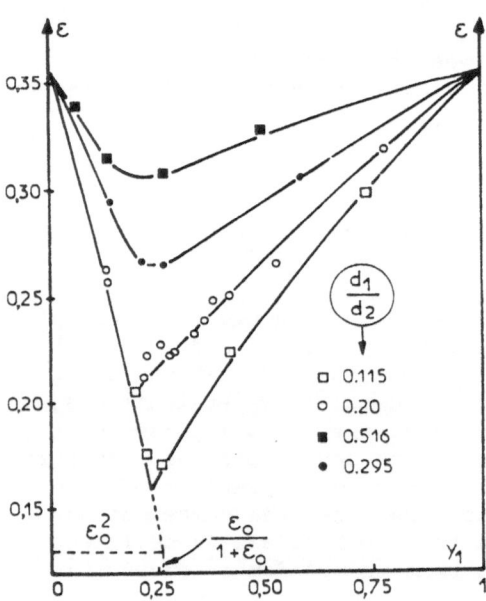

Fig. 4 Porosity vs volumetric fraction.

- The second domain corresponds to the introduction of separated
big spheres in a packing of small ones. These big beads create zones of zero porosity which will be partially conteracted by the
wall effect which they create at their vecinity with respect to
the small spheres. Ben Aim and Le Goff[4] showed experimentaly that
when $d_1/d_2 < 0.45$, the increase of volume V_2' brought about by a
sphere of volume V_2 is

$$V_2' = V_2 \left(1 + 0.9 \frac{d_1}{d_2}\right) \tag{8}$$

It can be seen that this relation may be explained theoretically
by supposing that the pertubation caused by a big sphere is limited around this sphere to a zone of thickness $d_1/2$ where the porosity would be $\varepsilon' = 0.56$ (Fig. 2).

When $d_1/d_2 < 0.2$, the corresponding value of V_2' leads to the
volumetric fraction of small particules corresponding to the Entectic state :

$$y_E = \frac{1 - \left(1 + 0.9 \dfrac{d_1}{d_2}\right)(1 - \varepsilon_o)}{2 - \left(1 + 0.9 \dfrac{d_1}{d_2}\right)(1 - \varepsilon_o)} \tag{9}$$

The volumetric fraction of an ordinary mixture is

$$y_1 = \frac{N_1 V_1}{N_1 V_1 + N_2 V_2} \tag{10}$$

where $N_1 N_2$ are the respective numbers of beads of diameters d_1
and d_2 and V_1, V_2 the corresponding volumes of one sphere. If the
mixture presents an excess of big spheres ($y_1 < y_E$), its mean porosity is

$$\bar{\varepsilon} = \frac{\varepsilon_o - y_1}{1 - y_1} \tag{11}$$

On the contrary, the porosity of a mixture presenting an excess of small spheres $(y_1 > y_E)$ is :

$$\varepsilon = \frac{(\frac{\varepsilon_o}{1 - \varepsilon_o} - 0.9 \frac{d_1}{d_2}) y_1 + 0.9 \frac{d_1}{d_2}}{(\frac{\varepsilon_o}{1 - \varepsilon_o} - 0.9 \frac{d_1}{d_2}) y_1 + 0.9 \frac{d_1}{d_2} + 1} \qquad (12)$$

In the general case where $d_1/d_2 > 0.2$ and where the two components in pure packing do not have the same porosity (this occurs when grains are not rigorously spherical), the mean porosity can be calculated by the following formulae

When $y_1 < y_E$
$$\varepsilon = \frac{(\frac{V_1'}{V_1} - \frac{1}{1 - \varepsilon_2}) y_1 + \frac{\varepsilon_1}{1 - \varepsilon_2}}{(\frac{V_1'}{V_1} - \frac{1}{1 - \varepsilon_2}) y_1 + \frac{1}{1 - \varepsilon_2}} \qquad (13)$$

When $y_1 > y_E$
$$\varepsilon = \frac{(\frac{1}{1 - \varepsilon_1} - \frac{V_2'}{V_2}) y_1 - \frac{V_2'}{V_2} - 1}{(\frac{1}{1 - \varepsilon_1} - \frac{V_2'}{V_2}) y_1 - \frac{V_2'}{V_2}} \qquad (14)$$

V_1' is the mean increase in volume brought by a small bead. These relation are in good agreement with experimental results.

The model describing the perturbation produced by obstacle in a random packing has been also used by Ben Aim and Le Goff[5], to determine contact-co-ordination number of a big sphere mixed with small ones. For random close packing the mean co-ordination number can be calculated by

$$C_c = 3 + 3.16 (\frac{d_2}{d_1})^2 \qquad (15)$$

Experimentally, relation (15) gives good agreement when $1 < \frac{d_1}{d_2} < 100$

Mixture of spheres and fibre-shaped bodies. Binary mixtures of spheres always produce a decrease in porosity. This is particularly the case when big spheres are added to small ones. On the contrary filter aids increase the porosity of a filter cake. The action was explained by Ben Aim and Le Goff[8] by means of the previous model where big spheres are replaced by cylinders of length a and diameter b. Each cylinder immersed in a bed of small spheres of diameter d is surrounded by a d/2 thick layer of disturbed porosity ε' (Fig. 5). The influence of this disturbed layer is more important as the shape of the cylinder differs greatly from that of a sphere. Two cases are interesting, the first one corresponds to thin disks (a < d < b), the second one corresponds to long cylinders (a > d > b). In these two cases an increase of porosity can be observed. (Fig. 6 and 7).

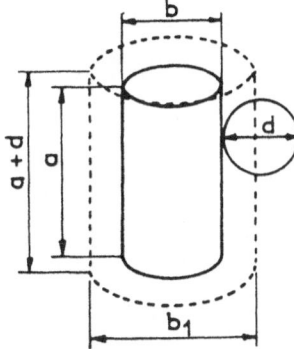

Fig. 5 Disturbed zone created by addition of cylinder in a bed of spheres.

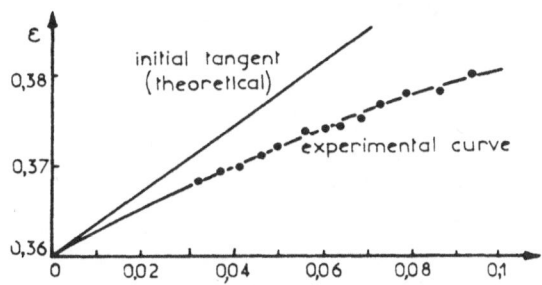

Fig. 6 Mixture of spheres and disks porosity vs volumetric fraction of disks (α = 0.22).

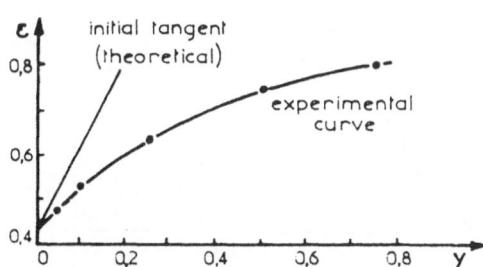

Fig. 7 Mixture of spheres and fibres. Porosity vs volumetric fraction of fibres (d = 82 μm, a = 500 μm, b = 14 μm)

164

In particular, the initial increase of porosity $(\frac{d\varepsilon}{dy})$ $y = 0$ can be determined as a fonction of the ratio of disk thickness or fibre diameter to the diameter of the spheres ($\alpha = \frac{a}{d}$ for disks or $\beta = \frac{b}{d}$ for fibres) (Fig. 8).

- For disks : $(\frac{d\varepsilon}{dy})_{y=0} = (1 - \varepsilon_o)\left[\frac{1}{\alpha}(\varepsilon' - \varepsilon_o) - \varepsilon_o\right]$ (16)

Supposing that $\varepsilon_o = 0.36$ and $\varepsilon' = 0.56$ an increase of porosity is observed if $\alpha < 0.55$, that is to say when the thickness of the disk is smaller than the sphere radius.
- For fibres considered as rigid cylinders, if $\beta > 0.152$

$$(\frac{d\varepsilon}{dy})_{y = 0} = (1-\varepsilon_o)\left[4(\varepsilon' - \varepsilon_o)(1 - \frac{1}{\beta} - \sqrt{1+\frac{1}{\beta}}) - \varepsilon_o\right]$$ (17)

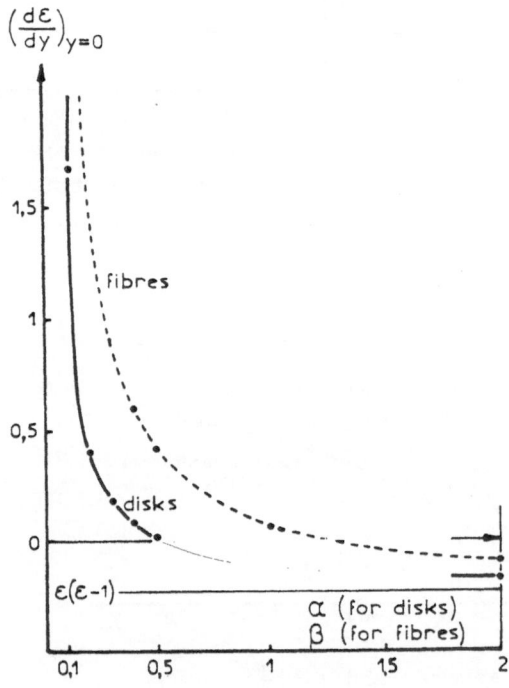

Fig. 8 Initial variation of porosity of a bed of spheres as a fonction of reduced dimension α or β of additional bodies.

Equation (17) gives an initial increase of porosity for $\beta<1.25$ if ε_o = 0.36 and ε' = 0.56. β = 0.152 corresponds to the largest diameter of cylinder which can penetrate in a constriction formed by three touching spheres. If β < 0.152 the previous model is no longer satisfactory but it may be assumed that the disturbed layer is as important as for β = 0.152. In this case

$$\left(\frac{d\varepsilon}{dy}\right)_{y=0} = \frac{1}{\beta^2}\,(1-\varepsilon_o)\left|-\beta^2+0.445(\varepsilon'-\varepsilon_o)+0.023(1-\varepsilon_o)\right| \qquad (18)$$

These results resumed in figure 9 show the importance of the shape of added grains on the increase of porosity. This should be useful in choosing filter aids. One of the purposes of filter aids is to increase the mean porosity of filter cakes with the minimal quantity of additive. Industrial filter aids are made of particules having various shapes always very different from a spherical symmetry.

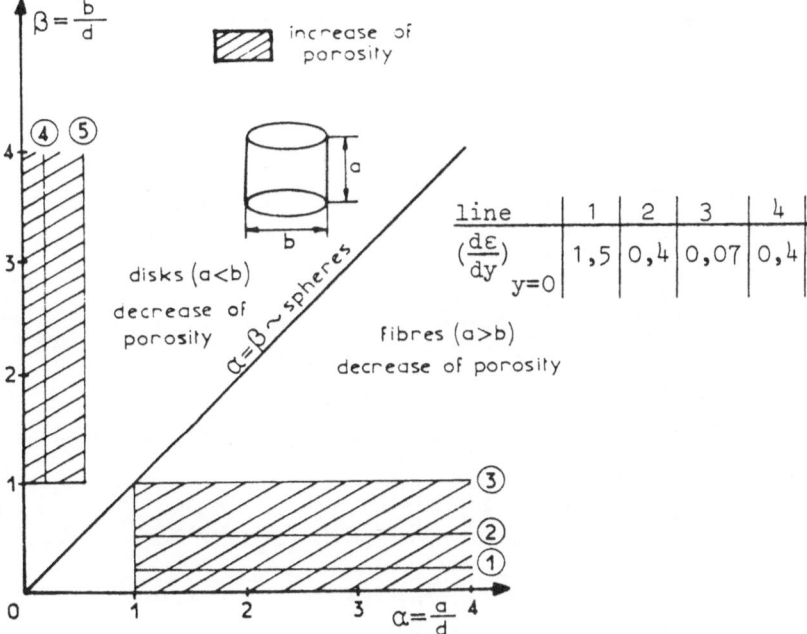

Fig. 9 Recapilatory diagram for the effect of α and β on porosity variation.

REFERENCES

1. Tiller, F.M., Haynes, S. Jr., and Wei-Ming Lu, The role of porosity in filtration VII. Effect of Side-Wall friction in Compression-permeability cells, AIChE Journal, 18, 13, 1972.

2. Allen, T., Particle size measurement, Chapman and Hall Ltd, London, 1968, chap. 14 and 15.

3. Orr, C. Jr., and Dallavalle J.M., Fine Particle measurement, The Macmillan Company, New York, 1959, chap. 8.

4. Ben Aïm, R., and Le Goff, P., Effet de paroi dans les empilements désordonnés de sphères et application à la porosité de mélange binaire, Powder Technol., 1, 281, 1967/68.

5. Ben Aïm, R., and Le Goff, P., La coordinance des empilements désordonnés de sphères. Application aux mélanges binaires de sphères, Powder Technol., 2, 1, 1968/69.

6. Bernal, J.D., The geometry of the structure of liquids, in liquids, structure, properties, solid interactions, Hughel, T.J., Ed, Elsevier, Amsterdam, 1965, 26.

7. Delachambre, Y., and Le Goff, P., Etude sur modèle du colmatage d'un milieu filtrant. Ecoulement d'une suspension de microsphères à travers un empilement de macrosphères, Rev. Fran. Corps Gras, 12, 3, 1965.

8. Ben Aïm, R., and Le Goff, P., Porosité des mélanges binaires de sphères et d'objets à symétrie cylindrique, Powder Technol., 2, 169, 1968/69.

PERMEABILITY OF FILTER MEDIA

D. Leclerc

Département Génie Chimique. Ecole Nationale Supérieure
des Industries Chimiques 1, rue Grandville
54000 Nancy France

NOTATION LIST

a	diameter of internal cylinder	L
a_p	particle specific area	L^{-1}
a_1, a_2	specific area of particle diameter d_1 or d_2	L^{-1}
b	diameter of external cylinder	L
b'	local permeability	L^2
B	permeability	L^2
B_1, B_2	permeability of packing of particles diameter d_1 or d_2	L^2
C, C_0	electrical conductivity	Ω^{-1}
d_c	capillary diameter	L
d_p, d_1, d_2	particle diameter	L
E	dispersion variable	$-$
h_k	Kozeny constant	$-$
k, k_n	constants	$L^2 T^{-1}$
N	number of spheres per unit volume of packing	$-$
N_c	number of capillaries in Kozeny's model	$-$
N_1, N_2	number of particles of diameter d_1 or d_2	$-$
P	pressure	$ML^{-1}T^{-2}$
$\bar{P}$	mean pressure	$ML^{-1}T^{-2}$
ΔP	differential pressure	$ML^{-1}T^{-2}$
Re_p	Reynolds number	$-$
T	tortuosity coefficient	$-$
u	local fluid velocity	LT^{-1}
u_m	superficial fluid velocity	LT^{-1}
V	volume of porous media	L^3
V_1, V_2	volume of particle of diameter d_1 or d_2	L^3
y, y_1	volumetric fraction of particles of diameter d_1	$-$

y_2	volumetric fraction of particles of diameter d_2	-
Y_1, Y_2	apparent volumetric fraction of particles of diameter d_1 or d_2	-
Z	thickness of porous media	L
Z_c	capillary length	L
α	angle between flow direction and capillaries	rad
γ	circularity factor	-
ε	porosity	-
ε_0	porosity of random close packing	-
η	dynamic viscosity	$ML^{-1}T^{-1}$
ρ	density	ML^{-3}
$\phi = u \times p$		$M.T^{-3}$
ω	cross sectional area of pores	L^2
Ω	cross sectional area of porous media	L^2

A very important characteristic of filter media is its permeability. This parameter can be easily defined as the proportionality coefficient B in Darcy's equation

$$u_m = \frac{B}{\eta} \frac{\Delta P}{\Delta Z} \tag{1}$$

u_m is the superficial liquid velocity and ΔP the pressure drop through a layer of depth ΔZ. Equation (1) can be generalized in vectorial form by

$$\vec{u} = - \frac{b'}{\eta} \overrightarrow{\mathrm{grad}\ P} \tag{2}$$

Although numerous studies have been made on local permeability b' and resumed in a recent bibliography[1], the present lecture only deals with overall permeability for unidirectional flow. We will successively present the limit of Darcy's equation, relations between permeability and overall characteristics of filter media. Such relations are based on models which will be discussed in the cases of monosized bead packs and binary mixtures.

DARCY'S EQUATION

Equation (1) applies only for liquid flow in the laminar regime. Experimentally this can be confirmed by the fact that superficial velocity is inversly proportional to viscosity η. In practice if there is a linear relation between ΔP and u_m, it can be assumed that the flow is laminar. For packing of monosized particles of diameter d_p the criterion for the laminar flow condition is the Reynolds number $Re_p < 1$

$$Re_p = \frac{u_m \rho d_p}{\eta} \qquad (3)$$

It is easy to verify that the dimensions of permeability are L^2. A unit frequently encountered is the Darcy which is defined as the permeability of a 1 cm cube through which flows a fluid of viscosity 1 centipoise under a differential pressure of 1 atmosphere. This unit corresponds approximatly to 1 μm^2.

Darcy's equation is not satisfactory for compressible fluids since volumetric flow varies from one point to an other. However it can be used by substituting $\phi = u.P$ for u_m, then for a medium of thickness dZ

$$\phi = \frac{B}{\eta} P \frac{dP}{dZ} \qquad (4)$$

This when integrated between two values P_1 and P_2 at the extremities of the porous medium, gives :

$$\phi = \frac{B}{\eta} \frac{P_1^2 - P_2^2}{2\Delta Z} = \frac{B}{\eta} \bar{P} \frac{\Delta P}{\Delta Z} \qquad (5)$$

In reality intermolecular collisions create an additional velocity at every point of flow and Darcy's equation must be modified to

$$\phi = (B \frac{\bar{P}}{\eta} + k) \frac{\Delta P}{\Delta Z} \qquad (6)$$

Contrary to the case for liquid flow, in gas flow the velocity at the wall of the pores is not zero. For that reason this type of gas flow is called slip flow.

When pressure is sufficiently low, each molecule flows individually without intermolecular shocks. This phenomena corresponds to a different flow regime. In this case

$$\phi = k_n \frac{\Delta P}{\Delta Z} \qquad (7)$$

where k_n mainly depends on the ratio of pore diameter to mean free path of the molecules. Gas flow measurements are frequently used to determine permeability of porous media. It is necessary to point out that they are not immediatly transposable to liquid flow.

PERMEABILITY MODELS

Although the direct measurement of permeability is the more convenient method for filtration practice, it is also necessary to predict permeability from the elementary characteristics of

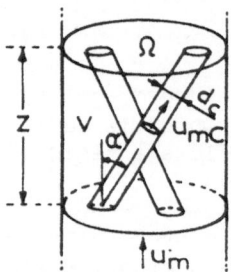

Fig. 1 - Kozeny's model

filter media such as grain sizes and positions, porosity, etc. It is not possible to take into account all these parameters in their integrality and models must be used. Of course these models do not represent the geometrical complexity of real media but they take into account some of the main characteristics such as specific surface a_p and porosity ε.

KOZENY'S MODEL

This model describes the void space as a bundle of N_C identical cylindrical, nonconnected capillaries of diameter d_C and length Z_C (fig. 1). The number and diameter of the capillaries is such that porosity and area are the same in the model as in the real media. For a media of porosity ε, in which the particle specific area is a_p

$$d_C = \frac{4\,\varepsilon}{a_p(1-\varepsilon)} \tag{8}$$

and

$$N_C = \frac{\Omega\,a_p{}^2(1-\varepsilon)^2}{4\,\pi\,\varepsilon\,T}$$

T is called the tortuosity factor, it takes into account the fact that the fluid flow path through the media is longer than the thickness of the media.

$$T = Z_C/Z = 1/\cos\alpha \tag{9}$$

Applying Poiseuille's equation to fluid flow through each capillary, we obtain the Kozeny Carman equation :

$$\frac{\Delta P}{Z} = h_k\,\eta\,u_m\,a_p{}^2\,\frac{(1-\varepsilon)^2}{\varepsilon^3} \tag{11}$$

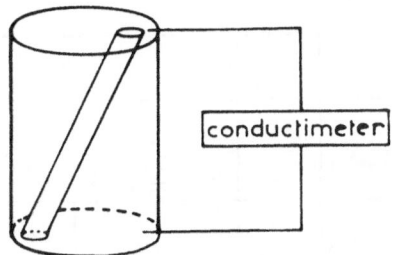

Fig. 2 - Conductivity measurement of tortuosity

where the Kozeny constant is

$$h_k = 2 \gamma T^2 \qquad (12)$$

γ is a circularity factor introduced to take into account the shape of capillary sections. For normal granular $\gamma = 1 \pm 0,15$. By
comparaison of equations (1) and (11), a relation between permeability and the main characteristics of filter media can be established :

$$B = \frac{\varepsilon^3}{h_k \, a_p^2 (1-\varepsilon)^2} \qquad (13)$$

Kozeny constant

Use of equation (13) presents difficulties. Firstly the determination of the circularity factor, secondly the determination of
the tortuosity coefficient which can be approached by electrical
or diffusional conductivity measurements.
If filter media was really constituted of cylindrical capillaries, conductivity measurements would give

$$\frac{C}{C_0} = \frac{\varepsilon}{T^2} \qquad (14)$$

where C_0 is the conductivity of a volume V of electrolitic solution and C the conductivity of the same volume of filter media
full of electrolitic solution (fig. 2). In reality pores are not
only tortuous but also the area ω of their sections vary. Then
the conductivity ratio would be approximately

$$\frac{C}{C_0} = \frac{\varepsilon}{T^2} \left[\overline{\omega} \cdot \overline{(\frac{1}{\omega})} \right]^{-1} \qquad (15)$$

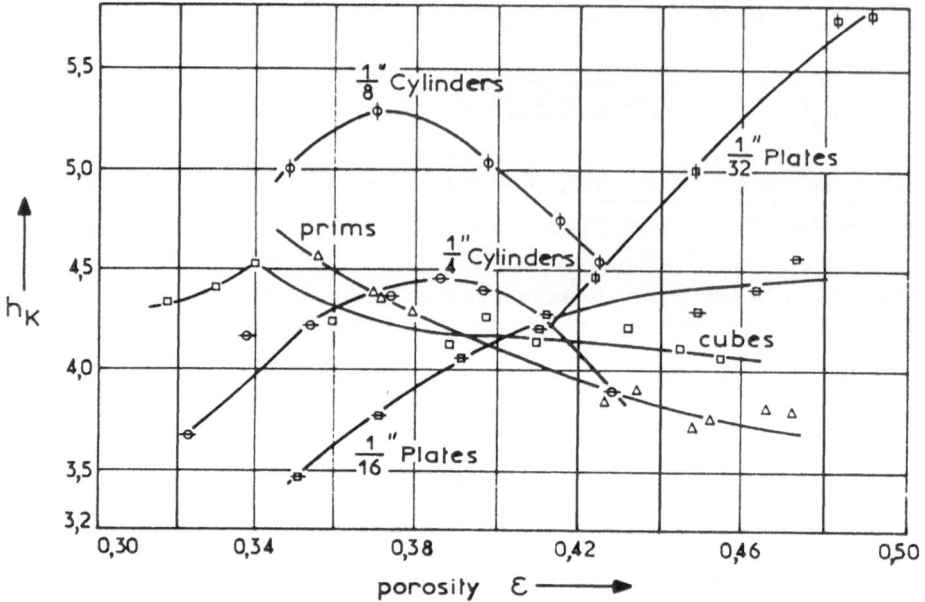

Fig. 3 - Kozeny constant vs porosity for packing of different
shaped particles

For a porous media of depth Z

$$\bar{\omega} = \frac{1}{Z} \int_0^Z \omega \, dZ \qquad (16)$$

$$\left(\frac{\bar{1}}{\omega}\right) = \frac{1}{Z} \int_0^Z \frac{1}{\omega} \, dZ$$

These factors are difficult, even impossible to determine. Finally, equation (15) is only significant if streamlines are identical
to the path followed by diffusing molecules or ions. This is not
necessarily so in the general case of tortuous pores. In conclusion, all relations proposed in the litterature to calculate tortuosity coefficient should be used with precaution. They remain
empirical relations, only useful in precise conditions.

Fortunately, it may be noted that for numerous granular media,
h_K = 4,5 ± 1 (fig. 3). This value of 4,5 is the most commonly
adopted for beds of spherical grains. This experimental result explains the interest of Kozeny's equation which is however not valid for media of high porosity such as fibrous media. For example

media composed of randomly packing fibres give :

ε	0,77	0,94	0,977	0,989
h_k	5,2	8,8	13,8	22,0

Such cases will be studied further on.

An important restriction on the use of Kozeny's model is the distribution of pore sizes. Equation (13) will give poor results if the media presented a wide variation in pore size. Small pores would give a large specific area which would have only a small effect on the fluid flow which would occur mainly in the large pores. In this case some improvments have been proposed to Kozeny's model. A review of the different equations which have been proposed has been made by Marshall[2] and Scheidegger[3].

Swarm of particles models[4]

Kozeny's equation does not apply for high porosity media ($\varepsilon > 0,7$). In such cases, particles can be considered as being completely independent. Then the fluid does not flow in capillary space but around the particles.

Swarm of spherical particles. The friction force corresponding to the flow of fluid around each particle of diameter d_p is given by Stokes'law

$$F = 3 \pi u_m d_p \eta \tag{18}$$

Since the number of spheres in a unit volume is

$$N = \frac{6(1-\varepsilon)}{\pi d_p^3} \tag{19}$$

the pressure through the bed will be

$$\frac{\Delta P}{Z} = \frac{18 \eta u_m (1-\varepsilon)}{d_p^2} \tag{20}$$

by comparison with Kozeny's equation (11) it can be seen that

$$h_k = \frac{\varepsilon^3}{2(1-\varepsilon)} \tag{21}$$

In reality interactions between spheres must be considered and equation (21) applies only for $\varepsilon > 0,95$. Brinkman has calculated a corrective term. He supposes that each spherical grain is surrounded by a porous aggregate to which he applies Darcy's equation. This gives :

$$\frac{\Delta P}{Z} = \frac{18\ \eta\ u_m}{d_p^2} \frac{1}{\frac{1}{1-\varepsilon} + \frac{3}{4}\left[1 - \left(\frac{8}{1-\varepsilon} - 3\right)^{1/2}\right]} \tag{22}$$

then

$$h_k = \frac{\varepsilon^3}{.2} \left\{ 1-\varepsilon + \frac{3}{4}(1-\varepsilon)^2 \left[1 - \left(\frac{8}{1-\varepsilon} - 3\right)^{1/2}\right] \right\}^{-1} \tag{23}$$

Equations (22) or (23) do not apply for porosities lower than 0,5.

   Swarm of cylindrical fibres. These models have been the subject of numerous studies dealing with
   - the flow through circular section cylinders disposed in the flow direction.
   - the flow through a model of two concentric cylinders.
The internal cylinder of radius a corresponds to the fibres of the porous media and the external cylinder of radius b corresponds to the void space around each fiber such that model porosity is the same as the real porosity.
   When flow is parallel to the cylinders, resolution of the Navier-Stokes equation gives

$$B = \frac{1}{8b^2}(4a^2b^2 - a^4 - 3b^4 - 4b^4 \text{Log } \frac{b}{a}) \tag{24}$$

with a hydraulic diameter

$$d_p = \frac{2(b^2 - a^2)}{a} \tag{25}$$

Comparison with equation (13) gives

$$h_k = \frac{2\varepsilon^3}{(1-\varepsilon)\left[2 \text{ Log } \frac{1}{1-\varepsilon} - 3 + 4(1-\varepsilon) - (1-\varepsilon)^2\right]} \tag{26}$$

   When flow is perpendicular to the axis of the cell, resolution of the Navier-Stokes equation gives

$$B = \frac{b^2}{4}\left[\text{Log }\left(\frac{b}{a}\right) - \frac{1}{2}\left(\frac{b^4 - a^4}{b^4 + a^4}\right)\right] \tag{27}$$

or

$$h_k = \frac{2^3}{(1-\varepsilon)\left[\text{Log }\frac{1}{1-\varepsilon} - \frac{1-(1-\varepsilon)^2}{1+(1-\varepsilon)2}\right]} \tag{28}$$

   Random packing of fibres can be represented by a model where 2/3 of the fibres are perpendicular to the flow and 1/3 parallel

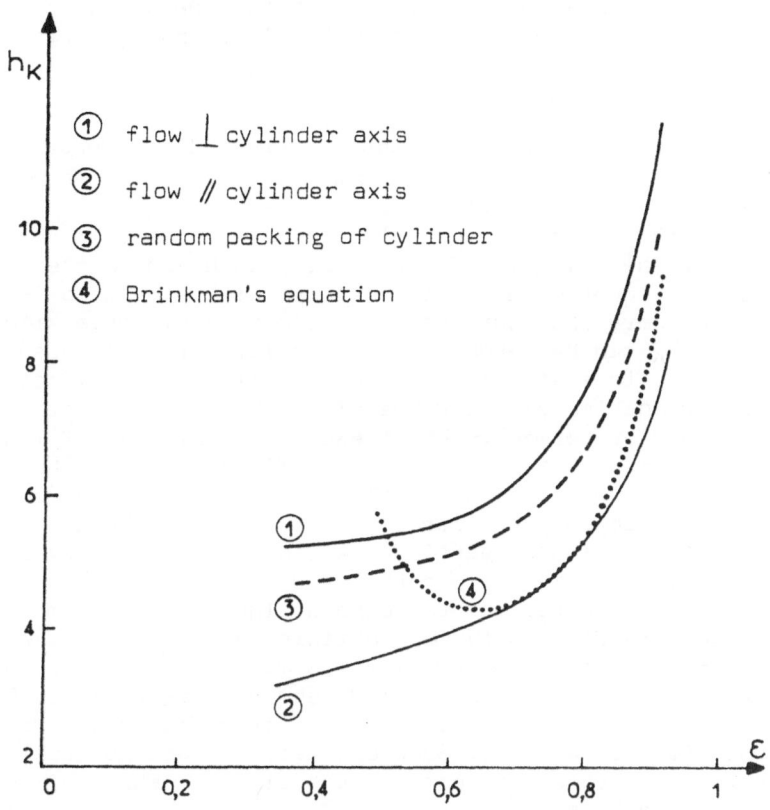

Fig. 4 - Theoretical values of h_k vs porosity

to the flow. Table 1 and fig. 4 give a comparison of the different

Porosity	0.4	0.5	0.6	0.7	0.8	0.9	0.99
Flow parallel to the cy-linder axis	3.4	3.7	4.0	4.4	5.2	7.3	31.1
Flow perpendicular to the cylinder axis	5.3	5.4	5.6	6.2	7.5	11.0	53.8
Flow through random packing of fibres	4.7	5.0	5.1	5.6	6.7	9.8	46.2
Brinkman' equation		5.5	4.3	4.4	5.4	8.8	61

Table 1 - h_k values vs porosity

equations as a fonction of porosity. These equations cannot be used when the porosity is lower than 0.7. In this case the Kozeny equation will generally give a sufficiently close approximation to permeability.

Determination of specific area by means of permeability measurements.

The basis of the method is the relation between specific area and permeability. The validity of permeametry is based on the validity of Kozeny's equation and the Knowledge of the value of Kozeny constant. We have seen that for an estimation of media with normal porosity, h_k can be taken as 4.5. For more precise measurements corresponding to a standard production, it can be sufficient to make one determination of the value of h_k. Permeametry methods have been described by Carman[5]. They present the interest of measuring the real surface which intervenes in fluid flow. On the contrary, adsorption methods measure the total surface. Thus permeability is well adapted for filtration problems. Practically permeametry can be applied to particles whose dimensions vary betwen 0.1 μm and 1 mm. Air is preferentially used for particles smaller than 10 μm and water for the others. The adsorption method can only be used with difficulty for particles greater than 1 μm because the surface is then too small for a measurable adsorption. However the adsorption method does not depend on pore texture. On the other hand permeametry can only be used for sufficiently disordered and uniform size media. So when grain dimensions are widely dispersed it is better to separate the big from the small grains before permeability measurements.

PERMEABILITY OF BINARY MIXTURES

We have seen previously that Kozeny's equation applies correctly to media constituted of monosized grains but that some difficulties can arise for non homogeneous media. Ben Aïm, Le Goff and Le Lec[6] studied binary mixtures and distinguished two cases. Firstly, maximally dispersed mixtures corresponding to low volumetric fraction y_2 of big particles homogeneously dispersed in a bed of small particles of diameter d_1. Secondly, minimally dispersed mixtures corresponding to low volumetric fraction y_1 of small grains. Then the big particles of diameter d_2 form aggregates around which small particles are in sufficient number to maintain the same properties as that of a random packing of particles.

Permeability of maximally dispersed mixtures

We have seen in a previous lecture on the characteristics of
unconsolidated media that for a volumetric fraction of small
beads

$$y = \frac{N_1 V_1}{N_1 V_1 + N_2 V_2} \qquad (29)$$

the porosity was

$$\varepsilon = \frac{\left(\dfrac{\varepsilon_0}{1-\varepsilon_0} - 0.9\,\dfrac{a_2}{a_1}\right)y + 0.9\,\dfrac{a_2}{a_1}}{\left(\dfrac{\varepsilon_0}{1-\varepsilon_0} - 0.9\,\dfrac{a_2}{a_1}\right)y + 0.9\,\dfrac{a_2}{a_1} + 1} \qquad (30)$$

The mean specific area of solid is

$$a = (a_1 - a_2)\, y + a_2 \qquad (31)$$

where a_1 and a_2 are the respective specific areas of grains of
diameter d_1 and d_2. Then permeability can be calculated as a function of new values of porosity and specific area, and compared with
the permeability of a random packing of small particles

$$B_1 = \frac{1}{h_k} \frac{\varepsilon_0^3}{(1-\varepsilon_0)^2 \, a_1^2} \qquad (32)$$

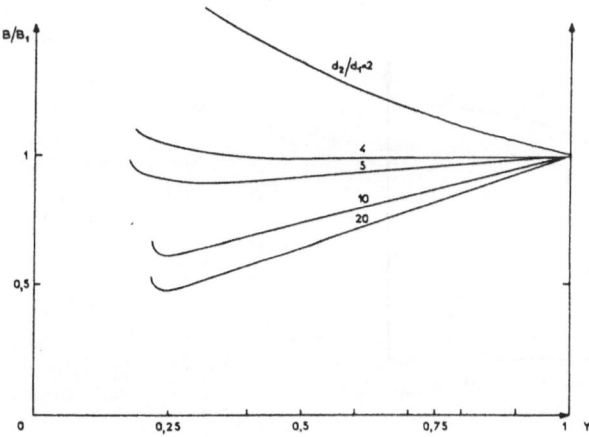

Fig. 5 - Permeability of binary mixture vs volumetric fraction
of small beads

$$\frac{B}{B_1} = \frac{(1-\varepsilon_0)^2 \left[\left(\frac{\varepsilon_0}{1-\varepsilon_0} - 0.9 \frac{a_2}{a_1} \right) y + 0.9 \frac{a_2}{a_1} \right]^3}{\varepsilon_0^3 \left[\left(\frac{\varepsilon_0}{1-\varepsilon_0} - 0.9 \frac{a_2}{a_1} \right) y + 0.9 \frac{a_2}{a_1} + 1 \right] \left[\left(1 - \frac{a_2}{a_1} \right) y + \frac{a_2}{a_1} \right]^2} \qquad (33)$$

It can be seen that the introduction of big spheres in a packing of small ones decreases the permeability when $d_2 > 3.5 \, d_1$ (fig.5) On the contrary, introduction of fibre or disk shaped particles increases permeability (fig. 6).

Permeability of minimally dispersed mixtures

In this case, direct use of kozeny's equation can led to erroneous results because of preferential flow paths through the media. Such media can be considered as a combination of two types of aggregates. One of permeability B_1 occupying an apparent volumetric fraction Y_1 and the other of permeability B_2 occupying the remaining fraction

$$Y_2 = 1 - Y_1 \qquad (34)$$

By electrical analogy, permeability of the mixture can be deduced from the respective disposition of the aggregates. If they

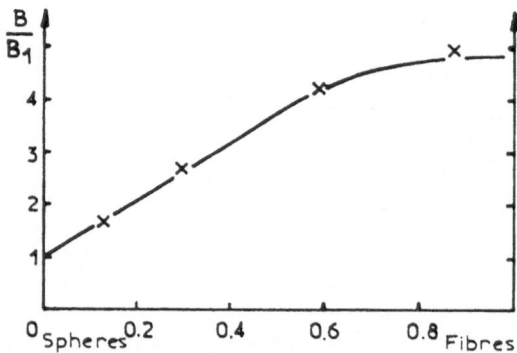

Fig. 6 - Permeability of fibres and spherical grains mixture
sphere diameter 82 μm - fibre : length 500 μm
diameter 14 μm
x experimental points ── Kozeny's equation

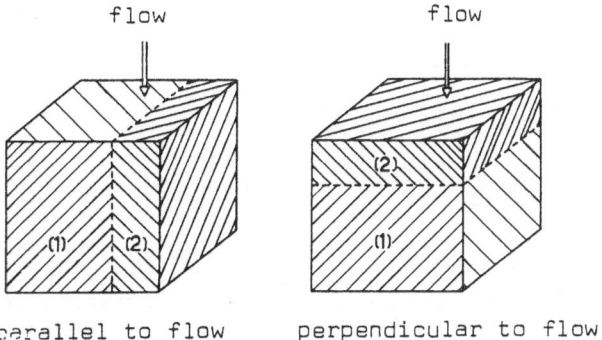

flow   flow

parallel to flow   perpendicular to flow

Fig. 7 - Disposition of aggregates

are disposed parallel to the fluid flow (fig. 7)

$$B = B_1 Y_1 + B_2 Y_2 \qquad (35)$$

on the contrary if they are disposed successively perpendicular to the fluid flow

$$\frac{1}{B} = \frac{Y_1}{B_1} + \frac{Y_2}{B_2} \qquad (36)$$

These two equations delimit a domain inside which the permeability of a binary mixture should be encountered.

A more representative model can be imagined. It is formed by superposition of a lattice with a random disposition of aggregates (fig. 8). For two successive lattices the probability for the fluid to cross two aggregates of the same permeability is Y^2 for permeability B_1, $(1-Y)^2$ for permeability B_2, and the probability

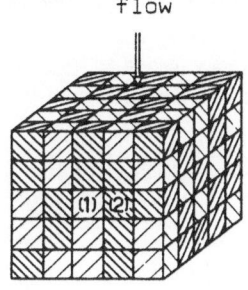

flow

Fig. 8 - Random disposition of aggregates

of crossing aggregates of different permeability is 2Y (1-Y). Then the permeability of such a model is

$$B = Y^2 B_1 + 4Y(1-Y) \frac{B_1 B_2}{B_1 + B_2} + (1-Y)^2 B_2 \qquad (37)$$

Equation (36) corresponds to the case of a flow without any interconnection between fluid paths. On the contrary, equation (37) corresponds to total interconnection of paths. In this case each aggregate receives fluid from all the aggregates of the upper strata and sends it to all the aggregates of the next strata below. So, if there is no preferential flow, permeability should lie between equation (36) and (37). This is the same as the empirical formulae theoreticaly explained by Matheron[7]

$$\text{Log } B = Y \text{ Log } B_1 + (1-Y) \text{ Log } B_2 \qquad (38)$$

Kozeny's equation, if applied to the model of minimal dispersion whose porosity remains equal to $\varepsilon_0 = 0.36$, also gives a result between the two limiting equations. Its specific area is given by

$$a = a_1 Y + a_2 (1-Y) \qquad (39)$$

Then substituing (39) and (32) in (13), we get

$$B = \frac{1}{\dfrac{Y^2}{B_1} + \dfrac{2}{B_1 B_2} Y (1-Y) + \dfrac{(1-Y)^2}{B_2}} \qquad (40)$$

As mentioned this equation gives results between those of equations (36) and (37)

All these equations can be represented by means of a dispersion variable E (fig. 9) which is a fonction of composition Y and ratio B_1/B_2.

$$E = \frac{B - B_1}{B_2 - B_1} \qquad (41)$$

In conclusion, we have pointed out some unrealistic characteristics of Kozeny's model. However the Kozeny's equation remains

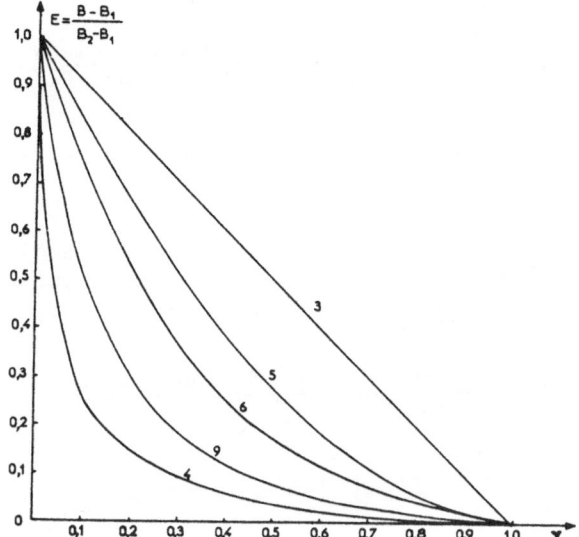

Fig. 9 - Theoretical variation of permeability for binary mixture of spheres (B_2/B_1 = 25)

3 equation (35)
4   "   (36)
5   "   (37)
6   "   (38)
9   "   (40)

a useful tool for industrial estimations of permeability when the
media is of normal porosity even if it is composed of particles
of different shapes.

REFERENCES

1. Rice, P.A., Fontugne, D.J., Latini, R.G., Barduhn, A.J., Ani-
   sotropic Permeability in Porous Media, in Flow Through Porous
   Media, American Chemical Society Publications, Washington, D.C.
   1970, 47

2. Marshall, T.J., Permeability equations and their models, in
   Proceeding of the Symposium on the Interaction Between Fluids
   and Particles, The Institution of Chemical Engineers, London,
   1962, 299

3. Scheidegger, A.E., The Physics of Flow Through Porous Media,
   University of Toronto Press, 1957, chap. 6.

4. Happel, J., Brenner, H., <u>Low Reynolds Number Hydrodynamics</u>, Prentice-Hall, Inc, Englewood Cliffs, N.J., 1965, chap. 8.

5. Carman, P.C., <u>Flow of Gas Through Porous Media</u>, Butterworths Scientific Publications, London, 1956, chap. 4.

6. Ben Aïm, R., Le Goff, P., Le Lec, P., La permeabilité des Milieux Poreux Formés par Empilement de Mélanges Binaires de Grains Sphériques, <u>Powder Technol.</u>, 5, 51, 1971/72.

7. Matheron, G., <u>Elément pour une Théorie des Milieux Poreux</u>, Masson et Cie, Paris, 1967, chap. 6.

CAPTURE MECHANISMS IN FILTRATION

K.J. Ives

Professor of Public Health Engineering
University College London

NOTATION

Symbol	Definition	S.I. Dimensions
B	Stokes-Einstein diffusion coefficient	m^2/s
C	concentration of suspension	-
$\triangle C$	change in concentration of suspension	-
d	particle diameter	m
d_i	diameter of particle of type i	m
d_j	diameter of particle of type j	m
D	grain diameter	m
E	inertia mechanism dimensionless group	-
g	gravitational acceleration (9.81)	m/s^2
G	velocity gradient	1/s
I	interception mechanism dimensionless group	-
k	Boltzmann's constant (1.38×10^{-23})	$J/^{o}K$
k_1	initial pressure drop through a strainer	N/m^2
k_2	Boucher filtrability index	1/m
l	initial thickness of a strainer	m
$\triangle L$	thickness of an elementary filter layer	m
n	arbitrary exponent of I	-
n_i	number of particles of type i per unit volume	$1/m^3$
n_j	number of particles of type j per unit volume	$1/m^3$
N	number of pores per unit strainer area	$1/m^2$

N_{ij}	number of collisions of i- and j-type particles per unit volume	$1/m^3$
m	arbitrary exponent of R	-
P	Peclet Number (diffusion mechanism)	-
ΔP	pressure drop across a filter	N/m^2
r	radial distance measured in a pore	m
r_o	radius of a clean pore	m
R	Reynolds Number (hydrodynamic mechanism)	-
R_G	Reynolds Number for shear gradient	-
R_{vp}	Reynolds Number for particle velocity	-
R_ω	Reynolds Number for flow pulsation	-
R_Ω	Reynolds Number for particle rotation	-
S	gravity mechanism dimensionless group	-
t	time during flow through filter	s
T	thermodynamic (absolute) temperature	^oK
U	velocity of liquid at infinite distance from sphere	m/s
v	approach velocity of filtration	m/s
v_i	interstitial velocity of liquid	m/s
v_p	velocity of particle relative to liquid	m/s
v_r	velocity of liquid at r in a pore	m/s
v_s	Stokes' settling velocity of a particle	m/s
V	volume of liquid strained per unit face area	m
α	exponent of interception group	-
β	exponent of diffusion group	-
γ	exponent of gravity group	-
δ	exponent of hydrodynamic group	-
ϵ	porosity of filter medium	-
ϵ_p	porosity of deposited particles	-
λ	filter coefficient (deep bed)	$1/m$
Λ	filter efficiency	-
μ	dynamic viscosity of liquid	$kg/m\,s$
ρ	density of liquid	kg/m^3
ρ_s	density of particles	kg/m^3
τ	shear stress at liquid-deposit interface	N/m^2
ω	flow pulsation frequency	$1/s$
Ω	angular velocity of rotating particles	rad/s

INTRODUCTION

It is generally accepted that the capture of fine particles in suspension by filtration through a porous medium may be divided into two principal steps: transport and attachment. A third step, detachment, may take place, possibly during filtration, but principally during cleaning of the filtration medium.

Attachment mechanisms have already been described in the lectures by J. Gregory and will not be considered here in any detail. However, the capture mechanism must be considered as a whole, so the two subjects should not be considered in isolation when considering fundamental filtration performance.

It is also convenient to consider filtration as taking place either as a cake (surface filtration) or as deposition within pores (depth filtration). In the former, straining mechanisms are dominant, whereas several other mechanisms interact in the latter. Although surface filtration does take place on deep filters, the nature of the cake is usually different to normal cake filtration as described in the lecture by P.M. Heertjes. Due to the large pore size of the medium in deep filters, the surface cake is discontinuous, with holes in it. This allows penetration of some suspension particles into the pores of the deep filter. This will be described later.

In all cases of filtration the flow is laminar, that is pressure drop is proportional to flow rate (Darcy's Law). Cleasby and Baumann at Iowa State University showed that even with filters considerably clogged with deposits, causing pressure drops many times greater than the clean medium value, Darcy's Law was still obeyed. Also experiments at University College London visualising flow with dye streams, round 5 mm grains showed no disturbance of the streamline flow at rates considerably higher than those encountered in practice.

STRAINING

Basically, straining takes place when a particle in suspension flowing through a pore is larger than the pore opening. Early descriptions of chemical engineering filtration described this as 'complete blocking filtration'. Other operations were also recognised and labelled: interception of particles by the walls of pores at the surface led to 'standard law filtration'; the accumulation of particles on the surface to form a permeable layer was called 'cake filtration'; a mathematical law of exponential rise of pressure drop with volume filtered was also listed, known as 'intermediate law filtration'. No physical basis for this was suggested but a mathematically identical relationship was discovered for the straining of suspensions through fine woven stainless steel mesh (pore opening approx. 35 μm) in water treatment technology, where it was known as Boucher's Law

$$\Delta P = K_1 \, exp\left(K_2 V\right) \tag{1}$$

where ΔP is the pressure drop, V is the volume strained per unit face area.

A modern review of these 'filtration laws' is given in the lecture by P.M. Heertjes, including the development of the case of several particles arriving simultaneously at a pore to block it. However, the 'intermediate' or 'Boucher's Law' has not been analysed. Development of this type of surface filtration depends on two simultaneous mechanisms: deposition in the surface pores, uniformly coating the internal surfaces and narrowing the flow channels, and deposition on the surface between the pore openings, thereby increasing the length of the surface pores. Consequently, the surface filtration forms a discontinuous cake, with holes in it corresponding with the pore openings immediately below.

Analysis of the geometry of the deposition in the surface pores, and assuming Poiseuille flow leads to equation (2).

$$K_2 V = 2\left[1 - \left(\frac{\Delta P}{K_1}\right)^{-\frac{1}{2}}\right] \tag{2}$$

Expansion of the right hand side into a logarithmic series:

$$K_2 V = \ell n \, \frac{\Delta P}{K_1} - \frac{1}{4}\left(\ell n \, \frac{\Delta P}{K_1}\right)^2 + \frac{1}{24}\left(\ell n \, \frac{\Delta P}{K_1}\right)^3 - \cdots \tag{3}$$

In the initial stages of straining (i.e. when $\Delta P/k_1 < 2$) the second and subsequent terms in the series can be neglected and another form of equation (1) results.

$$K_2 V = \ell n \left(\frac{\Delta P}{K_1}\right) \tag{4}$$

For greater values of $\Delta P/k_1$ the other terms in the series become significant.

Analysis of the build-up of deposits on the surface between the pores, gives equation (5), which is a linear relation between pressure drop and volume strained.

$$K_2 V = \frac{2\left(1 - \pi N r_o^2\right)}{\pi N r_o^2}\left[\frac{\Delta P}{K_1} - 1\right] \tag{5}$$

The value of k_2, sometimes called the Boucher filtrability index, from these analyses is given in equation (6)

$$k_2 = \frac{2 \Delta C /(1 - \epsilon_p)}{\pi N \ell r_o^2} \tag{6}$$

ΔC is the volume/volume concentration of particles removed from the flow, ϵ_p is the porosity of the deposited particles, N is the number of pores per unit strainer area, ℓ is the initial thickness of the strainer, and r_o is the initial radius of the surface pores. This defines k_2 in terms of properties of the suspension and the straining medium, but not all of equation (6) has been confirmed experimentally, and k_2 is usually determined empirically. The value k_1 is the initial pressure drop, and may be determined by clean water flow through the clean strainer. In theory, it can be described by the Kozeny equation, but in practice it is difficult to measure the factors (porosity, specific surface) in that equation.

Equation (1), written for strainers, has been found to describe the phenomenon of surface deposition on deep filters. It is an undesirable feature in deep bed filtration as it leads to an exponential rise in pressure drop. Unfortunately, none of the foregoing equations enables any predictions to be made whether surface deposition is likely, and what fraction of the inflow concentration will be removed by it.

A statistical approach has been made to the problem by Bodziony and Litwiniszyn in Krakow by considering a filtering layer (or strainer) having a distribution of pore sizes, through which flows a suspension having a distribution of particle sizes. Particles approaching a pore of equal or smaller size will be retained, whereas particles approaching a pore of larger size will pass through. For a constant inflowing suspension this leads to an asymptotic progression towards complete blocking of the holes, with time, expressed as an exponential function. The rate at which this complete blocking is approached depends the characteristics of both the pore size distribution and the particle size distribution.

INTERCEPTION

Transport mechanisms imply that in the laminar flow conditions found in filtration, the particles uniformly distributed through the liquid must be moved across the streamlines. Those that are transported to a position adjacent to a pore wall then come under the influence of surface forces as described in the lecture by J. Gregory.

An exception to this transport requirement are those particles whose centres are in streamlines near to the pore wall so that their radii cause them to touch the wall. This mechanism is called interception, and could be considered as the final mechanism before contact in all cases.

Interception has been regarded as a filtration mechanism for over 30 years, but more recently it has been characterised by the dimensionless ratio of equation (7)

$$I = \frac{d}{D} \tag{7}$$

where d and D are the particle and grain (or pore) diameters respectively. It is obvious that greater values of I will lead to more efficient particle capture. When I approaches 1.0, straining becomes the dominant mechanism. In the filtration of water and wastewater values of I lie between 2×10^{-4} and 1×10^{-1}, greater values leading to a straining effect.

Because interception leads to the final contact when the mechanisms are operative, it is difficult to separate its effect, although certain research workers claim to have done this.

INERTIA

Streamlines approaching a filter pore have to converge as the flow passes through it. If particles have sufficient inertia they maintain a trajectory which causes them to collide with the pore surface. Using the equations of motion of the liquid and a particle for flow past an isolated spherical grain, a dimensionless group characterises the efficiency of collection of particles on the grain. This efficiency is the ratio of the number of particles striking the grain to the number approaching it at an infinite distance upstream. This inertial efficiency group is given in equation (8).

$$E = \frac{\rho_s d^2 U}{18 \mu D} \tag{8}$$

where ρ_s is the density of the particle, U is the velocity of fluid at infinite distance relative to the spherical grain and μ is the dynamic viscosity of the fluid.

It can be seen that E is independent of the density of the fluid, and inversely proportional to the grain size D. For most packed beds of filter media, the pore size openings are proportional to the grain size, so E increases as the pore openings are smaller. The most significant parameters are the velocity U and viscosity μ. In water filtration, where U is approximately 2 mm/s and μ is 10^{-3} kg/m s , the value of E will

be between 2×10^{-9} and 1.5×10^{-3} which gives a negligible
collection efficiency. This may be contrasted with air
filtration, where significantly higher velocities are used
(e.g. 100 mm/s) and the viscosity is much lower (2×10^{-5} kg/m s).
This gives E values 10^5 to 10^7 times greater, and so inertial
transport, sometimes called impaction, is important in air
filtration.

Due to the low value of E in liquid filtration, it will not
be considered further.

GRAVITY

The effect of gravity, or sedimentation, on liquid
filtration was suggested over 70 years ago, when Hazen described
the pores of slow sand water filters as miniature settling basins.
However, for a long time sedimentation was dismissed as significant
in more rapid rates of filtration because it was argued, the
Stokes' settling velocities of the small particles to be filtered,
particularly flocs, were negligible compared with the filtration
velocities.

This can be illustrated with a simple numerical example,
using Stokes' Law, equation (9)

$$v_s = \frac{g(\rho_s - \rho)d^2}{18\mu} \qquad (9)$$

where $g = 9.81$ m/s^2 , and ρ is the density of the liquid (for
water 1000 kg/m^3). For a clay particle 10 μm diameter, density
2500 kg/m^3 in water with a viscosity of 10^{-3} kg/m s (approximately
20°C), $v_s = 0.82 \times 10^{-4}$ m/s or approximately 0.1 mm/s.

The mean interstitial velocity v_i is the approach velocity
of filtration v , divided by the porosity ϵ . In a typical
granular water filter $v = 2$ mm/s and $\epsilon = 0.4$, giving
$v_i = 5$ mm/s. Consequently, the ratio of settling velocity to
mean interstitial velocity is:

$$\frac{v_s}{v_i} = \frac{0.1}{5} = 0.02 \quad or \quad 2\%$$

This would indicate a low gravity effect, even for a quite dense
particle. (Many flocs have densities as low as 1005 kg/m^3.)

However, analysis of the viscous flow equations round a
sphere (approximating the filter grain) indicate that the
tangential velocity rapidly diminishes to zero at the surface.
For example, for flow round a fixed sphere, in the gravity
direction, at a location 20 μm distance from the surface of a

500 μm diameter sphere, at an angle of 10° from the flow axis at the upstream face the following components of flow velocity can be calculated.

velocity in direction of flow $\quad = \quad 0.10 \quad$ U
velocity perpendicular to flow $\quad = \quad 0.019 \quad$ U
velocity along streamline $\quad\quad = \quad 0.102 \quad$ U
velocity parallel to surface $\quad\quad = \quad 0.036 \quad$ U

where U is the velocity of the liquid relative to the sphere, at an infinite distance. This, of course, idealises the flow round grains in a porous medium, for the flow field cannot develop fully, nor can a proper meaning be attached to U. However, as an indication, if U = v_i , then for v_i = 5 mm/s , the velocity parallel to a tangent at the surface in the calculated example, would be:

v(tangential) = 0.036 x 5 = 0.18 mm/s

In this case, and position relative to the grain, the tangential velocity tending to impel the particle past the surface is of the same order as the Stokes' settling velocity (0.1 mm/s) and the ratio is:

$$\frac{v_s}{v(\text{tangential})} = \frac{0.1}{0.18} = 0.55 \text{ or } 55\%$$

All the liquid velocities of interest are proportional to U, anywhere in the flow field, and therefore proportional to v , the approach velocity. Consequently, the ratio v_s/v is a useful characteristic of the gravity mechanism although its numerical value should not be equated with collection efficiency. So the gravity transport mechanism is represented by the dimensionless group of equation (10).

$$S = \frac{g(\rho_s - \rho)d^2}{18\mu v} \qquad\qquad (10)$$

In water and wastewater filtration S has values lying between 0 and 1.4. The lower limit is set by particles that may be neutrally buoyant, having the same density as water. Cases could arise of S values being negative (oil or fat particles less dense than water), but this would represent a rather special situation.

There is a superficial similarity between the gravity group S and the inertia group E. However, the most significant difference is that E increases with velocity, whereas S diminishes as flow velocity increases. This supports Hazen's opinions of 70 years ago that sedimentation action was important in water filtration, as he was referring to slow sand filtration where the water velocities are very low (about 0.05 mm/s).

Theoretical and experimental investigations of the gravity mechanism by Ison at University College London have shown that in downflow through granular porous media the particles deposit as domes on the upper surfaces of the grains. In upflow they also deposit on the upper surfaces, but in much more limited extent, as small caps near the stagnation point.

DIFFUSION

It has long been observed that very small particles in liquids exhibit a random movement, due to the thermal energy of the water molecules (Brownian motion). Higher liquid temperatures cause greater movement because of the increased thermodynamic energy of the molecules, and the decreased viscous drag on the particles. The viscous drag force is given by the Stokes' drag $3\pi\mu dv_p$, where v_p is the velocity of the particle relative to the fluid. For particles greater than about $1\,\mu m$ in diameter the viscous drag and inertia of the particle restrict the Brownian movement, and the mean free path of the particles is at most one or two particle diameters, and so diffusive movement is not important. For particles less than $1\,\mu m$ the movement becomes increasingly significant with decreasing sizes.

An important parameter is the Stokes-Einstein diffusion coefficient, given in equation (11)

$$B = \frac{\kappa T}{3\pi\mu d} \tag{11}$$

where κ is Boltzmann's constant (energy per degree) and T is absolute temperature. This diffusion coefficient has dimensions of velocity times distance, so dividing B by D, gives the mean velocity imparted by Brownian motion over a distance of one grain diameter. This velocity can be expressed as a ratio with the velocity of the particle due to liquid movement (advective velocity) which is proportional to v.

$$\frac{\text{Brownian velocity}}{\text{advective velocity}} = \frac{B}{Dv}$$

This ratio is the reciprocal of the Peclet Number P, familiar in diffusion processes.

$$\frac{1}{P} = \frac{B}{Dv} = \frac{\kappa T}{3\pi\mu d v D} \tag{12}$$

In water filtration the range of values of $1/P$ is from about 10^{-8} to 0.5×10^{-5}. Experimentation with such small particles (sub-micronic) is very difficult but there is evidence from Sholji at University College London and Yao at University of North Carolina that equation (12) represents the removal characteristic for sub-micronic particles, and that smaller particles are collected more efficiently due to their greater Brownian motion.

Another aspect of diffusive motion is that of particle movements down concentration gradients studied by Litwiniszyn in Krakow. In special experimental circumstances a concentration front was established in a filter by injecting a concentrated suspension into clean liquid flow to establish significant concentration gradients. In normal filtration situations concentration gradients are too small to provide significant diffusive movement.

HYDRODYNAMIC ACTION

In a uniform liquid shear field, derived from a constant velocity gradient across the streamlines, a spherical particle will experience a greater liquid velocity on one side than on the other. This difference in drag on each side causes the particle to rotate and so create a spherical flow field. This in turn produces a pressure difference laterally to the flow direction, and the resultant force causes the particle to move across the flow field to the region of higher velocity.

In a non-uniform shear field, as in a filter pore, such lateral forces also exist, but they are not constant with position. In a non-stationary, non-uniform shear field as exists in a collection of interconnected pores of different sizes, the lateral forces are complex and time-dependent.

If the particle is not spherical, and its centre of mass and hydrodynamic centre do not coincide, the particle will experience further out-of-balance forces and will rotate at a non-uniform rate in a geometrically non-uniform manner.

As most particles to be filtered are non-spherical, suspended in a time-dependent, non-uniform laminar flow field, their movements due to the effects mentioned, will appear as a random drifting motion across the streamlines. Such migrations or drifts of particles have been observed in straight capillaries, but not directly in filter pores. Indirect evidence has been obtained by Ison at University College London from experiments in which all the other transport mechanisms were either negligible or held constant, and only the Reynolds Number for flow through the filter (equation (13)) was varied. This caused a change in

filtration efficiency which could only be attributed to the
hydrodynamic effect of a change in Reynolds' Number causing a
change in flow pattern, and presumably a change in the shear
field configuration.

$$R = \frac{v\,D\,\rho}{\mu} \qquad (13)$$

In water filtration R has a value of approximately 1.0.

Although this Reynolds' Number is a convenient expression
for the filter, using grain size D as the characteristic length
and approach velocity v as the characteristic velocity, it does
not reflect the intrinsic dependence of the hydrodynamic
transport mechanism on particle size d , or shear gradient G .
Other factors which may have significance in this mechanism are
the velocity of the particle relative to the liquid v_p , the
angular velocity of a rotating particle Ω , and the frequency ω
of a pulsating fluid flow due to pore size sequences.
By appropriate choice of the length and velocity terms four more
Reynolds' Numbers can be defined:

$$R_G = \frac{G\,d^2\rho}{\mu} \qquad (14a)$$

$$R_{v_p} = \frac{v_p\,d\,\rho}{\mu} \qquad (14b)$$

$$R_{\Omega} = \frac{\Omega\,d^2\rho}{\mu} \qquad (14c)$$

$$R_{\omega} = \frac{\omega\,d^2\rho}{\mu} \qquad (14d)$$

Due to obvious experimental difficulties, there has been no
assessment of the real significance of these various Reynolds'
Numbers. With the exception of the influence of particle size,
they can be expressed in terms of the simple Reynolds' Number
(equation (13)) and it may be that a parameter of the form of
equation (15) is more significant. This suggestion, however,
is entirely speculative

$$I^n R^m = \left(\frac{d}{D}\right)^n \left(\frac{v\,D\,\rho}{\mu}\right)^m \qquad (15)$$

There is another hydrodynamic phenomenon which is completely unconnected with the shear field drifting. This is the viscous resistance experienced by a particle during very close approach to a pore wall. When the separation of the particle and the wall is very small, less than the particle diameter, a displacement of the liquid in the gap is necessary for the particle to make contact. This displacement is a form of radial viscous flow, and due to the viscous resistance the particle experiences a force retarding its approach to the wall. This effect, however, is scarcely affecting transport mechanisms, but comes into action when surface forces are affecting a particle. Consequently, it enters as an extra in the force equations which include molecular and coulombic forces at very small particle-wall separations.

COMBINED MECHANISMS

The various transport mechanisms are shown diagrammatically on Figure 1. It is probable that all these will act simultaneously, although with varying degrees of relative importance depending on the nature of the suspension and filter medium.

As straining is not strictly a transport mechanism, and particle inertia is known to be insignificant in liquid filtration they will not be considered further. It could be argued that interception is also not a transport mechanism, but it enters as the limit condition for other transport mechanisms so cannot be neglected. In the limit as d/D approaches 1.0, interception becomes straining, and so straining may be regarded as a limiting condition of interception.

Inspection of the interception, gravity and diffusion mechanisms reveals that filtration efficiency will increase with some positive powers of I , S and $1/P$. In the case of the hydrodynamic mechanism, which is obscure in detail, represented by the simple Reynolds' Number R , its relationship with filtration efficiency is not so obvious. As the mechanism is attributed to the viscous flow fields in the pores, it would seem that the dominance of viscous effects would produce more random drift. Therefore, an inverse power relationship with R would be indicated as $1/R$ is the ratio of the viscous to the inertial fluid forces. This was verified by Ison in the one set of experiments where R was varied while all other mechanism parameters were held constant.

The efficiency of particle retention may be conveniently expressed as Λ , the fraction of suspension retained in a filter layer one grain diameter thick.

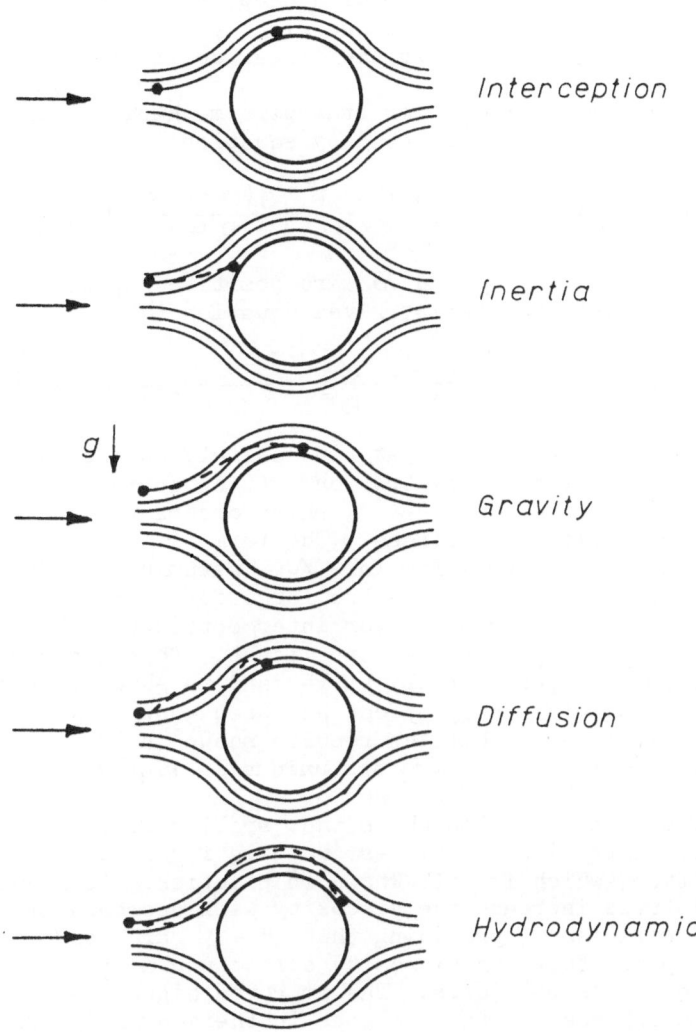

Figure 1. Simplified diagrams of particle transport mechanisms.

$$\Lambda = - \frac{\Delta C}{C} \frac{D}{\Delta L} \qquad (16)$$

where $-\Delta C/\Delta L$ is the concentration change per unit filter thickness and C is the concentration flowing into the layer ΔL. Comparison of equation (16) with deep bed theory (see lecture on Mathematical Models of Deep Bed Filtration by K.J. Ives) indicates that $\Lambda = \lambda D$ where λ is the filter coefficient.

In terms of the filter transport mechanisms, the filter efficiency may be represented by equation (17)

$$\Lambda = const. \left(\frac{d}{D}\right)^{\alpha} \left(\frac{\kappa T}{3\pi\mu\, dv D}\right)^{\beta} \left(\frac{g(\rho_s - \rho)d^2}{18, \mu v}\right)^{\gamma} \left(\frac{\mu}{v D\rho}\right)^{S} \qquad (17)$$

where α, β, γ and δ are positive exponents. Collecting together terms, gives equation (18).

$$\Lambda = const. \frac{d^{\alpha - \beta + 2\gamma}}{\mu^{\beta+\gamma-\delta} D^{\alpha+\beta+\delta} v^{\beta+\gamma+\delta}} (\kappa T)^{\beta} \frac{(\rho_s - \rho)^{\gamma}}{\rho^{\delta}} \qquad (18)$$

Equation (18) indicates that, in general, an increase in particle size d will improve filtration. An exception is where β is large compared to $\alpha + 2\gamma$, which occurs for very small particles affected by diffusion forces. The form of the exponent of d would give a minimum value for Λ, assuming all other factors to be constant, when d is too large for effective diffusion (small β), but too small for interception (small α) or gravity (small γ) to be significant. This has been demonstrated experimentally by Yao and is shown on Figure 2. This minimum efficiency effect has been known for some time in aerosol filtration, but the results shown in Figure 2, giving a minimum around $d = 1 \,\mu m$ are unique in liquid filtration.

Equation (18) also shows that smaller grain sizes D, and smaller filtration velocities v give improved efficiency of collection, which is well-known in practice. As liquid temperatures increase the viscosity will decrease leading to better efficiency, providing that $\beta + \gamma$ are large relative to δ. This implies that the viscosity effect would be most marked for small or dense particles. The improved higher temperature effect is also enhanced by the increase in thermodynamic energy kT.

It is possible to formalise the derivation of equation (17) by means of dimensional analysis using Buckingham's Theorem. This does not yield any additional insight into the problem, but does ensure that the variables are properly grouped without any ambiguous or redundant entries. One of the principal sets of experiments made by Ison to evaluate filtration transport mechanisms used the dimensionless groups of equation (17) to define the experiments, varying only one group at a time.

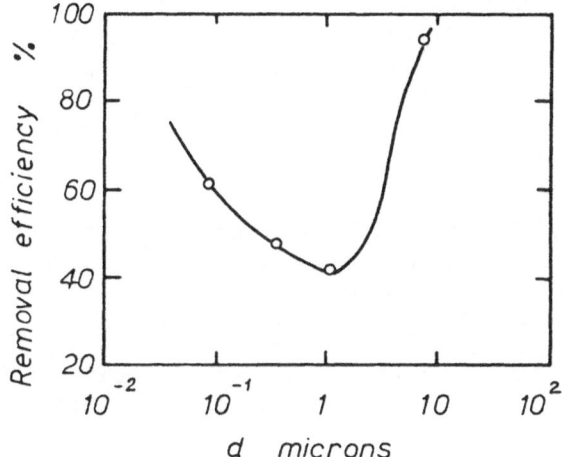

Figure 2. Filter efficiency against particle size, showing
           minimum at 1 micron. (Yao).

However, it can be seen from equation (18) that for a suspension
of particles of given density ρ_s , over a practical temperature
range (say, 5 to 25°C) where variation in liquid density ρ and
thermodynamic energy kT would be small, a set of observations
could be made with the four variables d , μ , D and v.
The power dependence of this form would yield four simultaneous
equations allowing α , β , γ and δ to be evaluated.
This would then define the relative importance of the four
mechanisms for the suspension under study.

    The form of equation (18) shows why there is so little
agreement among published results of the dependence of filtration
efficiency upon various operating variables: particle size,
grain size, flow rate, temperature, etc. A number of different
transport mechanisms may operate simultaneously; their relative
significances depend on the nature of the suspension as well as
the operation of the filter.

Flocculation

    In view of the importance of particle size d , and to a
lesser extent particle density ρ_s , it is worth considering
the changes in these characteristics that would be achieved by
flocculation within the filter pores. This is not a consideration
of the flocculation that might take place as a pretreatment
because that takes place in circumstances quite unaffected by the
filtration process.

In the filter pores the laminar flow conditions create shear gradients (velocity gradients), consequently, orthokinetic flocculation is possible. The rate of particle collision, leading to particle aggregation, in a velocity gradient G is given by the Smoluchowski equation (19).

$$\frac{dN_{ij}}{dt} = \frac{G}{6} n_i n_j \left(d_i + d_j\right)^3 \tag{19}$$

where n_i and n_j are the numbers of particles per unit liquid volume of diameters d_i and d_j. Time t is residence time during the passage of the suspension through the filter pores, so that Δt may be written as $\Delta L/v_i$ or $\epsilon \Delta L/v$.

In a non-uniform shear field, the velocity gradient can be replaced by the Camp-Stein function of power dissipation, equation (20).

$$G = \left(\frac{power\ dissipated}{volume\ x\ viscosity}\right)^{1/2} \tag{20}$$

In a filter the power dissipated by fluid motion through the pores is evinced as pressure drop ΔP and equation (20) becomes:

$$G = \left(\frac{v\,\Delta P}{\epsilon \mu \Delta L}\right)^{1/2} \tag{21}$$

The Kozeny equation can be used to give the pressure gradient.

$$\frac{\Delta P}{\Delta L} = 5\mu v \frac{(1-\epsilon)^2}{\epsilon^3}\left(\frac{6}{D}\right)^2 \tag{22}$$

$$G = \frac{13.4\,v\,(1-\epsilon)}{D\,\epsilon^2} \tag{23}$$

Using equation (23) and the transform $\Delta t = \epsilon \Delta L/v$, the Smoluchowski equation becomes:

$$\frac{dN_{ij}}{dL} = 2.23 \frac{(1-\epsilon)}{\epsilon D} n_i n_j \left(d_i + d_j\right)^3 \tag{24}$$

Equation (24), so far not tested experimentally, raises some interesting speculations. It indicates that the rate of particle aggregation in the pores, with respect to depth, is independent of flow rate and that it is directly proportional to the internal pore surface per unit pore volume, $6(1-\epsilon)/\epsilon D$.

Flocculation within filter pores is a phenomenon known and used in practice. The factors which affect is and how it may be controlled are still unknown, and equation (24) requires verification before it can be accepted.

DETACHMENT

There is experimental and practical evidence that increasing the flow in a deep bed filter, when deposited particles are present in the pores, leads to detachment of some of these particles causing a locally increased suspension concentration. It is the opinion of Mints of U.S.S.R. that such detachment takes place even at constant flowrate, because deposits in the pores cause local increases in interstitial velocity. This subject is discussed more fully in the lecture Mathematical Models of Deep Bed Filtration by K.J. Ives.

Such detachment, whether at constant or increasing flowrate, is primarily due to increased liquid shear stress at the deposit surface. Rupture of part of the deposit at a plane of weakness causes a particle, or more usually an aggregate of particles, to detach and be entrained in the flow. Thereafter, the behaviour of the detached particle is the same as any other in suspension, and it may be subsequently redeposited. The shear stress at the liquid-deposit boundary is given by Newton's equation (25)

$$\tau = \mu \left(\frac{dv_r}{dr} \right)_{r_{max}} \tag{25}$$

where v_r is the local velocity at a radial position r measured from the centreline axis of the pore. As v_r will be linearly proportional to the interstitial velocity v_i($= v/\epsilon$)and r_{max} will depend on pore size (a function of grain size, porosity and quantity of deposited particles), it is obvious that increased velocities and smaller pore sizes will give greater shear stresses. It is obvious because such conditions also lead to greater pressure drops, which are another manifestation of shear stresses at the boundaries, shown in its simplest form in the derivation of Poiseuille's equation in capillary flow.

These factors do not give any guidance however to the fundamental property which affects detachment: shear strength of the deposits. It is claimed that one of the principal advantages of using polyelectrolytes in filtration is that they increase the shear strength of the deposits, thus allowing less detachment. There is not at the present time an adequate method of measuring shear strength, particularly of deposits in situ.

Other modes of detachment have been observed visually at University College London in deep bed filter models, when large accumulations of deposit have built up in filter pores. Such accumulations appear to be unstable and a particle impinging on the upper part of such a deposit may create an "avalanche" with many aggregates breaking away into the flowing liquid. This phenomenon is even less amenable to analysis than the shear rupture of deposits, but once again it must depend to a large degree on the shear strength of the deposits.

CONCLUSION

Although it is possible by careful experimentation to study the various filtration transport mechanisms, and it is possible to put numerical values to the dimensionless parameters which characterise the mechanisms, it is not possible to predict filter performance from prior knowledge of the physical constituents of the process.

This inability to predict filter performance arises principally from the following factors:
(i)   lack of knowledge of the detailed structure of fluid flow in the filter pores;
(ii)   lack of knowledge of the characteristics of the particles in suspension;
(iii)   insufficient knowledge of the physics of the system: pore-particle-liquid;
(iv)   insufficient knowledge of the geometry of the porous medium;
(v)   complexities arising due to deposited particles changing the geometry of the pores.

Consequently, filter theory and filter design can rarely use this understanding of the transport mechanisms except to predict semi-quantitatively changes that might take place in a known filtration system when an operating variable is altered. Modern filter theory accepts that there is a certain interaction between a given suspension and a given filter medium, leading to clarification of the suspension. This interaction is designated by the filter coefficient λ ($= \Lambda/D$) which is subsequently elaborated as given in the lecture Mathematical Models of Deep Bed Filtration.

BIBLIOGRAPHY

The following publications deal with the subject more extensively and contain reference lists.

1. Herzig, J.P., Leclerc, D.M., Le Goff, P., Flow of suspensions through porous media, Ind. Eng. Chem., 62, 8, 1970.

2. Ison, C.R. and Ives, K.J., Removal mechanisms in deep bed filtration, Chem. Eng. Sci., 24, 717, 1969.

3. Ives, K.J., Theory of filtration, in International Water Supply Association Eighth Congress Vienna, Volume 1, I.W.S.A., Park Street, London, 1969.

4. Ives, K.J., Rapid filtration, Wat. Res., 4, 201, 1970.

5. Ives, K.J., Filtration of water and wastewater, Critical Reviews in Environ Contr., 2, 293, 1971.

6. Mints, D.M., Modern theory of filtration, in International Water Supply Association Seventh Congress Barcelona, Volume 1, I.W.S.A., Park Street, London, 1966.

7. Yao, K.M., Habibian, M.T., O'Melia, C.R., Water and waste water filtration : concepts and applications, Environ. Sci. Technol., 5, 1105, 1971.

MATHEMATICAL MODELS OF DEEP BED FILTRATION

K.J. Ives

Professor of Public Health Engineering
University College London

NOTATION

Symbol	Definition	S.I. Dimensions
a_1	filter coefficient constant (Ives)	$1/m$
a_2	filter coefficient constant (Ives)	$1/m$
A	inlet face area of filter	m^2
b	packing constant	-
C	concentration of suspension, vol/vol	-
C_o	inlet concentration of suspension	-
D	grain diameter	m
D_o	grain diameter at inlet face	m
F	function of	-
g	gravitational acceleration (9.81)	m/s^2
H	head loss	m
H_d	head loss due to pore deposition	m
H_o	head loss of clean filter	m
H_s	head loss due to surface deposition	m
j	gradient of grain size/distance relationship	-
k	head loss constant	-
k_s	initial head loss at surface	m
k_t	rate constant of surface head loss	$1/s$
l	length of a capillary per unit depth	-

L	distance from inlet face of filter	m
m	$2(p_1+j)$	-
m_1	exponent of grain size	-
m_2	exponent of velocity	-
n	number of contact points on a grain	-
N	number of capillaries per unit face area	$1/m^2$
p_1	filter coefficient constant (Diaper-Ives)	-
p_2	filter coefficient constant (Diaper-Ives)	-
ΔP	pressure drop across a filter	N/m^2
Q	volumetric flow rate of suspension	m^3/s
r	radius of a capillary	m
R	radial distance from axis of radial filter	m
R_o	radial distance of inlet face	m
$\bar{R}$	R/R_o	-
S	specific surface of filter pores	$1/m$
S_o	initial specific surface	$1/m$
t	elapsed time of filtration	s
T	time group (Diaper-Ives)	-
v	approach velocity, Q/A	m/s
v_c	critical interstitial velocity	m/s
v_i	interstitial velocity	m/s
v_o	velocity at inlet face of radial filter	m/s
V	volume of a coated grain	m^3
V_o	volume of a clean grain	m^3
x	exponent of velocity term	-
y	exponent of spherical specific surface term	-
z	exponent of capillary specific surface term	-
α	scour coefficient	$1/s$
β	bulking factor for deposits	-
e	porosity of filter medium	-
e_o	porosity of clean filter medium	-
θ	thickness of deposit coating a capillary	m
λ	filter coefficient	$1/m$
λ_o	initial filter coefficient	$1/m$
λ_{oR_o}	inlet face filter coefficient for radial filter, $\lambda_o v_o$	$1/s$
ν	kinematic viscosity of filtering liquid	m^2/s
ρ	density of filtering liquid	kg/m^3
σ	specific deposit, vol. of deposit/unit filter vol.	-
σ_a	absolute specific deposit	-
σ_o	specific deposit at inlet face	-
σ_u	saturation value of specific deposit	-

INTRODUCTION

Deep bed filters are normally operated to clarify dilute
suspensions. Like all filters they exhibit a pressure drop,
which rises as the filter becomes clogged. Consequently,
mathematical models of deep bed filtration must represent both
the clarification process and the pressure drop changes.
Both of these have initial conditions when the filter is clean,
and both change in a time-dependent manner during the process
of clogging.

The clarification process is measured by changes in
suspension concentration C , depending on the quantity of deposit
per unit filter volume σ (specific deposit) and the independent
variables of distance from the inlet face of the medium L , and
elapsed time of filtration t. Pressure drop ΔP variation will
depend on the same variables. Mathematical models are produced
in order to predict concentration and pressure drop at any values
of L and t , in terms of the physical parameters and operating
variables of filtration (inlet concentration C_o , filtration
approach velocity v, grain size D, initial porosity ϵ_o , water
temperature, suspension particle characteristics and surface
chemistry of the system). Such prediction should produce better
designs and aid optimisation studies.

CLARIFICATION IN A UNIFORM FILTER

In spite of extensive knowledge of the surface chemistry of
filtration systems, and of transport mechanisms in filter pores,
there is still insufficient understanding of the nature of
suspensions, and the physics of filtration, to enable removal
efficiencies to be predicted from the physical characteristics
of the components of the system. Consequently, an empirical
measure of the interaction between a uniform suspension and a
uniform filter medium has been used as the basis for filtration
clarification theory. This is the filter coefficient λ .

The basic theoretical hypothesis is that the removal of
suspension with respect to distance (depth) in the filter is
first order.

$$- \frac{\partial C}{\partial L} = \lambda C \tag{1}$$

This defines the filter coefficient as the proportion of suspension
concentration removed per unit depth. Equation (1) assumes that
initially, when the filter medium is clean (i.e. contains no
deposited suspension particles), every layer of the filter is
equally efficient at removing particles from suspension. Also,
it assumes that in every layer the suspension entering it and

206

leaving it is uniformly dispersed. At the commencement of
filtration, when the elapsed filtration time is zero, equation (1)
can be integrated to yield equation (2)

$$C = C_o \exp(-\lambda_o L) \qquad (2)$$

The initial filter coefficient is λ_o at t = o , and C_o is the
inlet concentration at L = O. This initial exponential decline
in suspension concentration is presented on Figure 1.

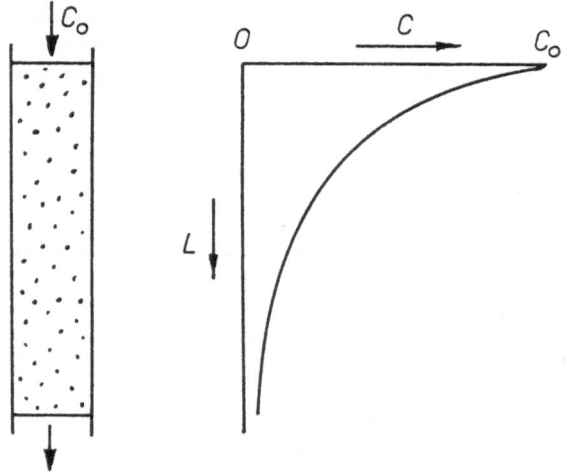

Figure 1. Exponential decline of concentration of suspension
with depth at time t = o.

Another basic equation is the mass balance for the suspension
particles. In principle it is simply a statement that particles
removed from suspension are deposited in the filter pores. In an
element of the filter medium, face area A and thickness ΔL ,
the suspension experiences a loss of concentration, volume by
volume, of $-\Delta C$. The inflowing suspension is carried by a
volumetric flowrate Q , and the flowthrough takes time Δt.
During this time the specific deposit (volume of deposited
particles per unit filter volume) has increased by $\Delta\sigma_a$.

Volume of particles removed from suspension $= -\Delta C Q\, \Delta t$
Volume of particles increased in deposits $= \Delta\sigma_a A\, \Delta L$

$$-\Delta C\, Q\, \Delta t = \Delta\sigma_a A\, \Delta L \tag{3}$$

In differential form equation (3) becomes equation (4)

$$-\frac{\partial C}{\partial L} = \frac{A}{Q}\frac{\partial\sigma_a}{\partial t} \tag{4}$$

Here σ_a represents the absolute specific deposit, being the solid volume of the particles deposited per unit filter volume. It is useful to consider the volume effectively occupied in the pores, which includes the self-porosity of the deposited particles (some measured values have given about 60%). So the effective specific deposit $\sigma = \beta\sigma_a$, where β is a bulking factor. A consequence of this redefinition is that the local instantaneous porosity is given by equation (5).

$$\epsilon = \epsilon_o - \sigma \tag{5}$$

The balance equation (4) does not take into account low order terms in C and $\partial C/\partial t$, and the effects of diffusional gradients. In the dilute suspensions which are clarified by deep bed filtration (less than 500 mg/l) such terms are negligible; more complete discussions are available in the papers by Herzig et al[1] and Ives and Horner.[2]

The approach velocity v is defined as Q/A, so equation (4), incorporating the effective specific deposit, becomes equation (6).

$$-\frac{\partial C}{\partial L} = \frac{1}{\rho v}\frac{\partial\sigma}{\partial t} \tag{6}$$

During the process of clarification and deposition, leading to progressive clogging of the filter pores, the removal efficiency of the filter changes and the exponential decline of equation (2) is no longer valid. Consequently, some modification of equation (1) is required, which must take account of the changes in the pores that are both depth and time dependent.

The complexity of these changes even for an initially uniform filter can be readily demonstrated by a numeric example. Assume that an inlet concentration of 1000 units flows through a filter that is 70% efficient per unit layer.

Time interval 1.

	Initial deposit	Concentration	Final deposit
Inlet	-	1000	-
Layer 1	0	300	700
Layer 2	0	90	210
Layer 3	0	27	63
Layer 4	0	8	19

During time interval 2, layer 1 will have an efficiency modified by the presence of 700 units of deposit, layer 2 will be modified by 210 units, layer 3 by 63 units, and so on. Therefore each layer will remove a different proportion of the inflowing suspension, and modify even further its efficiency and accentuate the differences between the layers.

Two principal approaches have been made to this problem, one known as the deposition and scour theory, the other known·as the geometric or modified filter coefficient theory.

Deposition and scour

The deposition and scour hypothesis propounded by Daniel Mints of Moscow, relies on the fact that the deposits in the pores narrow the channels for flow causing a rise in interstitial velocity (at constant flow, which is normal). This increased local velocity causes some of the deposits to be scoured back into suspension, the quantity scoured being proportional to the amount of deposit present.

Equations (1) and (6) can be combined to form equation (7)

$$\frac{\partial \sigma}{\partial t} = \beta v \lambda C \qquad (7)$$

This, however, only gives the rate of increase·of specific deposit at the initial condition when $t = o$ and $\sigma = o$. As soon as deposit appears, some fraction of it is scoured back into suspension, to diminish the rate of increase of σ .

$$\frac{\partial \sigma}{\partial t} = \beta v \lambda C - \alpha \sigma \qquad (8)$$

The scour coefficient is α . Dividing equation (8) by βv and substituting equation (6):

$$-\frac{\partial C}{\partial L} = \lambda C - \frac{\alpha \sigma}{\beta v} \qquad (9)$$

If equation (9) is differentiated with respect to t , and again equation (6) is substituted, equation (10) is derived.

$$-\frac{\partial^2 C}{\partial t \, \partial L} = \lambda \frac{\partial C}{\partial t} + \alpha \frac{\partial C}{\partial L} \qquad (10)$$

$$\frac{\partial^2 C}{\partial t \, \partial L} + \lambda \frac{\partial C}{\partial t} + \alpha \frac{\partial C}{\partial L} = 0 \qquad (11)$$

Boundary conditions for equation (11) are $C = C_o$ at $L = 0$, and
equation (2) when $t = o$. The solution to equation (11) is:

$$\frac{C}{C_o} = \exp -(\lambda L + \alpha t) \sum_{i=0}^{\infty} \left(\frac{\alpha t}{\lambda L}\right)^{i/2} I_i \left[(\lambda L \alpha t)^{1/2}\right] \tag{12}$$

where $I_i \left[(\lambda L \alpha t)^{1/2}\right]$ is a Bessel function of order i.

In his earlier publications, Mints gave a different series
solution to equation (11); it was however an approximation and
equation (12) is correct.

Similar equations to (11) and (12) can be derived for the
variation of σ with L and t. The forms of the solutions are
shown graphically on Figure 2, where it can be seen that there is
a limit condition which starts at the inlet face and progresses
through the depth of the filter. In this limit condition the
specific deposit has reached its saturation value σ_u, and
according to Mints the rate of deposition equals the rate of scour.
Consequently, the suspension concentration is unchanged and $C = C_o$.
It follows that $\partial C / \partial L = 0$, so equation (9) becomes:

$$0 = \lambda C_o - \frac{\alpha \sigma_u}{\beta v} \tag{13}$$

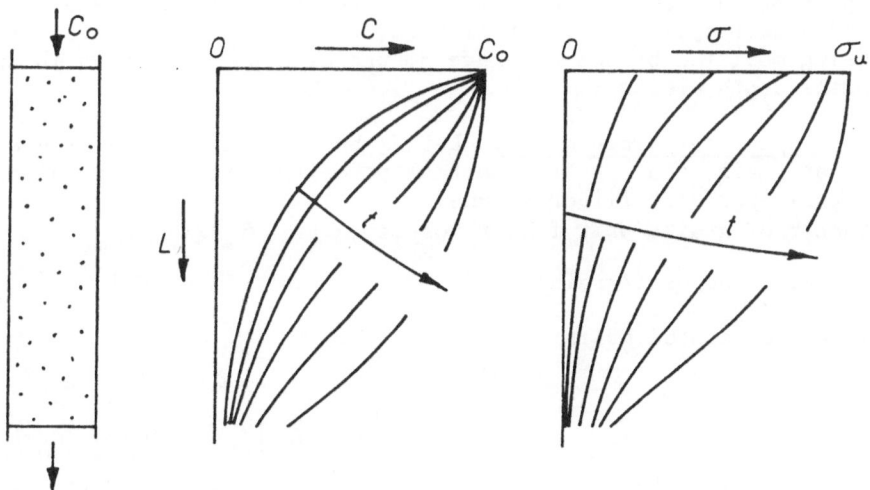

Figure 2. Variation of concentration C and deposit σ with
depth L and time t in a deep bed filter.

Equation (13) becomes a means of defining the scour coefficient α, which is a constant of the process.

$$\alpha = \frac{\beta v \lambda C_o}{\sigma_u} \qquad (14)$$

As β, v and λ are also constants, it follows that the ratio C_o / σ_u must be constant. This raises a problem, because C_o can be increased to any value, whereas σ_u cannot exceed the initial porosity ϵ_o (the deposit volume cannot be greater than the pore volume). It seems likely, therefore, that α cannot be a constant, but some function of concentration C.

Modified filter coefficient

A variety of mathematical models has been presented in which the filter coefficient λ of equation (1) is modified to be a function of the specific deposit σ. These models assume that the changing efficiency of the filter is due to the changes in pore geometry caused by the deposits. They are sometimes referred to as the geometric theory. The most general of these takes account of changes in specific surface in the pores, and the increased interstitial velocity due to narrowing of the pore flowpaths.

This general model will be considered under three headings: the porous medium being represented by an assembly of individual spheres, the porous medium being represented by an assembly of individual cylindrical capillaries, the interstitial velocity being modified by the average amount of deposit present in any small depth element.

Spherical grain model. Let V_o be the volume of a single clean grain, and ϵ_o the initial porosity, and σ the volume of deposit per unit filter volume.

Number of grains/unit filter volume $= (1 - \epsilon_o)/V_o$

Volume of deposit/grain $= \sigma V_o/(1 - \epsilon_o)$

Volume of coated grain $= V$

$= V_o + \sigma V_o/(1 - \epsilon_o)$

$= V_o \left[1 + \sigma /(1 - \epsilon_o)\right]$

$1/(1 - \epsilon_o) = \left[\epsilon_o/(1 - \epsilon_o)\right]\left[1/ \epsilon_o\right] =$ packing constant$/\epsilon_o = b/\epsilon_o$

Therefore, $V = V_o (1 + b\sigma/ \epsilon_o)$

The surface per unit filter volume of clean, and deposit-containing filter (specific surface) are respectively S_o and S. The ratio of specific surfaces is proportional to the ratio of the volumes to 2/3 power.

$$\frac{S}{S_o} = \left(\frac{V}{V_o}\right)^{2/3} = \left(1 + b\sigma/\epsilon_o\right)^{2/3} \tag{15}$$

Capillary model. Let there be N capillaries per unit face area, length l per unit depth, internal radius r.

Initial porosity $\epsilon_o \quad = \quad \pi r^2 Nl$

Specific surface $S_o \quad = \quad 2\pi rNl$

The inside of each capillary is coated with a thickness θ.

Specific deposit $\sigma = \pi r^2 Nl - \pi(r - \theta)^2 Nl$

Coating thickness $\theta = r\left[1 - (1 - \sigma/\epsilon_o)^{\frac{1}{2}}\right]$

Specific surface $S \quad = \quad 2\pi(r - \theta)Nl$

$$= \quad S_o - 2\pi\theta Nl$$

$$= \quad S_o - 2\pi rNl + 2\pi rNl(1 - \sigma/\epsilon_o)^{\frac{1}{2}}$$

$$S \quad = \quad S_o(1 - \sigma/\epsilon_o)^{\frac{1}{2}}$$

$$\frac{S}{S_o} = (1 - \sigma/\epsilon_o)^{1/2} \tag{16}$$

Combined specific surface model. Initially, deposits on the grains will cause the sperical model to dominate but, as deposits become contiguous, side spaces will be filled in and flow will be through channels approximating tubes or capillaries.

Combining the two models:

$$S = S_o\left(1 + b\sigma/\epsilon_o\right)^{2/3}\left(1 - \sigma/\epsilon_o\right)^{1/2} \tag{17}$$

Clearly, the geometry of the pores is not that of the ideal geometries assumed here, so the power functions are generalised.

$$S = S_o\left(1 + b\sigma/\epsilon_o\right)^{y}\left(1 - \sigma/\epsilon_o\right)^{z} \tag{18}$$

This is the model presented by Mackrle et al.[3] This reaches the limit $S = 0$ only at $\sigma = \epsilon_o$, that is when the pores are completely blocked by deposit. In practice, this is not the limiting case, since in deep bed filtration the removal of suspension effectively stops before the pores are completely blocked, while there is still flow. So a limiting factor other than specific surface must be incorporated.

Interstitial velocity. The approach velocity of filtration is $v = Q/A$. The local interstitial velocity v_i is v/ϵ. The critical velocity at which no further deposition can take place due to the high shear gradient at the pore boundary is v_c and occurs when the specific deposit has reached its saturation value σ_u. Using equation (5) for this limit condition:

$$v_c = v/(\epsilon_o - \sigma_u)$$

It is generally accepted that deep bed filtration clarification efficiency is an inverse function of velocity (see lecture on Capture Mechanisms in Filtration by K.J. Ives). Assuming, therefore, that λ is proportional to some power x of the difference between the reciprocals of the interstitial and critical velocities:

$$\lambda = \text{const.} \left(\frac{1}{v_i} - \frac{1}{v_c} \right)^x$$

$$= \text{const.} \left(\frac{\epsilon_o - \sigma}{v} - \frac{\epsilon_o - \sigma_u}{v} \right)^x$$

$$= \text{const.} \left(\frac{\sigma_u - \sigma}{v} \right)^x$$

For the initial state $\lambda = \lambda_o$ when $\sigma = 0$,

$$\lambda_o = \text{const.} \left(\frac{\sigma_u}{v} \right)^x$$

$$\frac{\lambda}{\lambda_o} = \left(\frac{\sigma_u - \sigma}{\sigma_u} \right)^x = (1 - \sigma/\sigma_u)^x \qquad (19)$$

This is similar to the development by Maroudas and Eisenklam.[4]

Filter coefficient. It is assumed that the filter coefficient is simultaneously a function of specific surface and interstitial velocity.

$$\frac{\lambda}{\lambda_o} = \frac{S}{S_o} \times \text{velocity term}$$

$$\lambda/\lambda_o = (1 + b\sigma/\epsilon_o)^y (1 - \sigma/\epsilon_o)^z (1 - \sigma/\sigma_u)^x \qquad (20)$$

This approaches the correct limit $\lambda \to 0$ as $\sigma \to \sigma_u$.

Models of various investigators. Most of the mathematical models proposed for the variation of λ with σ can be expressed by equation (20) with appropriate selection of values for x , y and z.

Ives[5] The original model proposed by Ives analysed dome-like deposits on filter grains, and utilised the velocity term in a modified form. The result was equation (21).

$$\lambda = \lambda_o + a_1\sigma - a_2\sigma^2/(\epsilon_o - \sigma) \tag{21}$$

If x = y = z = 1 in equation (20) then the expansion of equations (20) and (21) become the same.

Mackrle[3] If x = 0 , the general equation reduces to Mackrle's equation (22)

$$\lambda/\lambda_o = (1 + b\sigma/\epsilon_o)^y (1 - \sigma/\epsilon_o)^z \tag{22}$$

Maroudas[4] If y = z = 0 and x = 1 , the general equation becomes Maroudas' equation (23)

$$\lambda/\lambda_o = (1 - \sigma/\sigma_u) \tag{23}$$

Shekhtman,[6] and Heertjes and Lerk[7] The model proposed by Shekhtman assumed arbitrarily that due to increases in σ and interstitial velocity, λ declined linearly with σ to zero. Heertjes and Lerk based their model on a unit cell concept, in which particles close to the pore boundary experienced a force balance between viscous drag and London-van der Waals' surface force. In both cases, mathematically the models are represented by the general equation where x = y = 0 and z = 1 , resulting in equation (24).

$$\lambda/\lambda_o = (1 - \sigma/\epsilon_o) \tag{24}$$

An advantage of equation (24) is that, combined with equations (1) and (6), explicit solutions can be obtained for C and σ as functions of L and t.

$$\frac{C}{C_o} = \frac{\exp(\lambda_o \beta v C_o t/\epsilon_o)}{\exp(\lambda_o L) + \exp(\lambda_o \beta v C_o t/\epsilon_o) - 1} \tag{25}$$

$$\frac{\sigma}{\epsilon_o} = \frac{\exp(\lambda_o \beta v C_o t/\epsilon_o) - 1}{\exp(\lambda_o L) + \exp(\lambda_o \beta v C_o t/\epsilon_o) - 1} \tag{26}$$

Filtration is thus characterised by the dimensionless groups $(\lambda_o L)$ and $(\lambda_o \beta v C_o t/\epsilon_o)$.

Deb[8] The model of a coated sphere was used by Deb, but with a refined approach allowing for the contact points between spherical grains, where the geometry of the deposit would be modified. An analysis of Deb's equations shows that for small values of σ ($\sigma \ll (1 - \epsilon_o)$) the general equation (20) can represent his model, if x = o, y = y (signified by symbol A in Deb's paper), and z = 1-y. Also, the ratio b/ϵ_o in equation (20) is equal to $F(n)/3(1 + F(n))(1 - \epsilon_o)$, where $F(n)$ is a nonlinear function of the average number of contact points of each grain.

Herzig[1] Refinements of the differential equations (1) and (6) were made by Herzig et al, who did not define the functional relationship $\lambda = \lambda_o F(\sigma)$. Therefore, their solutions were even more general than those using equation (20). A most useful relationship, in the form of equation (27), was found to be general, and independent of the form of $F(\sigma)$.

$$\frac{C}{C_o} = \frac{\sigma}{\sigma_o} \qquad\qquad (27)$$

Here σ_o is the value of the specific deposit in the surface layer at L = 0, but which is a variable with filter run time t. Equation (27) considerably simplifies mathematical solutions of filter clarification equations, and also allows separate evaluation of on-surface and in-surface deposition in experimental systems.

The various modified filter coefficient theories are illustrated diagrammatically on Figure 3. It can be seen that several are non-monotonic, the initial rise in filter coefficient representing an increase in filter efficiency during the early stages of deposition. The exact forms of these curves depends on the values of x , y , z and b (see, for example, Ives and Horner[2]).

Trajectory models. There is a growing interest in computing trajectories of particles in suspension as they approach a filter grain. Such trajectory calculations have been made previously, principally for single collector fibres represented as cylinders, in aerosol filtration. Mathematical models of the initial collection efficiency of a sphere in a liquid suspension have been proposed by Spielman and Goren,[9] representing a bridge between the analysis of capture mechanisms and mathematical models of the process dynamics. Current research is extending these trajectory models to consider the progress of collection efficiency during filtration, and to employ more sophisticated packed bed models than the single spherical collector.

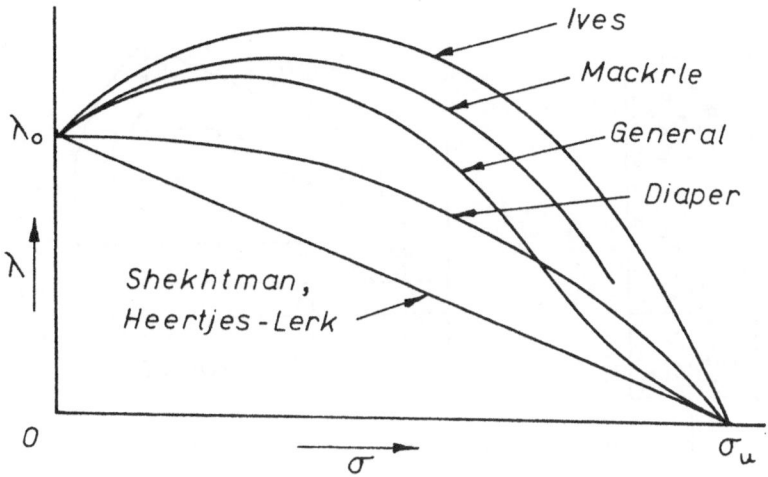

Figure 3. Diagram of the function of the modified filter
           coefficient λ according to various authors.

CLARIFICATION IN A NON-UNIFORM FILTER

     Three basic cases can be considered where the initial
conditions are not uniform at all distances from the filter
inlet face. One is a filter containing mixed sizes of grains
but homogeneously dispersed through all the filter (as in slow
sand water filters). This could be treated theoretically as a
uniform filter, for even with uniform grains there is a
distribution of pore sizes. The second case is of grains,
graded in size, with different sizes at different distances
where $D = F(L)$. This function might be continuous, with a
continuous size grading caused by the fluidisation-washing
process, or it might be discontinuous as in multilayer filters.
The third case is where the local velocity varies with distance,
$v_i = F(L)$. This occurs in radial flow filters where L is
replaced by R , the distance from the axis of the filter.
These two latter cases are shown diagrammatically in Figure 4.

216

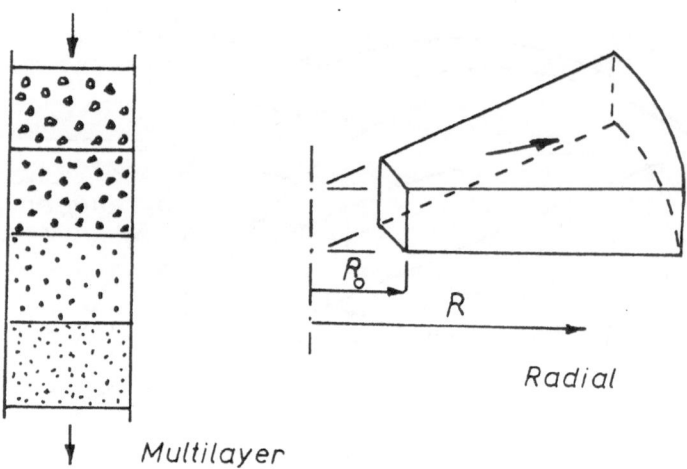

Figure 4. Non-uniform filters: (a) grain size variation with
distance L ; (b) velocity variation with distance R.

Size-graded media

Both from the study of filter transport mechanisms and the
mathematical models dealing with specific surface it is evident
that the filter coefficient varies as an inverse function of
grain size D.

$$\lambda = \text{const.} D^{-m_1}$$ (28)

The value of m_1 has been variously reported between 1 and 3,
presumably depending on the dominant transport mechanism. If it
is determined empirically equation (28), (1), (6) and (20) can be
combined to form a mathematical description of the concentration
and deposit variations, provided that the variation of size D with
distance L is known. It is likely, however, that the exponents
x , y and z , and constant b are also functions of D , which
complicates the model further. Such an analysis and computation,
based on experiments with a five-layer filter, have been
presented by Mohanka.[10]

The only explicit solutions for a size-graded filter are in the paper by Diaper and Ives,[11] in which they assumed that $m_1 = 1$, and a linear variation of grain size with distance $D = D_o + jL$. The grain size at the inlet surface $L = 0$ is D_o, and j can be positive or negative according to whether the grain size is increasing or diminishing in the direction of flow. This enabled Diaper and Ives to compare analytically the processes of downflow and upflow filtration through continuously size-graded media. Their system of equations and solutions is as follows.

$$\lambda D = p_1 - p_2 \sigma^2 \tag{29}$$

Equation (29), together with equation (1) and the balance equation (6) and the linear relation $D = D_o + jL$ gives equation (30)

$$\frac{c}{c_o} = \left[\text{sech}^2 T \left(\frac{D_o + jL}{D_o} \right)^{m/j} + \tanh^2 T \right]^{-p_1/m} \tag{30}$$

$$T = \beta v c_o t (p_1 p_2)^{1/2} / D_o \tag{31}$$

$$m = 2(p_1 + j) \tag{32}$$

It will be noted that equation (29) does not appear to be a particular case of the general equation (20). It is, in fact, an approximation of equation (21), to avoid a non-monotonic function for mathematical simplicity.

Velocity variation (radial flow)

Study of the transport mechanisms shows that the filter coefficient is an inverse function of flow velocity, that is lower velocities lead to improved filtration efficiency.

$$\lambda = \text{const. } v^{-m_2} \tag{33}$$

Values of m_2 have been variously quoted between 0.7 and 4, presumably depending on the dominant mechanism.

Normally, constant rate deep bed filters operating as beds of fixed cross-section normal to flow cannot take any advantage of equation (33). However, in radial flow filters of cylindrical form where the suspension enters at the axis and filters radially outwards to be collected at the periphery, local filtration velocity varies inversely with radial distance. Hence filtration efficiency increases in the direction of flow.

This[6] has been treated theoretically and experimentally by
Shekhtman[6] using equation (24) as a basis, and by Mackrle et al[3]
using equation (22). The most complete study has been made by
Ives and Horner[2] who assumed $m_2 = 1$ and transformed equations (1)
and (20) into cylindrical coordinates.

$$-\frac{\partial C}{\partial \bar{R}} = \frac{R_0 \bar{R}}{v_0} \lambda_{oR_0} \left(1 + \frac{b\sigma}{\epsilon_0}\right)^y \left(1 - \frac{\sigma}{\epsilon_0}\right)^z \left(1 - \frac{\sigma}{\sigma_u}\right)^x C \qquad (34)$$

Here $\bar{R}$ is R/R_0 where R_0 is the radial distance from the axis to
the inlet face, and λ_{oR} is a radial filter coefficient equal
to $\lambda_0 v_0$, where v_0 is the inlet face velocity. Together
with a form of the balance equation (6) in cylindrical coordinates,
equation (34) has been solved in an implicit form.

Some comparison of linear and radial flow filtration can be
made using the mathematical models. The simple relationship of
equation (35) emerges for equal performance of the two systems
based on the number of bed volumes filtered to a given pressure
drop limit, to provide a given standard of filtrate.

$$L = R_0 (\bar{R}^2 - 1)/2 \qquad (35)$$

From equation (35) for example, if a linear filter is 1m deep,
and a radial filter has a face radius of 100mm, $\bar{R}$ becomes 4.6.
Hence R is 460mm, and the filter has a radial thickness of 360mm.
So a radial filter 0.36m thick can produce equivalent performance
to a linear filter 1m deep.

PRESSURE DROP

If filter media are clarifying suspensions as they flow
through, it follows that the pores of the media accumulate
deposits which cause a loss of permeability, on increased flow
resistance.

The filter media exhibit a resistance to flow, even to clear
liquid which can be calculated from the Carman-Kozeny equation (36).

$$\left(\frac{\partial H}{\partial L}\right)_0 = \frac{5 \nu v S_0^2}{g \epsilon_0^3} \qquad (36)$$

Here $(\partial H/\partial L)_0$ is the hydraulic gradient at t = o , and ν is
the kinematic viscosity of the liquid. The head loss H ($= \Delta P/\rho g$)
is more convenient than pressure drop due to its use in pressure
diagrams (see Figure 6).

In the case of a filter layer containing specific deposit σ ,
both the specific surface S and the porosity ϵ are modified,
according to equations (18) and (5) respectively.

Consequently, equation (36) becomes:

$$\frac{\partial H}{\partial L} = \frac{5 \gamma v S_o^2}{g \epsilon_o^3} \left(1 + \frac{b\sigma}{\epsilon_o}\right)^{2y} \left(1 - \frac{\sigma}{\epsilon_o}\right)^{2z-3} \tag{37}$$

$$\frac{\partial H}{\partial L} = \left(\frac{\partial H}{\partial L}\right)_o \left(1 + \frac{b\sigma}{\epsilon_o}\right)^{2y} \left(1 - \frac{\sigma}{\epsilon_o}\right)^{2z-3} \tag{38}$$

In the special case $y = z = 1$

$$\frac{\partial H}{\partial L} = \left(\frac{\partial H}{\partial L}\right)_o \frac{\left(1 + b\sigma/\epsilon_o\right)^2}{\left(1 - \sigma/\epsilon_o\right)} \tag{39}$$

Expansion of equation (39) yields:

$$\frac{\partial H}{\partial L} = \left(\frac{\partial H}{\partial L}\right)_o \left[1 + (2b+1)\frac{\sigma}{\epsilon_o} + (b+1)^2\left(\frac{\sigma}{\epsilon_o}\right)^2 + (b+1)^3\left(\frac{\sigma}{\epsilon_o}\right)^3 + \cdots\right] \tag{40}$$

So, to a first approximation the head loss per unit depth is proportional to the local specific deposit, particularly when $\sigma \ll \epsilon_o$. This was shown to be true for several other mathematical models by Herzig et al[1] and for several empirical formulations reported by Mints.[12] Consequently, a convenient approximation to equation (40) is frequently quoted.

$$\frac{\partial H}{\partial L} = \left(\frac{\partial H}{\partial L}\right)_o + \kappa\sigma \tag{41}$$

The value of σ, as a function of L and t, must be obtained from solutions of the clarification equations, either analytically as in equation (26) or by digital computation. Integration of equation (41) with respect to L gives the total head loss through depth L at any instant of time t.

$$H = \int_o^L \left(\frac{\partial H}{\partial L}\right)_o dL + \int_o^L \kappa\sigma \, dL \tag{42}$$

$$H = H_o + \kappa \int_o^L \sigma \, dL \tag{43}$$

Using the simple analytic solution for σ given in equation (26)

$$H = H_o + \kappa\epsilon_o\left[L - \frac{1}{\lambda_o} \ln\left(\exp(\lambda_o L) + \exp(\lambda_o \beta v C_o t/\epsilon_o) - 1\right) + \beta v C_o t/\epsilon_o\right] \tag{44}$$

It should be noted that $\exp(\lambda_o L)$ at significant depths (e.g. L greater than 150mm) is numerically much greater than $\exp(\lambda_o \beta v C_o t / \epsilon_o)$ for normal times of filter operation. This is due to the presence of the inlet concentration value C_o, which is usually in parts per million (vol/vol), so that a typical value is 100 parts per million, or 100×10^{-6}. In such cases this reduces $\exp(\lambda_o \beta v C_o t / \epsilon_o)$ to be approximately 1.0 and the logarithmic term becomes $\exp(\lambda_o L)$.

$$H = H_o + \kappa \epsilon_o \left[L - \frac{1}{\lambda_o} \ln \left(\exp(\lambda_o L) \right) + \beta v C_o t / \epsilon_o \right] \qquad (45)$$

$$H = H_o + \kappa \beta v C_o t \qquad (46)$$

Equation (46) shows that, for deposition in the pores, the head loss is linear with time for most practical cases. Equation (46) is really general and independent of the form of the function relating σ to L and t, providing that the filtrate concentration is low compared with C_o (e.g. 0.05 C_o). This means that all the suspension flowing in up to time t ($v C_o t$) has been retained in the filter, which is another statement of the integral $\int_o^L \sigma \, dL$.

In addition to the head loss (pressure drop) created by deposits in the pores, there may be an extra contribution due to deposits on the surface. As shown in the lecture Capture Mechanisms in Filtration by K.J. Ives, such deposits form a discontinuous cake at the inlet surface of a deep bed filter. The rise in head loss is exponential with time, sometimes known as Boucher's Law.

$$H_s = \kappa_s \exp(\kappa_t t) \qquad (47)$$

Here k_s is the initial head loss through the surface layer, usually extremely small, and k_t is a rate constant of the surface deposition process. In deep bed filtration it is desirable to minimise or eliminate H_s as it considerably shortens the length of the filter run.

The three components of the head loss in a deep bed filter: H_o the initial head loss, H_d the head loss due to deposition in the pores, and H_s the surface deposit head loss, are shown in Figure 5.

These models of head loss (pressure drop) changes in the filter have implicitly assumed a uniform filter. However, it is not difficult to integrate numerically the hydraulic gradient equation (41) for size-graded media (see, for example, Mohanka[10]) or for radial flow (see Ives and Horner[2]) providing the functional variations with L are specified either mathematically or numerically.

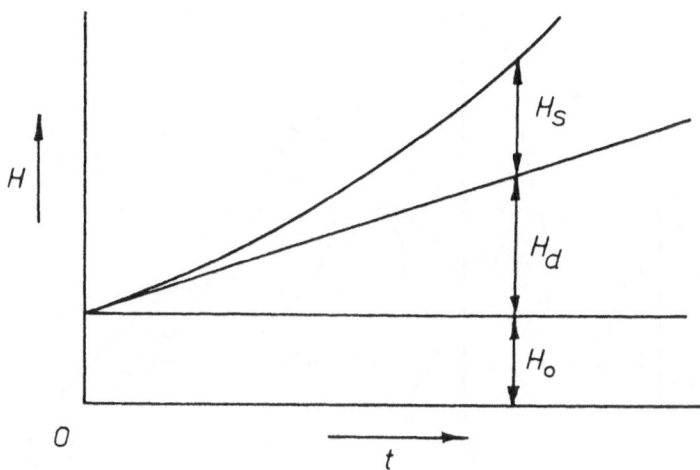

Figure 5. Head loss variation with time in a deep bed filter.

A useful method of presenting both empirical and theoretical information on head loss through a linear flow deep bed filter is by means of a pressure diagram as shown in Figure 6, for a uniform filter. In the static case (no flow) 1 metre of pressure head is gained for every metre increment of depth. A pressure drop is experienced through the media, increasing linearly with depth, when flow starts, according to the Carman-Kozeny equation (36). This linearity is not maintained, however, as clarification proceeds with consequent deposition in the pores. The quantity of deposit varies from layer to layer as illustrated previously in the numeric example following equation (6). Consequently, the local hydraulic gradient increases according to equation (41) and the pressure line becomes distorted. When the pressure line touches the atmospheric pressure value it is desirable to end the filter run to avoid air being drawn out of solution by sub-atmospheric pressure conditions. The pressure diagram is a useful guide to experimentalists, filter operators, designers and those concerned with optimisation studies. The diagram of Figure 6 does not include a component due to surface deposition, although it can easily be incorporated in such diagrams.

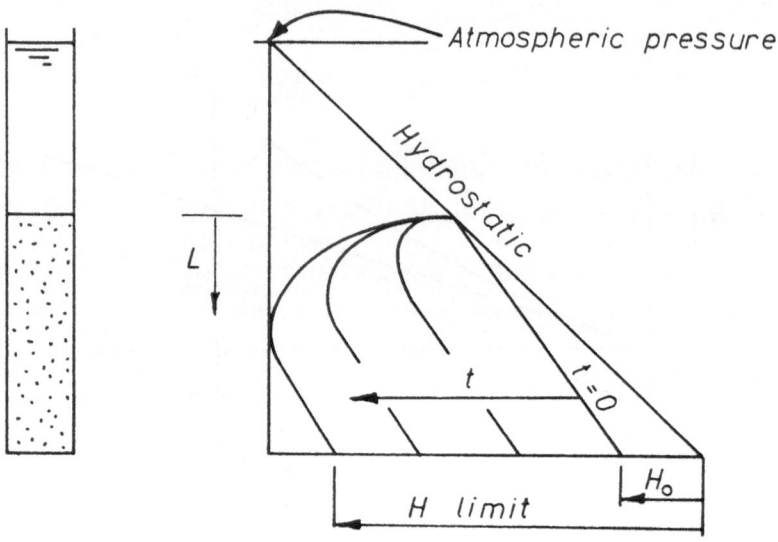

Figure 6. Pressure diagram for a uniform deep bed filter.

CONCLUSION

Mathematical modelling of the clarification and pressure drop changes in deep bed filtration continues to become more sophisticated with the digital computer as a very necessary tool. It should, however, progress in step with empirical tests of the models to check their validity.

The models should have as their objectives not only better understanding of the processes, but also to achieve better filter design. To this end, mathematical models have already made the following contributions:

(i)   indicated the value of pressure probes through filter
        walls to check operation by pressure diagrams;
(ii)   allowed extrapolation of experimental and plant data in a
        rational manner;
(iii)   suggested the observations to be made on experimental
        models, e.g. surface head loss, in-depth sampling, in situ
        porosity observations;
(iv)   allowed comparisons of different designs such as upflow
        and downflow, or linear and radial filtration;
(v)   provided a basis for optimisation studies.

In spite of these merits none of the models is entirely predictive and recourse to experimentally determined constants or coefficients is still necessary.

REFERENCES

1. Herzig, J.P., Leclerc, D.M., Le Goff, P., Flow of suspensions
    through porous media, Ind. Eng. Chem., 62, 8, 1970.

2. Ives, K.J., Horner, R.M.W., Radial filtration, Proc. Inst.
    Civ. Engrs., 55, 2, 229, 1973.

3. Mackrle, V., Dracka, O., Svec, J., Hydrodynamics of the
    Disposal of Low Level Liquid Radioactive Wastes in Soil,
    International Atomic Energy Agency Contract Report No. 98,
    Czech. Academy of Sciences Institute of Hydrodynamics,
    Prague, 1965.

4. Maroudas, A., Eiskenklam, P., Clarification of suspensions :
    a study of particle deposition in porous media, Chem. Eng. Sci.
    20, 867, 1965.

5. Ives, K.J., Rational design of filters, Proc. Inst. Civ. Engrs.,
    16, 189, 1960.

6. Shekhtman, Yu. M., Filtration of Suspensions of Low Concentration,
    Publishing House of the U.S.S.R. Academy of Sciences, Moscow,
    1961. (In Russian).

7. Heertjes, P.M., Lerk, C.F., The functioning of deep bed filters,
    Part II, The filtration of flocculated suspensions, Trans. Inst.
    Chem. Engrs., 45, T138, 1967.

8. Deb, A.K., Theory of sand filtration, J. San. Eng. Div.,
    Proc. Amer. Soc. Civ. Engrs., 95, SA3, 399, 1969.

9. Spielman, L.A., Goren, S.L., Capture of small particles by London forces from low-speed liquid flows, Environ. Sci. Technol., 4, 135, 1970.

10. Mohanka, S.S., Theory of multilayer filtration, J. San. Eng. Div., Proc. Amer. Soc. Civ. Engrs., 95, SA6, 1079, 1969.

11. Diaper, E.W.J., Ives, K.J., Filtration through size-graded media, J. San. Eng. Div., Proc. Amer. Soc. Civ. Engrs, 91, SA3, 89, 1965.

12. Mints, D.M., Modern theory of filtration, in International Water Supply Association Seventh Congress Barcelona, Volume 1, I.W.S.A., Park Street, London, 1966.

BIBLIOGRAPHY

Additional material will be found in the following references:

1. Ives, K.J., Theory of filtration, in International Water Supply Association Eighth Congress Vienna, Volume 1, I.W.S.A., Park Street, London, 1969.

2. Ives, K.J., Rapid filtration, Wat. Res., 4, 201, 1970.

3. Ives, K.J., Filtration of water and wastewater, Critical Reviews in Environ. Contr., 2, 293, 1971.

LEAST COST DESIGN — OPTIMIZATION OF DEEP BED FILTERS

E. Robert Baumann

Department of Civil Engineering and Engineering
Research Institute, Iowa State University, Ames, Iowa

INTRODUCTION

The design of a deep bed filter requires specification of
the filtration rate, the maximum pressure drop across the media,
the media size, and the media depth. For successful filtration,
of course, these depend on the quality of filter influent water
and the desired quality of the filter effluent. Equivalent
filter performance can be obtained from deep bed filters using
the same grade of filter media by operating shallow-depth filters
at lower rates or deeper-depth filters at higher rates. Conversely, if the filtration rate were held constant, the same
quality of filter effluent may be obtained from a shallow-depth
filter of fine media as from a deeper filter of coarse media.

Operationally optimum filter design occurs when the clarification capacity of the filter is exhausted at any given filtration rate, when all of the available pressure drop provided
across the filter media is exhausted. Many alternative filter
designs can provide operationally optimum filter operation.
However, only one can produce the filtrate at the least cost per
cubic meter of filtrate. Such a filter would be the least cost
filter design and would represent the ultimate objective in
optimization of the design of deep bed filters.

In order to design an operationally optimum, least cost
filter, it is necessary either to have extensive data for the
filtration of water through different thicknesses of media of
different grain sizes at many different rates, with respect to
filtrate quality, head loss, and cost per cubic meter of
filtrate, or to have reliable mathematical models of the

filtration process. In fact, two mathematical models are required:

1.   An operationally optimum model of the filtration process
to describe the relationship between the quality of the
filtrate and the total head loss and independent variables of
filter depth and time during the filter operation.

2.   An economic mathematical model to evaluate the effects of
the filter structure, hydraulic appurtenances, energy and
maintenance requirements, and size of the total filter plant on
the cost of operationally optimum filters.

Huang[1], Gur[2], Huang and Baumann[3], and Ives[4] have recently
worked in the area of optimization of deep bed filters. This
discussion is primarily taken from the papers by Huang and
Baumann[3] and Ives[4].

MATHEMATICAL MODELS

Assuming that the nature of the suspension and the water
temperature remain reasonably constant with time, the influence
of filtration velocity, grain size and grain size distribution
with depth must be known in a mathematical form, as they affect
filtrate quality and head loss. Hsiung and Cleasby[5] proposed a
method for prediction of deep bed filter performance by using
pilot plant operating data and the analogy of the filter performance to the statistical chi-square distribution. This
method is based on the fact that in filtration, suspended
solids are transported through the filter by a fluid which
changes direction in a random manner seeking a relatively un-
obstructed pathway. The particles are brought within the
range of the Van Der Waal's forces of the filter medium or previously deposited particles and attach themselves to surfaces
exhibiting such forces. The point at which a suspended particle
becomes attached in the filter bed is determined in a random
manner.

The similarity between filtration data collected from a
pilot plant and the chi-square distribution established a relationship between U, v, and P_C of the chi-square distribution and
L, t, and C/C_O, respectively, obtained from filtration experiments. Here, U is the random variable in the chi-square distribution, v is the degree of freedom, P_C is the cumulative
probability for a continuous random variable $f(U)dU = dP_C$, L is
the sand depth in filter measured from surface, t is the time
of filtration, and C/C_O is the ratio of the effluent suspended
solids concentration to the influent suspended solids concentra-

tion.

The development of the chi-square filter performance prediction technique was based on filtration data collected in the
laboratory using three thin-layer filters. The suspended
solid used in these tests resulted from the addition of ferrous
sulfate to aerated tap water. Water temperature was maintained
at 77 ºF throughout the filter runs. The tests were conducted
using several sand sizes, several sand depths, different flow
rates, and different influent iron concentrations (C_O). Both
effluent water quality (C) and head loss were measured at different filter depths and at different filtration times. Hsiung
and Cleasby concluded that a rational filter design could then
be achieved after performance curves (which depend on the influent suspension characteristics obtained from a compact pilot
unit consisting of three thin filters of varying depth) had
been developed.

Ives[4], on the other hand, followed a more fundamental approach to development of the required filtration model. He
concluded that the following first order relation between
concentration (C) and depth (L) proposed by Iwasaki[6] is valid
for the initial condition of filtration through a uniform bed
of media:

$$- \frac{\partial C}{\partial L} = \lambda C \qquad (1)$$

where λ is the filter coefficient. Also the continuity equation

$$- \frac{\partial C}{\partial L} = \frac{1}{v} \frac{\partial \sigma}{\partial t} \qquad (2)$$

where σ is the specific deposit (volume of deposit per unit
filter volume), t is filter run time, and v is the approach
velocity of filtration, relates the quantity of suspension
particles removed from the flowing liquid to the quantity of
deposit accumulating in the filter pores.

During the filter run the efficiency of the filter changes
due to the accumulation of deposits in the pores. There are two
principal theories describing these changes: one due to Mints[7]
which attributes changes to scour of deposits by the water
flowing through the pores, and one due to Ives[8] which attributes
changes to geometric and velocity changes in the pores. Mints
described optimization procedures based on his theory, and
these have been extended in a very detailed manner by Gur[2].
The relationship derived by Ives is

$$\frac{\lambda}{\lambda_o} = (1 + \frac{\beta\sigma}{f})^y (1 - \frac{\sigma}{f})^z (1 - \frac{\sigma}{\sigma_u})^x \qquad (3)$$

where λ_o is the filter coefficient at $t = 0$ (clean filter), f is the initial filter porosity, σ_u is the ultimate saturation value of specific deposit, and β, x, y, z, are empirically derived factors. It has been shown that Eq. (3) is general in that all other theories, except Mints', can be described by appropriate choices of x, y and z.

Equations (1), (2) and (3) describe clarification of a suspension by filtration at constant rate (v), in a uniform filter of constant grain size (d). If the filter is operated at a different rate, or with a different grain size, then the factors λ_o, β, σ_u, x, y, z will change. Although some theoretical investigations have been made of the effects of changing flow-rate and grain size (Ives and Sholji[9]; Ison and Ives[10]), it is necessary to obtain experimental information because the relationships depend on the nature of the suspension (Mohanka[11]).

These relationships may be written

$$\lambda_o = \Lambda/d^{m1} v^{n1}, \qquad (4.1)$$

$$\beta = Bd^{m2} v^{n2}, \qquad (4.2)$$

$$\sigma_u = S/d^{m3} (1 + v)^{n3}, \qquad (4.3)$$

$$x = X/d^{m4} v^{n4}, \qquad (4.4)$$

$$y = Y/d^{m5} v^{n5}, \qquad (4.5)$$

and

$$z = Z/d^{m6} v^{n6}. \qquad (4.6)$$

Taking granular material with a porosity (f) of 0.44, filtering ferric hydroxide floc suspension in water, Mohanka obtained the following results, in cm g min units:

$$\Lambda = 5.9 \times 10^{-3}, \quad m_1 = 1.35, \quad n_1 = 0.25,$$

$$B = 11.8, \quad m_2 = 0.75, \quad n_2 = 0,$$

$$S = 0.44, \quad m_3 = 0, \quad n_3 = 0.75,$$

$$X = 0.95, \quad m_4 = 0.61, \quad n_4 = 0.24,$$

$$Y = 1.5, \qquad m_5 = 0, \qquad n_5 = 0, \text{ and}$$

$$Z = 0.75, \qquad m_6 = 0, \qquad n_6 = 0.$$

Note that λ_o has dimensions of cm^{-1}, all other factors (β, σ_u, x, y, z) are dimensionless.

In addition to the action of clarifying the suspension, the filter shows a loss of head. The Kozeny-Carman equation is

$$\left(\frac{dH}{dL}\right)_o = 5 \, \frac{\mu}{p} \, \frac{v}{g} \, \frac{(1-f)^2}{f^3} \, \left(\frac{6}{d}\right)^2 \qquad (5)$$

where μ/p is the kinematic viscosity, g is the gravitational acceleration, and $(dH/dL)_o$ signifies the hydraulic gradient at t = 0. During the course of filtration the specific deposit occupies part of the pore space, causing an increase in head loss, shown by

$$\frac{\partial H}{\partial L} = \left(\frac{dH}{dL}\right)_o + k\sigma \qquad (6)$$

where k is a dimensionless head loss factor, dependent upon grain size (d) and flow rate (v) and where

$$k = Kv^{n7}/d^{m7} . \qquad (7)$$

Mohanka's experiments gave, in cm g min units:

$$K = 0.8, \; m_7 = 0.9, \text{ and } n_7 = 0.4.$$

Equation (6) has been reported empirically by several research workers (Mints[7]), but it was given a theoretical basis by Horner[12].

Integration of Eqs. (1) to (7) can be carried out by finite difference techniques on a digital computer.

For a uniform filter, with given values of grain size (d) and flow rate (v) kept constant, the resulting curves of concentration and head loss are as shown on Figure 1. Alteration of the values of d or v will lead to different, but similar, curves.

230

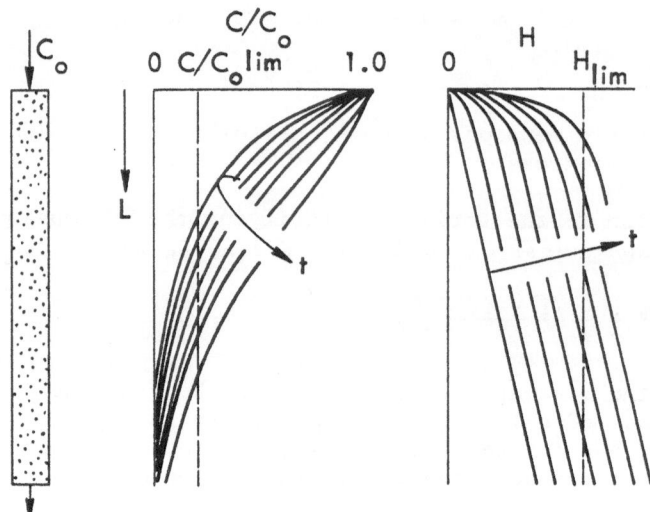

Figure 1. Curves of concentration ratio (C/C$_o$) and head loss (H) varying with depth in the filter (L), and with time of filter run (t).

OPERATIONAL OPTIMUM FOR UNIFORM FILTERS

As can be seen on Figure 1 the filtrate concentration changes during the filter run, rising in value as time proceeds. Also the total head loss rises with time.

Although the filter could be designed to produce satisfactory filtrate quality during the early stages of the filter run, there will come a time when the filtrate quality will become unsatisfactory, and the filter run will have to be terminated (time = t_c). Similarly the head loss will rise until it reaches a limit set by the hydraulic conditions of the design; when this limit head loss is reached the filter run must be stopped (time = t_H).

A filter will be in an operational optimum condition if these two times are equal; that is, when the clarification capacity of the filter has been exhausted simultaneously with the hydraulic capacity, as in

$$t_{OPT} = t_c = t_H .$$ (8)

This operational optimum can be determined simply by drawing on Figure 1 the criteria for the limiting filtrate concentration

(C/C$_O$ limit) and for the limiting head loss (H limit), as shown
in dashed lines. These dashed lines intercept the concentration
and head loss curves at depths and times where the limit values
are reached. The values of time, with corresponding values
for depth can be plotted as curves, as shown on Figure 2. Where
these two curves cross, Eq. (8) is satisfied and the optimum
filter run time, and the optimum media depth, are defined.

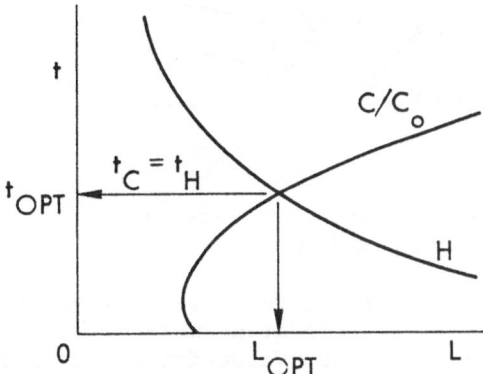

<u>Figure 2.</u> Curves of values of depths (L) and time (t) which meet
the criteria C/C$_O$ limit and H limit, for a uniform filter.

If the curves of Figure 1 are recalculated for a different
filtration velocity, then a different pair of depth-time curves
will result, with a different intersection for t_{opt}. This can
be repeated for several filtration velocities, with the result
shown on Figure 3. Any point on the locus of optima will define
the filtration velocity (v), filter depth (L) and run time (t)
for an operational optimum filter of uniform grains of a given
size.

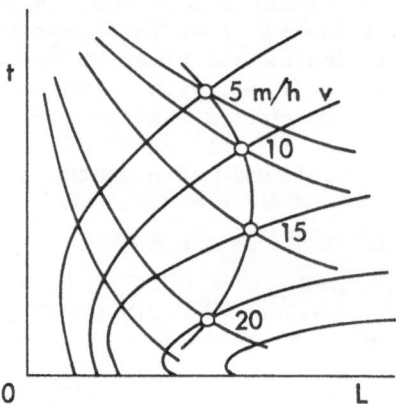

<u>Figure 3</u>. Variation of optimum run time (t_{OPT}) with filtration velocity (v) for a uniform filter.

Again, if the curves of Figure 1 are recalculated for a different grain size, then a different pair of depth-time curves will intersect at another value for t_{opt}. Repeating this for several different grain sizes will produce several different loci of optima as shown on Figure 4.

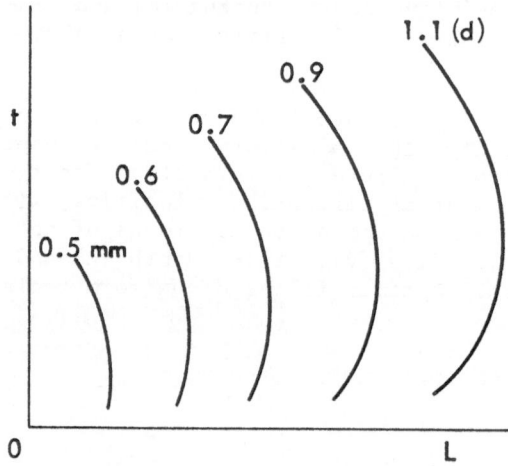

<u>Figure 4</u>. Loci of optima for different grain sizes of uniform filter medium.

The result of the interaction of these four variables: depth, time, velocity and grain size, is to produce a response surface as shown on Figure 5. Any point on this surface pro-'duces an operational optimum design, according to the constraints of limiting filtrate concentration (C/C_O limit) and limiting head loss (H limit). If either, or both of these constraints are changed, then a new response surface will be formed.

It can be seen that there are an infinite number of design solutions which conform to an operational optimum.

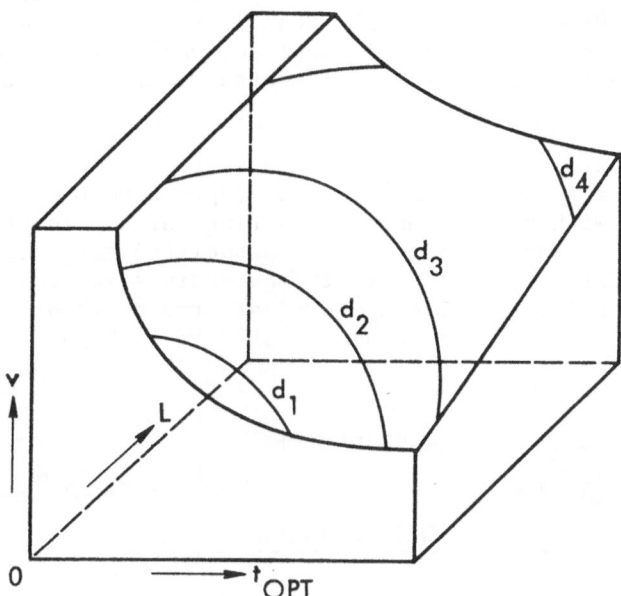

Figure 5. Response surface for operational optimum, depending on depth, time, velocity and uniform grain size.

OPERATIONAL OPTIMUM FOR GRADED-MEDIA FILTERS

Filters used in practice rarely contain media of uniform grain size. The media are either graded from fine to coarse in the direction of flow due to backwashing, or form a graded sequence of coarse to fine as in upflow and multilayer filters. Consequently, an additional variable is introduced into the design: the sequence of size in the depth of the filter.

In considering alternative designs for size-graded filters, the following variations are possible, either singly or in combination: 1. The total depth of the media can be changed; 2. the inlet face grain size can be changed; 3. the outlet face grain size can be changed; and 4. the size distribution within the depth can either be changed (a) as a continuous gradation or (b) as a series of discrete size jumps (i) monotonically increasing or decreasing, (ii) nonmonotonically changing. It is this range of possibilities which make formal approaches to optimization very difficult, so that a computer is indispensible.

The programming of optimization procedures for size-graded filters depends on the principle that if t_C is not equal to t_H, then a step change in design is applied, and t_C and t_H are again compared, and this process is reiterated until t_C and t_H converge.

For example, a sequence of grain sizes can be read into the program, covering a very long range of depth 0 to L max (e.g. 3 m depth). Within this range a starting value L initl and end value L end can be chosen, with associated grain sizes representing the inlet face and outlet face grain sizes respectively. The distance L initl to L end represents the depth of the filter. This may be represented diagrammatically as in Figure 6.

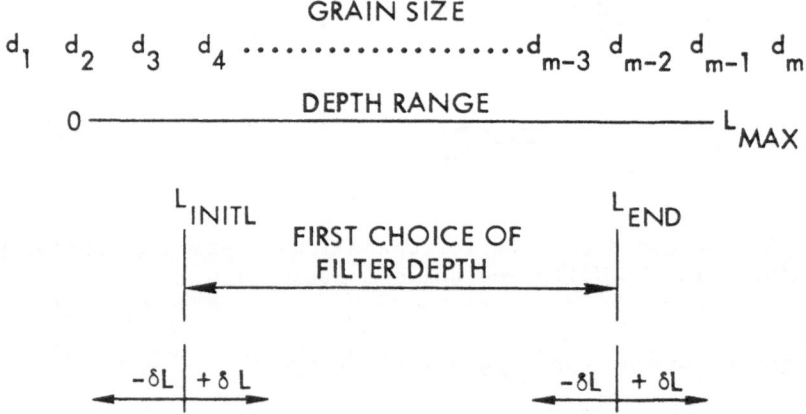

Figure 6. Diagram of array of grain sizes through filter depth.

The values ot t_C and t_H are calculated for the first choice of filter depth. If $t_C > t_H$ the bed must be made shorter by removing a layer of fine grains at the top (L'initl = L initl + δL) and a layer of coarse grains at the bottom (L'end = L end - δL). Conversely, if t_c < t_H the bed must be made longer by adding a layer of grains at top and bottom (L'initl = L initl - δL; L'end = L end + δL). The process is repeated until t_c = t_H. It is necessary to have a sufficient depth range (0 to L max) for the optimum to be selected. This can be readily achieved by digital computation with a deck of data cards, each of which represents a depth interval δL containing grains of size d_L.

By not varying either L initl or L end, it is possible to change the design by varying either only the outlet face grain size, or only the inlet face grain size.

With the data cards representing depth intervals containing given grain sizes, it is possible to calculate t_c and t_H for a full range 0 to Lmax. If t_c > t_H, the filter can be made shorter by multiplying Lmax by a factor less than 1.0. For example, L'max = 0.8 Lmax; the filter has been shortened by 20%. The depth interval value assigned to each data card is correspondingly shortened, δL' = 0.8 δL, but the grain size assigned to each card (depth interval) remains unchanged. This effectively contracts the bed depth, but does not alter the grain size distribution. This is shown as a diagram on Figure 7.

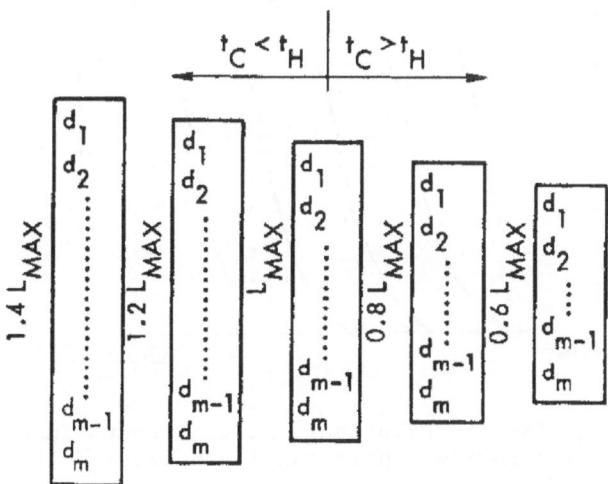

Figure 7. Contraction or extension of filter depth without changing the grain size distribution, to achieve an operational optimum.

The procedures described so far, deal with cases 1, 2 and 3 of the variations in design of size-graded filters, that is, changes in total depth, and inlet and outlet grain sizes. For case 4 where the size distribution can be changed, a different method has been used. As the number of size-distributions which may be used with either continuous or stepped size changes, is infinite, some choice of possibilities has to be established before computation is started. It is assumed, therefore, that a number of practical size distributions (depending on availability of media) are possible, as shown on Figure 8. One of these, marked 0 on Figure 8 is selected as the original grading, and with this providing the grain size data, the values of t_c and t_H are calculated, in a manner similar to that for Figure 6, but with a fixed depth 0 to Lmax. However, in this case if $t_c - t_H$ is negative then finer gradations are required, represented by the odd numbered distributions on Figure 8. If $t_c - t_H$ is positive then coarser gradations are required, represented by the even numbered distributions on Figure 8. Size distributions in an appropriate direction will be selected sequentially until $t_c = t_H$, that is an operational optimum is achieved.

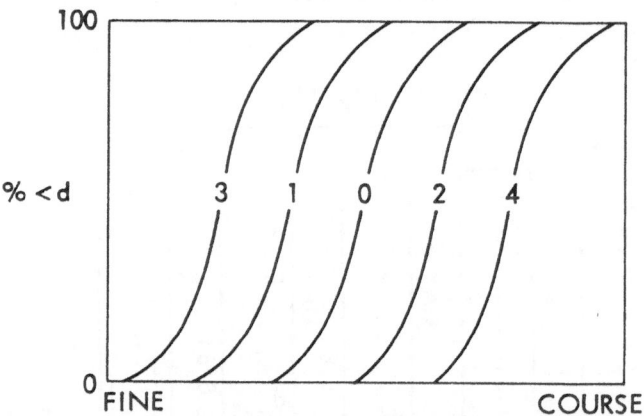

Figure 8. Grain size distributions, as continuous functions, available for a size-graded operationally optimum filter.

The process can also be used for stepped size distribution as in dual media (anthracite-sand) or multilayer filters.

It is obvious from the number of variations of size-graded media, that very many operationally optimum designs are possible.

EMPIRICAL APPROACH TO OPTIMUM DESIGN

According to Huang and Baumann[3], two filters may be said
to provide equivalent performance when they produce the same
quantity and quality of filtered water from the same water source
during the same time period, i.e., one day. The empirical filter
performance prediction model based on the analogy of the chi-
square distribution to the filter performance[5] was used in
developing the computer program for predicting equivalent performance of sand filter designs[1]. The computer program involves the following steps:

1.   The values of the variate U in the chi-square distribution are fed into the program as built-in input data in which
the variate U is related to the filtration data as $C/C_o = P_c$
(cumulative probability) and length of run T, in hours equals v,
degree of freedom, i.e., 1 hr = 1 degree of freedom. For a
given influent water quality and any desired effluent water
quality, the C/C_o ratio, the ratio of effluent suspended solids
concentration to the influent suspended solids concentration is
calculated by the computer and used as an index in picking up
the corresponding value of U in the chi-square distribution,
which has been built into the program.

2.   Two filter performance curves (Figures 9 and 10) are
programmed to predict the filtration rate, Q, and the terminal
head loss, H_t, at a given influent quality, C_o, and terminal
effluent quality, C, the given sand depth, L, and the given
length of run, T. The two performance curves (I, II) can be
represented by two mathematical models. One is used to predict
filtration rate and the second is used to predict terminal head
loss.

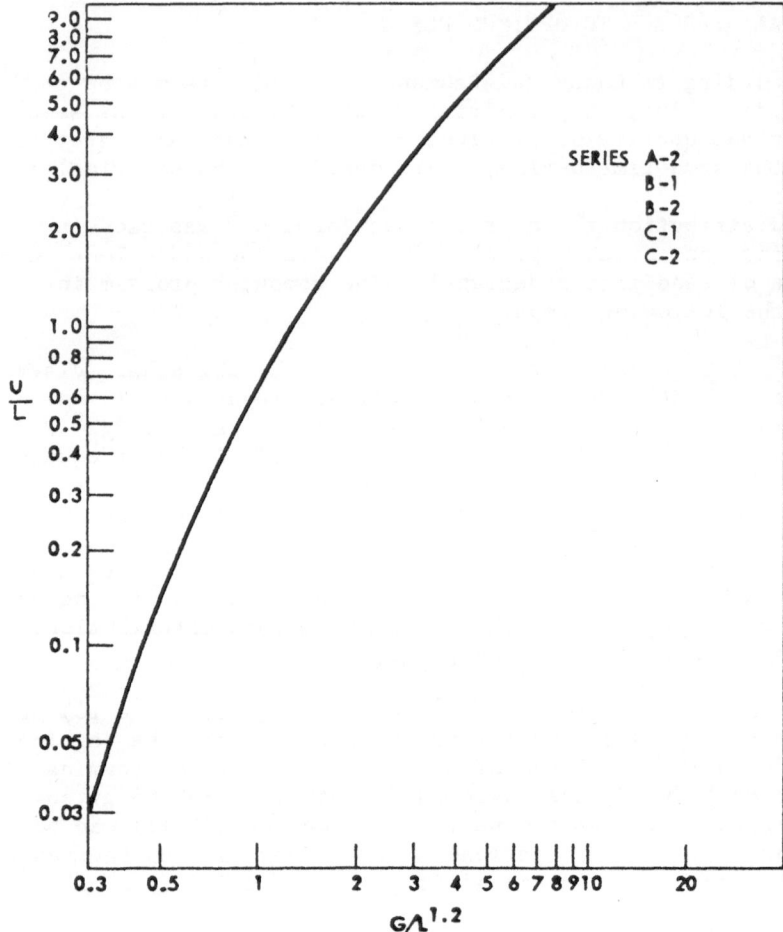

Figure 9. Performance curve I: U/L vs $G/L^{1.2}$, lumped data for various grain sizes, flow rates and influent concentrations.

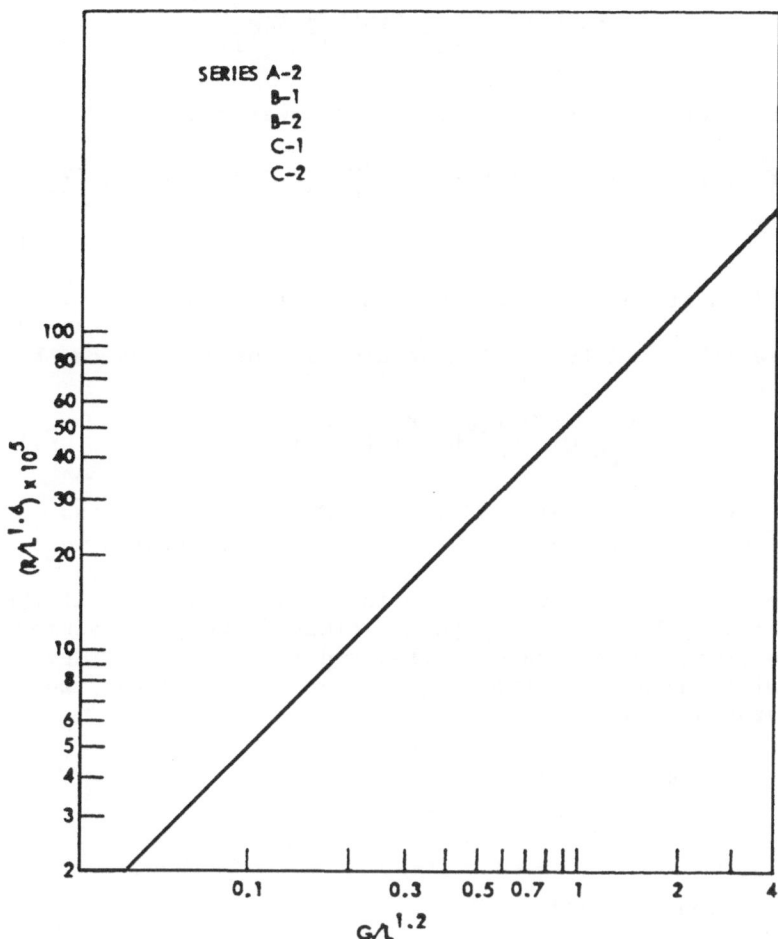

<u>Figure 10</u>. Performance curve II: $R/L^{1.6}$ vs $G/L^{1.2}$, lumped data for various grain sizes, flow rates and influent concentrations.

The first performance curve is expressed mathematically as

$$\log \left(\frac{U}{L}\right) = -0.208 + 1.950 \log \left(\frac{G}{L^{1.2}}\right) - 0.645 \left[\log \left(\frac{G}{L^{1.2}}\right)\right]^2$$

$$(9)$$

where U, L, are as defined previously, and G is a grouped term defined as $Q^{0.29}d^{0.62}T$ for this particular influent suspension of iron in water. In the computer program, the value of $G/L^{1.2}$ is calculated from Eq. (9) in which U/L is a known value. $G/L^{1.2}$ is set equal to $(Q^{0.29}d^{0.62}T)/L^{1.2}$. Therefore,

$$Q = [(\frac{G}{L^{1.2}})^{3.448} L^{4.1376}] / [d^{2.1377} T^{3.448}]. \tag{10}$$

The second performance curve is expressed mathematically as

$$\log \frac{R}{L^{1.6}} = -3.25 + 1.013 \log (\frac{G}{L^{1.2}}) - 0.036[\log (\frac{G}{L^{1.2}})]^2 \tag{11}$$

where R is a grouped term defined as $d^{2.5}(H_t - H_o)/Q^{1.2}C_o^{1.4}$.

Therefore, the head loss relationship for the filters can be related as

$$H_t - H_o = [\frac{R}{L^{1.6}} Q^{1.2} C_o^{1.4} L^{1.6}] / [d^{2.5}] \tag{12}$$

where H_t is the terminal head loss, and H_o equals head loss through the clean filter at the beginning of filtration.

The initial head loss, H_o, which is the head loss at the beginning of filtration through the clean filter, is assumed to be proportional to the rate of flow and filter depth. One empirical formula for estimating the head loss for the flow of water through clean sand is

$$H_o = [27Lq(73 - f)]/[10^5 d^{1.89}(t_w + 20.6)] \tag{13}$$

where

H_o = head loss (ft),

L = depth of sand (in.),

q = filtration rate (MGAD),

d = 50% sand size (mm),

t_w = water temperature (F), and

f = porosity of sand (%).

The total terminal head loss can be predicted using Eqs. (12) and (13).

A computer program was designed to predict the filtration rate and terminal head loss for any combination of C/C_o ratio, length of run, sand size and sand depth.

Of the variables affecting filtration performance, two · must be held constant in this approach to operationally optimum design: 1. Kind of influent suspended solids, and 2. Method of filter operation (influent constant-rate control). The performance prediction model assumes that the flow rate will remain constant during a filter run.

All of the other variables involved in this study are controllable. As a typical example of sand filter optimization, it was assumed a 4-cell single media filter was desired for the removal of iron from 3 MGD of an iron-bearing water. One influent-effluent quality condition studied was as follows:

Case 1 C_o = 5 mg/1

C = 0.3 mg/1.

With the values of these variables assumed, in each case the mathematical model was used to predict the filter design combinations which would give equivalent performance for each of several sand sizes (0.6 to 1.3 mm in 0.1 mm increments) for several lengths of filter run (10 to 50 hr in 4-hr increments) and several sand depths (10 to 40 in. in 1-in. increments). In each case, an effluent water quality safety factor of 2 was assumed. This means that the model predicted the performance as if C were 0.15 instead of 0.3 mg/1. The prediction model was used to determine the filtration rate and head loss for the particular combination of sand size, sand depth, length of run, and C/C_O ratio used in that case.

For example, in Table 1, alternate 10, we find the computer output for case 1, sand size of 0.60 mm, sand depth of 19.0 in., and a run length of 34 hr. In this case, the water quality at the end of the run would meet the case requirement when the filtration rate was 8.27 gpm/sq ft. In this run, the head loss through the sand would be 38.25 ft of water at the end of the run. The controlled variable in Table 1 was the sand depth. All alternate filter designs in Table 1 would produce the same quality of effluent in a 34-hr filter run using 0.6 mm sand media. Table 2 shows the effect of the length of run on the lowest cost alternate design. Table 3 shows the effect of sand size on the lowest cost alternate design.

Table 1. Unit cost, $/mg, of filtering iron bearing water on sand filters providing equivalent performance, at temperature 77 °F with a safety factor of 2, design life 20 yr, interest rate 5%, salvage value 15% of first cost.

C_o = 5.0 MG/L, C = 0.3 MG/L, C/CO = 0.06, T = 34 HR, DIAM = 0.60 MM

					FIRST	COST*OPERATION			COST	TOTAL
					EQUIP MENT	FILTER STRUCT	POWER	LABOR &MAIN	BACK WASH	COST
ALT	L IN	Q GPM/SF	A SF	H FT	$/MG	$/MG	$/MG	$/MG	$/MG	$/MG
1	10	2.17	1056	4.09	4.258	3.326	0.736	2.548	0.615	11.484
2	11	2.61	875	5.61	3.889	3.044	0.813	2.548	0.483	10.776
3	12	3.11	734	7.53	3.626	2.955	0.924	2.548	0.396	10.449
4	13	3.67	623	9.90	3.430	2.947	1.067	2.548	0.334	10.327
5	14	4.28	534	12.81	3.281	2.975	1.234	2.548	0.287	10.325
6	15	4.95	462	16.33	3.164	3.082	1.435	2.548	0.253	10.482
7	16	5.69	402	20.54	3.071	3.233	1.674	2.548	0.227	10.754
8	17	6.48	353	25.54	2.996	3.438	1.963	2.548	0.207	11.152
9	18	7.34	311	31.41	2.934	3.656	2.294	2.548	0.191	11.623
10	19	8.27	276	38.25	2.882	3.956	2.679	2.548	0.179	12.245
					FIRST	COST*OPERATION			COST	TOTAL

Table 2. Least cost of filter designs varying with run length, C_O = 5.0 mg/l of iron, C = 0.3 mg/l of iron, sand size = 0.6 mm.

LENGTH OF RUN HP	SAND DEPTH IN	FILTRATION RATE GPM/SF	HEAD LOSS FT	UNIT COST $/MG
10	12.1	8.61	10.78	12.175
14	11.9	6.26	9.15	11.276
18	12.3	5.62	9.97	10.774
22	12.6	4.91	9.96	10.519
26	12.9	4.49	10.33	10.377
30	13.2	4.16	10.64	10.314
31	13.8	4.09	10.91	10.346
32	13.1	4.01	10.68	10.285
33	13.8	4.02	11.31	10.324
34	13.6	4.03	11.58	10.299
35	13.9	3.82	11.10	10.340
36	13.4	3.83	11.27	10.293
37	13.7	3.65	10.87	10.354
38	13.9	3.83	12.04	10.329
42	13.2	3.34	10.56	10.349
46	13.6	3.29	11.41	10.403
50	14.0	3.25	12.31	10.476

Table 3. Recommended optimum sand filter designs varying with sand size, for iron removal C_0 = 5.0 mg/1, C = 0.3 mg/1, S.F. = 2.

SAND SIZE MM	SAND DEPTH IN	FILTRATION RATE GPM/SF	RUN LENGTH HR	HEAD LOSS FT	UNIT COST $/MG
0.6	13.1	4.01	32	10.7	10.29
0.7	16.4	4.10	36	10.5	9.90
0.8	19.3	4.40	36	9.5	9.70
0.9	22.8	4.44	40	9.2	9.50
1.0	26.3	4.60	42	8.9	9.30
1.1	29.6	5.00	42	8.6	9.19
1.2	32.9	5.28	42	8.3	9.07
1.3	36.0	5.48	42	7.9	8.95

Once all of the design combinations were predicted for each case, the cost of each filter was calculated and output by the computer. Table 1 also shows the cost of each of the 10 design alternatives. Alternate 10 in Table 1 indicates that this filter could be built and operated to produce water at a cost of $12.24 per million gallons. Comparison of the costs involved with the various sand depths in Table 1 indicates which sand depth is optimum for that sand size, run length, and C/C_0 ratio.

Ultimately, the data produced in each case can be used to plot graphs to show the following for Case 1: 1. The effect of the various cost components (filter equipment, filter structure, power, labor and maintenance, and backwashing) on total filtration cost at various sand depths for a fixed sand size, run length, and C/C_0 ratio (example: Figure 11). 2. The effect of sand size and sand depth on the filtration cost for a fixed run length and C/C_0 ratio (example: Figure 12). 3. The effect of filtration rate, sand depth, head loss, and length of run on the unit cost of filtration for each sand size (example: Figure 13). 4. The optimum combination of design variables (sand depth, filtration rate, head loss, and run length) for each sand size required to produce least cost filtration (example: Figure 14).

244

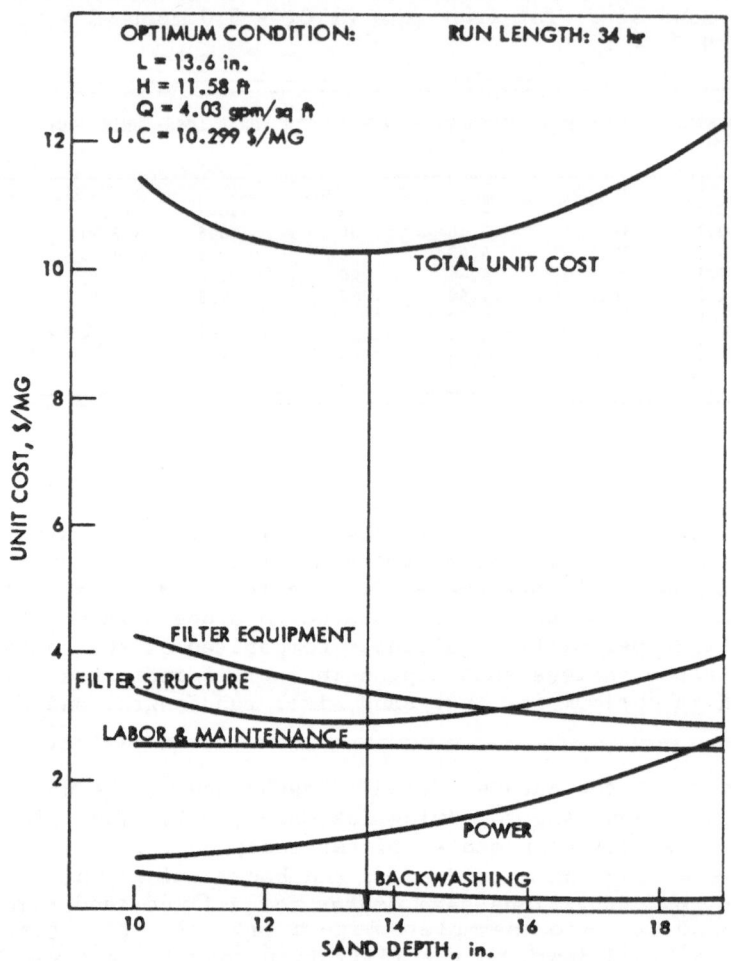

Figure 11. Unit filtration costs vs sand depth. Case 1, sand size = 0.60 mm, run length = 34 hr.

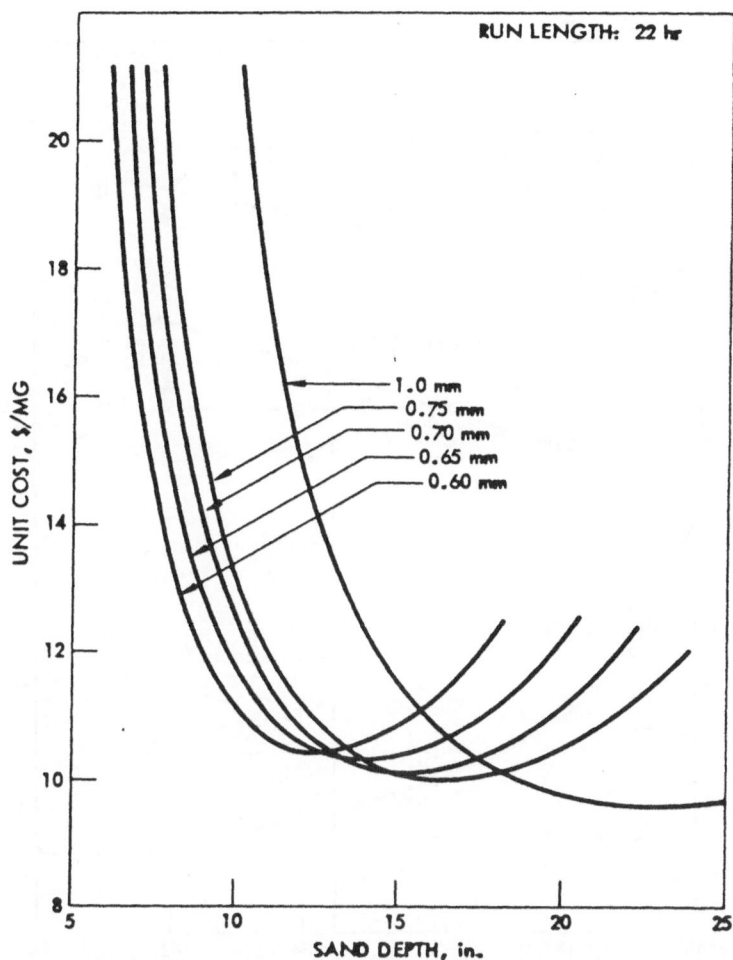

Figure 12. Total unit filtration cost for various sand sizes and sand depths, case 1, run length = 22 hr.

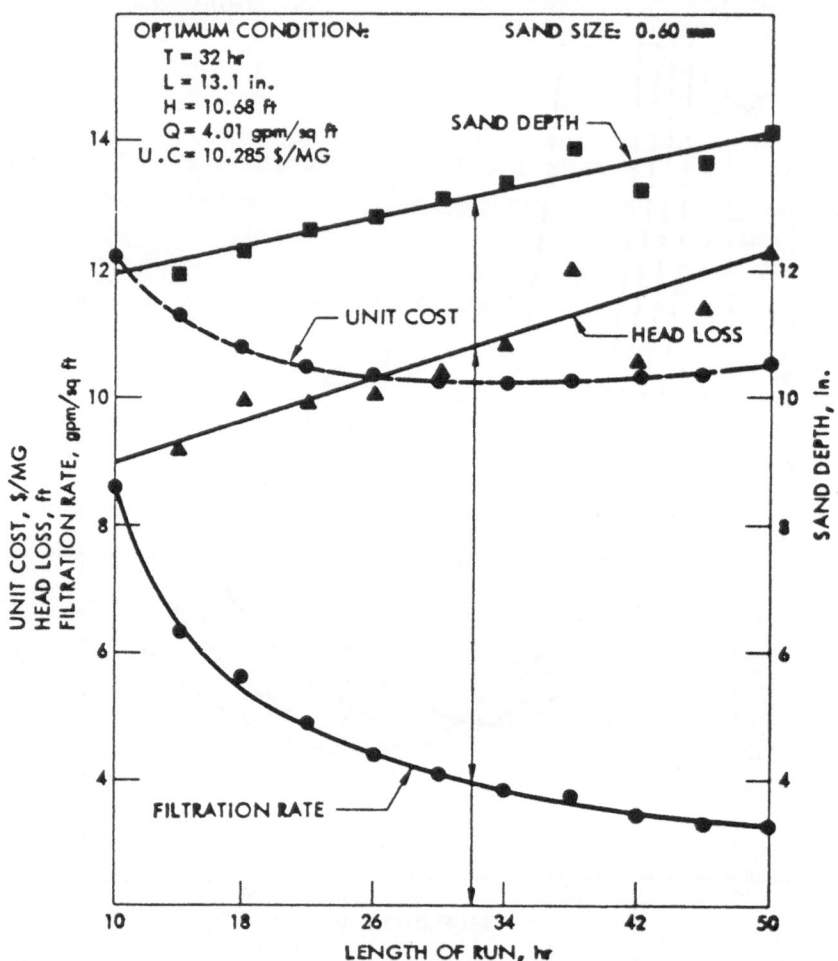

<u>Figure 13</u>. Effect of run length on filtration costs and the required filtration rate, head loss, and sand depth. Case 1, sand size = 0.60 mm.

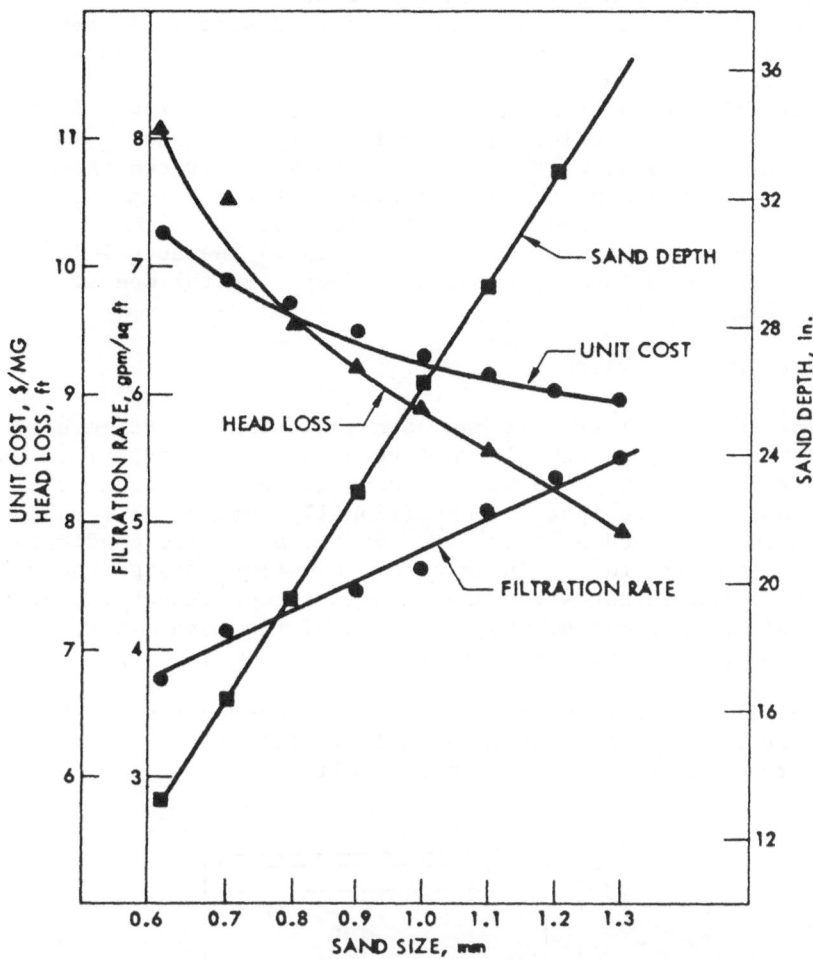

Figure 14. Effect of sand size on the optimum design characteristics of sand filters. Case 1.

As a result of such studies, more details of which are published elsewhere[3], Huang and Baumann calculated the cost of water per unit filtrate for a range of sand sizes, filtration rates, sand depths and limiting head losses. As the sand size increased (Figure 14), the unit cost decreased with increasing filtration rate and sand depth, but with lower head loss. In view of the increasing difficulty in cleaning sand media over 1.3 mm size, the calculations were not continued far enough to reach a minimum in unit cost. At the 1.3 mm sand size, the lowest cost occurred at a filtration rate of 14 m/h through 0.9 m of sand to a 2.5 m head loss. The actual least cost filter would involve a coarser, deeper filter operating at a still higher rate.

ECONOMIC OPTIMUM DESIGN

According to Ives[4], it has been shown that to achieve an operational optimum many alternative designs are possible. But of these designs only one will produce filtered water at the least cost. The designs for operationally optimum filters considered only the media, flow rate and the hydraulic conditions limiting the head loss. In an economic optimum design the effects of the filter structure, hydraulic appurtenances, energy and maintenance requirements, and size of the total filter plant also have to be considered.

The filter will be considered to be constructed as shown in Figure 15, and it is assumed that all filters in the treatment plant are identical, with no shared walls.

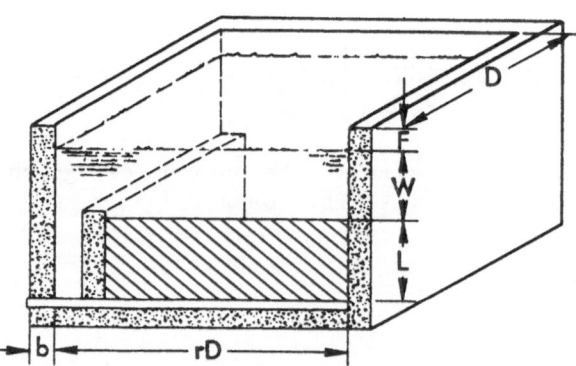

Figure 15. Sketch of filter giving principal dimensions.

The following items contribute to the cost of filtration, established as a cost per month (30 days):

1. Cost of walls

$$T_1 = C_1 ab[(2r + 3)LD + 2(r + 1)(WD + FD)]; \qquad (14.1)$$

2. Cost of floor including underdrains and piping

$$T_2 = C_2 arD^2\left(\frac{v}{v_s} + u\right); \qquad (14.2)$$

3. Cost of media assuming wash gully width $\ll rD$

$$T_3 = C_3 aLrD^2 ; \qquad (14.3)$$

4. Cost of energy due to head loss

$$T_4 = C_4 H_{1im} rD^2 vt\left(\frac{30 \times 24}{t + t_w}\right)\left(\frac{10^3 \ kgf/m^3}{3.68 \times 10^5 \ kgf - m/kWh}\right); \qquad (14.4)$$

and

5. Cost of backwashing

$$T_5 = C_5 v_w t_w rD^2\left(\frac{30 \times 24}{t + t_w}\right) . \qquad (14.5)$$

The values for Eq. (14) are as follows:

a = capital recovery factor per month (amortization) 0.0075,

b = wall thickness (m) 0.3,

D = width of filter unit (m) 8.0,

L = media depth (m),

r = ratio of length to width of filter unit 1.5,

W = water depth over media (m),

H_o = initial head loss (m),

H_{1im} = terminal head loss (m),

t = filter run period (h),

t_w = washing period (h) 0.12,

v = filtration velocity (m/h),

v_s = standard filtration velocity (m/h)   10,

v_w = washwater velocity (m/h)   50,

u = proportionality factor for underdrains   0.5,

C_1 = unit cost of finished concrete ($\pounds/m^3$)   43,

C_2 = unit cost of floor including underdrains
($\pounds/m^2$)   270,

C_3 = unit cost of filter media ($\pounds/m^3$)   28.5,

C_4 = unit cost of energy ($\pounds/kWh$)   0.006,

and C_5 = unit cost of treated water ($\pounds/m^3$)   0.0085.

It is assumed that

$$0.8(W + L) = H_{\ell im} , \qquad (15)$$

which is reasonable considering the shape of pressure profiles through filters (see Ives[8]), and allows 25% head loss in under-drains, piping, etc. Also it is assumed that $H_o \ll H_{\ell im}$ and, therefore, that H_o can be neglected. Some simplification can be made by assuming $t_w \ll t$, therefore, $t + t_w = t$; also draining and recharge times have been neglected.

The volume of filtrate produced per month is $vrD^2 \times 24 \times 30 m^3$. Therefore, the total cost per cubic metre filtered is

$$T = (T_1 + T_2 + T_3 + T_4 + T_5)/720 \ vrD^2 \qquad (16)$$

If a choice is made for r and F, it appears from inspection of Eqs. (14.1) to (14.5) that the variables are D, W, L, v and t. In an analysis of the effect of width D on monthly cost, Gur[2] found that for different flow rates v, the cost was not sensitive (varying only $\pm$ 7%) for a range of D from 3 m to 11 m. If an extra filter is provided for standby, then the monthly cost increases markedly with filter size, as would be expected. Assuming no standby, but that load sharing takes place during washing of a filter, the choice of D can be arbitrary, not affecting cost optimization.

The variables W, L, v and t (and grain size d of a uniform filter which is implicitly included) are related for operational optimum conditions as given in Figure 5, and other graphs similar to Figure 5 for different $H_{\ell im}$ values. The water depth W and media depth L, taken together, specify the head loss limit $H_{\ell im}$ Eq. (15). The value $C/C_o = 0.05$ is assumed constant.

So the procedure for calculating Eq. (16) is to specify values of W, L, and v. This fixes $H_{\ell im}$. From Figure 5 or analogues of it (see Gur[2]) the value of t (operational optimum run time) is read. This, with fixed data for costs, amortization, etc., enables T to be calculated.

For example, taking the cost and other values assigned in the symbol list, Eq. (14.1) to (14.5) and (16) simplify to:

$$T = [\frac{1}{v} (36.35L + 5.6W + 144 + 5100/t) + 1.3L + 1.3W$$

$$+ 28.3] \times 10^{-5} \; \pounds/m^3 \tag{17}$$

If W = 2m, L = 1m, then $H_{\ell im}$ = 2.4 and t = 17h,

$$T = 0.81 \times 10^{-3} \; \pounds/m^3 \text{ or } 0.081 \text{ New Pence}/m^3$$

Equation (17) has been calculated for an array of values W = 1, 2, 3; L = 1.0, 1.8, 2.5; v = 10, 20, 30. The resulting cost data are presented on Table 4. It can be seen that these do not include the optimum conditions (minimum cost). Up to a value of W = 3m the optimum is at L = 5.5m, v = 20 m/h, with corresponding values of t = 30h, $H_{\ell im}$ = 6.8m and d_3 = 1.5 mm, giving a cost of $0.66 \times 10^{-3} \; \pounds/m^3$.

Table 4. Values of cost of filtered water in 10^{-3} ₤/m^3 *.

$W^\dagger = 1$

L^{**} $v^\ddagger$	1.0	1.8	2.5
10	0.84	0.74	0.72
20	0.92	0.74	0.68
30	1.23	0.87	0.76

$W = 2$

L v	1.0	1.8	2.5
10	0.81	0.74	0.71
20	0.88	0.83	0.68
30	1.07	1.02	0.70

$W = 3$

L v	1.0	1.8	2.5
10	0.80	0.75	0.71
20	0.83	0.71	0.68
30	1.01	0.84	0.71

*Calculated from Eq. (17).

$\dagger$W = water depth over media, m.

**L = depth of media, m.

$\ddagger$v = filtration velocity, m/h.

As can be seen from Table 4, the cost is not very sensitive to variation in W, near the optimum. Extending W to 4m, produces a slight advantage to T = 0.65 $\times$ 10^{-3} ₤/m^5, at L = 4m, v = 20m/h, t = 25h Hℓim = 5.6m and d = 1.3 mm.

The response of cost, with respect to L and v has been calculated by Gur[2] using the theory of Mints[7], but with the same basic cost data. In this case, because of the nature of the cost function, the optimum (minimum cost) could be obtained by differentiation. The result was very similar to the present calculated optimum, with a depth $L = 3.8m$, velocity $v = 20m/h$ and cost 0.67×10^{-3} $£/m^3$. It may be argued that, therefore, it does not matter very much which mathematical model of the filtration process is used.

It is interesting to note that the standard conditions for rapid filtration in Europe and North America are $W = 2m$, $L = 1m$, $v = 5m/h$. This gives $H_{\ell im} = 2.4m$ and $t = 40h$ (if operational optimum conditions are achieved, which is rare). The corresponding cost is 0.96×10^{-3} $£/m^3$, and doubling the rate of filtration reduces the cost to 0.81×10^{-3} $£/m^3$, a saving of 15.6% in the cost of the water. This is similar to the figure given by Miller[14] who estimated a saving of 17% on the cost of filtering water by doubling the rate.

REFERENCES

1. Huang, J. Y. C.; _Least cost sand filter design for iron removal_; M.S. thesis; Iowa State University, Ames, Iowa; 1969.

2. Gur, A.; _Theory and optimization of water filtration_; Ph.D. thesis; University College, London; 1969.

3. Huang, J. Y. C., and Baumann, E. R.; Least cost sand filter design for iron removal; _Journ. Sanitary Engr. Division_, _ASCE_; SA2; 97; 171; April 1971.

4. Ives, K. J.; _Optimization of deep bed filtration_; First Pacific Chemical Engineering Congress; Part I; Session 2, Separation Techniques; 99-107; Society of Chemical Engineers, Japan and AIChE; Oct. 10-14, 1972.

5. Hsiung, K. Y., and Cleasby, J. C.; Prediction of filter performance; _Journ. Sanitary Engr. Division_, _ASCE_; SA6; 94; 1043; 1968.

6. Iwasaki, T.; Some notes on sand filtration; _Journ. Am. Wat. Wks. Assn._; 29; 1591; 1937.

7. Mints, D. M.; Modern theory of filtration; Special Subject No. 1D; International Water Supply Congress; Barcelona; Published: I.W.S.A.; London; 1966.

8. Ives, K. J.; Theory of filtration; Special Subject No. 7; International Water Supply Congress; Vienna; Published: I.W.S.A.; London; 1969.

9. Ives, K. J., and Sholji, I.; Research on variables affecting filtration; Journ. San. Engr. Division; SA4; 91; 1; 1965.

10. Ison, C. R., and Ives, K. J.; Removal mechanisms in deep bed filtration; Chem. Eng. Sci.; 24; 717; 1969.

11. Mohanka, S. S.; Theory of multilayer filtration; Journ. San. Engr. Division; SA6; 95; 1079; 1969.

12. Horner, R. M. W.; Water clarification and aquifer recharge; Ph.D. thesis; University College, London; 1968.

13. Hulbert, R., and Feben, D.; Hydraulics of rapid filter sand; Journ. Am. Wat. Wks. Assn.; 25; 19-65; 1933.

14. Miller, D. G.; Filtration: experimental developments; Journ. Inst. Wat. Engrs.; 25; 21; 1971.

BACKWASH OF GRANULAR FILTERS

John L. Cleasby, Appiah Amirtharajah,[*] & E. Robert Baumann

Department of Civil Engineering and
Engineering Research Institute
Iowa State University

INTRODUCTION

The backwashing of deep granular filters is one of the neglected areas of filtration research. At a time when such filters are assuming an increasingly important role in waste water filtration, many questions concerning backwashing remain unanswered. Waste water filters using new filter media sizes and gradations are being designed which prevent transfer of old "rule of thumb" design practices developed in water treatment to waste water treatment. Some of the unanswered questions regarding backwashing are:

1) What is the minimum degree of bed expansion necessary to ensure adequate cleaning of the filter media over a long period of service?

2) Can coarser media filters be successfully cleaned over a long period by air scour followed by backwash at rates below the level required to fluidize the bed?

3) If fluidization and bed expansion are essential to adequate cleaning, how can the degree of expansion be calculated from measured properties of the media and fluid, such as size gradation, density, fixed bed porosity, fluid temperature and viscosity?

4) Is some degree of intermixing of the media of dual or multimedia filters desirable from the standpoint of filtrate

*Former graduate student, now with Government of SriLanka

quality and head loss development? If so, how can the degree
of intermixing be predicted from measurable properties of the
media?

5) Is air scour or surface wash an essential requirement for
    waste water filters to achieve adequate cleaning? If so,
    which method is the most effective?

6) Will extended service with air scour result in abrasive loss
    of anthracite coal?

These questions are currently under study at Iowa State University. The purpose of this brief paper is to summarize our
findings to date.

SOME FLUIDIZATION FUNDAMENTALS

A number of fluidization fundamentals should first be re-
viewed since they form a foundation for discussion of any of
the questions previously raised.

The phenomenon of fluidization can best be visualized by
passing a fluid (gas or liquid) upward through a bed of solid
particles in which it encounters a resistance to flow and a resultant pressure drop Δp. As the flow rate V is increased there
is a linear relationship between Δp and V. As V is further increased a point is reached at which the pressure drop is sufficient to bear the weight of the solid particles. Any further
increase in flow rate causes the bed to expand and accommodate
the increased flow while maintaining the pressure drop Δp effectively the same. The fluidized bed thus formed closely re-
sembles that of a liquid[10]. The feature which distinguishes
the fluidized bed from other processes (fixed bed, filtration,
etc.) is the motion of the particles within the bed. The characteristics of an ideal fluidized bed and the distortions due
to real conditions are indicated in Fig. 1.

Within the last two decades a flowering of thought has occurred in the field of fluidization, fertilized by the necessity
of its use in the catalytic cracking of heavy hydrocarbons into
petroleum products. This has given rise to five books in English[8, 11, 12, 16, 17], six symposia and countless papers. The
above books and the reviews of recent work by Coulson and Richardson[7] and Botterill[3,4] form excellent general references.

Point of incipient fluidization or minimum fluidizing velocity-V_{mf}

This is the fluid velocity required for the onset of fluidization. It could be defined exactly as point A in Fig. 1 for an

ideal fluidized bed. For a real graded bed it is defined by
some as the intersection of the two linear sections of the
curve[8],[11] while alternate definitions are presented by others[13].

The bed is completely fluidized when the friction drag or
pressure drop across the bed is just enough to support the weight
of the filter media[7]. Mathematically, this relationship is given
by

$$\frac{\Delta_{pw}}{I_e} = (\rho_s - \rho_f)g(1 - \epsilon) \tag{1}$$

where

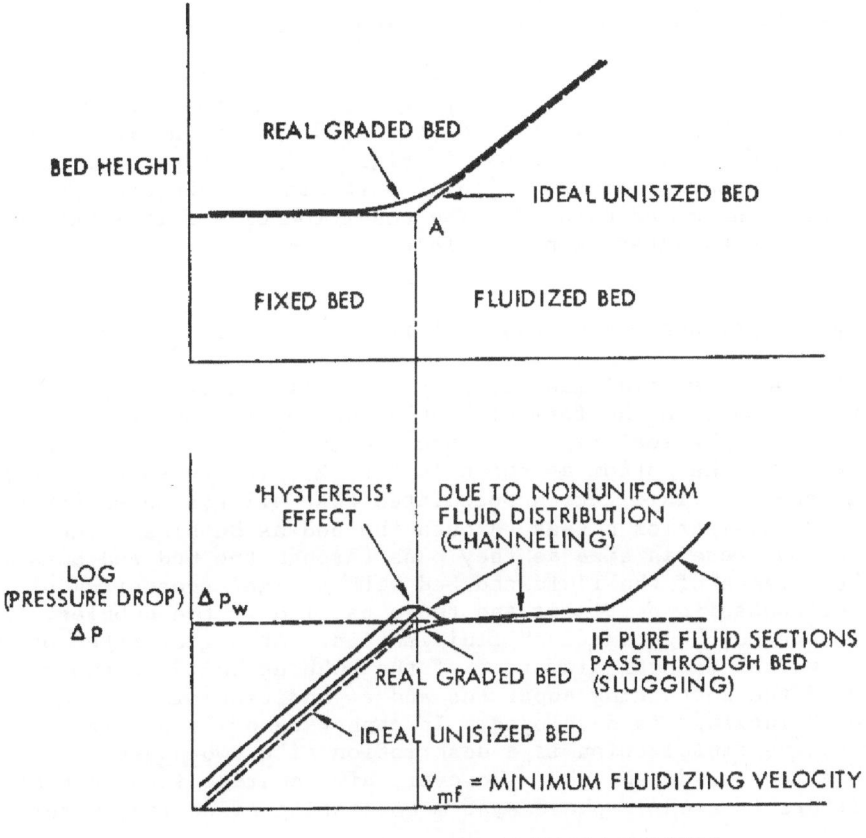

Fig. 1. Characteristics of fluidized beds.

Δ_{pw} = pressure drop across the fluidized bed
I_e = height of expanded bed
ϵ = porosity of expanded bed
g = acceleration due to gravity
ρ_s = particle mass density, and
ρ_f = fluid mass density.

The simplest bed expansion can be worked out by considering a bed which is fluidized from initial porosity ϵ_o at height I_o to a porosity ϵ and a height I_e. Since the volume of solids within the bed remains constant, then for a bed of constant cross section we have

$$I_o(1 - \epsilon_o) = I_e(1 - \epsilon). \qquad (2)$$

Particulate or homogeneous fluidization

In most liquid fluidized beds there is a uniform increase of bed height for velocities greater than V_{mf}, and the liquid passes smoothly and appears uniformly distributed within the interstices of the solid particles as shown in Fig. 2. This type of fluidization with a uniform distribution of particles was termed "particulate" by Wilhelm and Kwauk[14]. The condition of a filter bed while being washed by water is particulate.

Aggregative or nonhomogeneous fluidization

For most gas fluidized systems, part of the gas breaks through the bed in the form of bubbles which are considered shaped as a spherical cap or a sphere with a collection of solid particles at the bottom as shown in Fig. 2. The two-phase theory of aggregative fluidization postulates that all gas in excess of minimum fluidization passes through the bed as bubbles. The bubbles increase in size as they pass through the bed and burst at the surface of the fluidized bed with a light scattering of the surface solid particles and those carried by the bubbles. This is termed "aggregative" fluidization. At higher rates of gas flow, the frontal diameters of the bubbles build to the diameter of the containing apparatus and a condition of the bed called "slugging" is developed. It cannot be overemphasized that aggregative fluidization is a description of a two-phase system, composed of solids and gas. However, air scouring in backwashing of filters is a three-phase system consisting of solids, water and air.

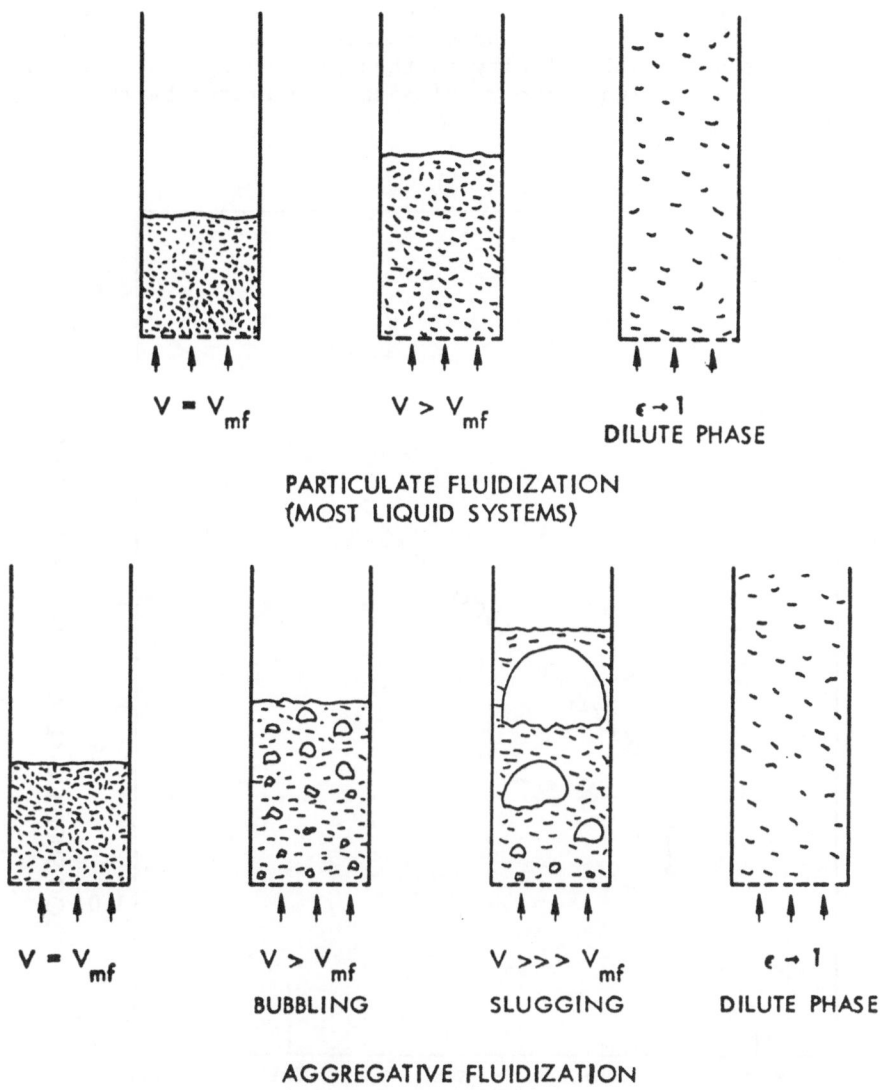

$V = V_{mf}$ $V > V_{mf}$ $\epsilon \to 1$
                                  DILUTE PHASE

PARTICULATE FLUIDIZATION
(MOST LIQUID SYSTEMS)

$V = V_{mf}$ $V > V_{mf}$ $V >>> V_{mf}$ $\epsilon \to 1$
                 BUBBLING   SLUGGING   DILUTE PHASE

AGGREGATIVE FLUIDIZATION
(MOST GAS SYSTEMS)

Fig. 2.   Fundamental behavior patterns.

Bed expansion

If the depth and porosity of a fixed bed are measured, and
then expansion of the bed is observed at various superficial
flow rates (upflow rate per unit gross filter bed area) the porosity at each flow rate can be calculated from Eq. 2. If the
log of the superficial velocity is then plotted against the log
of porosity, a straight line results as illustrated by the example
in Fig. 3.

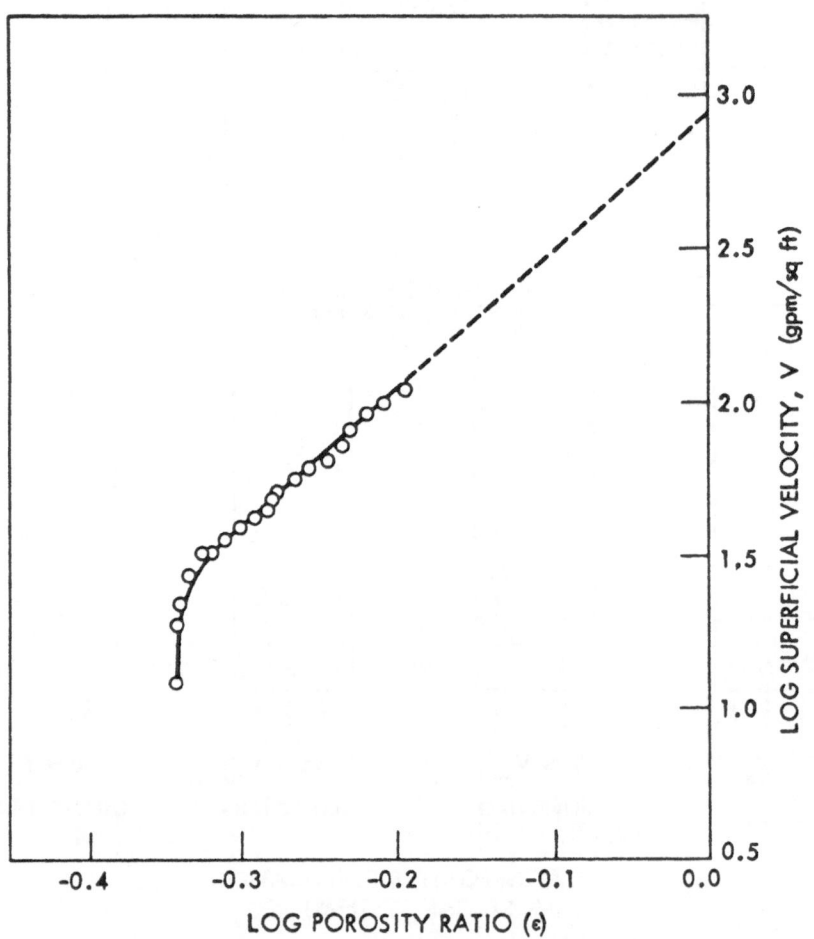

Fig. 3.   Log plot of V versus ε for garnet sand media passing 14
mesh, retained on 16 mesh, Run 1[15]. (gpm/sq ft × 0.6792 =
ℓ/sec.m²)

The line can be extended to a porosity (ϵ) of 1.0 ($\log \epsilon = 0$), the intercept velocity at this point is denoted V_i. For beds of uniformly sized spherical particles, V_i is equal to the unhindered subsiding velocity of the particles (V_s). However, for uniform nonspherical particles, V_i is somewhat higher than V_s.

The equation for the straight line (Eq. 3) is a fundamental equation applicable to all fluidized beds and is useful in various mathematical models for bed expansion.

$$\frac{V}{V_i} = \epsilon^n \tag{3}$$

where: V is superficial velocity
V_i is intercept velocity at $\epsilon = 1.0$
n is slope of the straight line.
For the special case of uniform spherical particles,

$$\frac{V}{V_s} = \epsilon^n \tag{4}$$

where: V_s is the unhindered subsiding velocity of the particles.

BED EXPANSION CORRELATIONS

Question 3 has been partially solved in a publication describing an expansion model for filter sand[2]. However, extension of the model to anthracite coal was not successful so an alternate model has been developed.

The revised model uses the following approach[15]:

1. Calculate V_i from measured properties of the media by using an empirical correlation between Re_i and the Galileo number (Ga)

$$Ga = \frac{d^3}{\mu^2} \rho(\rho_s - \rho)g \tag{5}$$

where: Ga = Galileo No. (dimensionless)
d = mean diameter of bed particles
μ = viscosity of fluid

$$Re_i = \frac{dV_i\rho}{\mu} = \text{Reynolds number based on the intercept velocity } V_i$$

For example, the correlation developed for garnet sand is:

$$Re_i = 0.0702 \; Ga^{0.823} \tag{6}$$

2. Calculate n from Re_i using an empirical correlation between n and Re_i

For example, the correlation developed for garnet sand is:

$$n = 5.768 \; Re_i^{-0.0541} \tag{7}$$

3. Calculate bed porosity at any desired superficial velocity using Eq. 3 and the values of V_i and n determined above.

The above approach has been successfully applied to graded and uniform garnet sand, and should be equally applicable to silica sand and crushed anthracite coal providing the appropriate empirical correlations are developed for each filter media. These correlations for both media are currently being developed.

OPTIMUM BED EXPANSION DURING WATER BACKWASH

The question of optimum expansion has been studied and published[1,6]. A number of interesting conclusions were reached in that study.

1. The cleaning of granular filters by water backwash alone to fluidize the filter bed is inherently a weak cleaning method because particle collisions do not occur in a fluidized bed and thus abrasion between the filter grains is negligible.

2. The cleaning which results in a water fluidized bed is due to the hydrodynamic shear at the water-filter grain interfaces. A simple mathematical model was developed to calculate the porosity of maximum hydrodynamic shear in a fluidized bed. Maximum hydrodynamic shear in a fluidized bed occurs at a porosity of 0.68 to 0.71 for sand sizes normally used in filtration. Optimum cleaning of the filter media at this porosity was demonstrated experimentally.

3. When backwashing with water alone, there is a slight economy in total washwater used by expanding the bed to the optimum porosity outlined in conclusion 2 above. Lower wash rates (anywhere above the rate for minimum fluidization) will result in nearly the same terminal washwater turbidity, but proportionately longer backwash times will be required. Therefore no economy of water use is achieved by use of low backwash rates.

Various studies in the literature had shown that several properties of fluidized beds maximized around a bed porosity of about 0.70. Those properties were turbulence, heat transfer, mixing, and mass transfer. It was this evidence that led to the hypothesis that backwashing effectiveness would also be optimized at a porosity of about 0.70. Using the classical equation for velocity gradients induced as a function of power input to a fluid body originally developed by Camp and Stein[5,9],

$$\frac{dV}{dz} = \frac{P}{\mu C} \tag{8}$$

where: $\frac{dV}{dz}$ = velocity or shear gradient

P = power dissipation
C = volume

and the basic headloss equation (Eq. 1) and expansion equation (Eq. 3), Amirtharajah was able to develop a simple equation for the porosity of maximum hydraulic shear,

$$\epsilon = \frac{n - 1}{n}. \tag{9}$$

He proceded to show experimentally that with water backwash alone, optimum cleaning was achieved at the porosity determined by the above equation. However, he also showed that the curve of shear versus ϵ around the optimum is very flat and thus hydraulic shear is quite insensitive to changes in ϵ around the maximum. For example, a change in backwash porosity from the optimum of 68% to a level of 60% resulted in a decrease in shear of only 2.5%. At 52% porosity, the shear was 7.8% lower than at the optimum.

Thus, while the porosity of maximum hydraulic shear is a noteworthy academic contribution, Amirtharajah was the first to admit that water backwashing alone is inherently a weak method of cleaning the filter grains because in a fluidized bed the grains do not collide.

THE BENEFIT OF AIR SCOUR IN BACKWASHING

Due to the inherent weakness of water backwashing cited above, auxiliary means of improving filter bed cleaning are generally desirable. A study of the benefits of air scour has been completed. The experimental supporting evidence will not be presented in detail due to space limitations but it is available upon request[6]. A brief summary of the work follows.

264

The objective of the experimental study was to determine the relative effectiveness of the following two backwashing techniques.

1. Air scour followed by water backwash. This is a common method of cleaning filters in European countries, and it has been in use for over 50 years.

2. Water backwash only. This is the common U.S. practice. This system is associated with some filter bed troubles which are not common where the air and water wash system are being used.

Pilot scale and plant scale studies were conducted. The pilot plant consisted of three identical filters of 6-in. (15.24 cm) diameter. Before backwashing these filters using different techniques, filters were dirtied under identical conditions. After the dirtying run, the three filters were backwashed differently. Usually one filter was backwashed with water only and the other filters were air and water backwashed. The following parameters were observed to determine the effectiveness of the various washing techniques: (1) washwater quality, (2) cleanliness of sand after backwashing, (3) effluent quality and headloss in the subsequent filter run.

Two different chemicals were added to Iowa State University (ISU) tap water to prepare the influent suspensions used to dirty the filters. The chemicals were ferrous sulphate during series I and sodium hydroxide during series II.

When using ferrous sulphate (series I), it was found that the preformed suspended iron flocs were apparently trapped in the bed due to physical mechanisms and had little tendency to stick on the sand grains. Therefore, the greater agitation of sand grains resulting from air scour did not improve cleaning of sand grains compared to water wash alone. The results obtained during series I are summarized as follows:

1. There was no apparent difference in effluent quality during the observation runs following different washing techniques.

2. There was no consistent difference in backwash water quality during the first seven runs of series I when the influent suspension was aerated. Small differences in initial wash water quality were observed during the later runs of series I when influent suspension was not aerated. There was no apparent difference in the terminal wash water quality for the various washing techniques studied, providing that the same volume of washwater was used up to the point of comparison.

3. No consistent results were observed regarding the cleanliness
    of sand, as measured by a sand abrasion test, after back-
    washing filters using different techniques.

When using sodium hydroxide to prepare the influent suspension (series II), the intent was to make water unstable, thus
encouraging precipitation of calcium carbonate within the filters
or on the sand grains. In this case, the precipitates were not
just physically trapped like the iron flocs, but the precipitate
actually crystallized on the sand grains and, therefore, adhered
quite strongly. The washing scheme for each filter after each
dirtying run was kept the same through series II in order to observe the cumulative effect of several identical filter cycles.
The results obtained are as follows:

1. There was no apparent difference in effluent quality even
    after the fifth run of the series.

2. Some interesting results were observed when washwater quality
    was compared up to any selected volume of washwater. For
    filters using air scour, the initial quality of washwater
    was more turbid when compared with filters which received
    water wash only as shown in Fig. 4. This indicates that air
    scouring was helpful in increasing the removal of suspended
    particles or precipitates retained on the sand grains. There
    was no appreciable difference in the terminal washwater
    quality for the first two or three runs, but the difference
    in terminal wash water quality was more pronounced during
    the last two runs, being better for the air and water-washed
    filters.

3. The best and most direct evidence of the benefit of air scour
    during the backwash was obtained using a standardized abra-
    sion test on a small sample of sand after each run in series
    II as shown in Table 1[6]. The filter sand which was subjected
    to air and water wash was found to be cleaner than the sand
    receiving water wash only as evidenced by the amount of sus-
    pended solids removed during the abrasion test. The dif-
    ference in the amount of removable suspended solids became
    more pronounced during the later runs of series II.

The plant scale studies conducted at the Ames, Iowa water
treatment plant added further evidence to support the above conclusions (series III). This lime-soda ash softening plant deli-
vers a water containing some suspended calcium carbonate crystals
to the filters. Polyphosphate is used to reduce the deposition
tendency of the water. Even under these conditions, the use of
air scour followed by water backwash was shown beneficial in

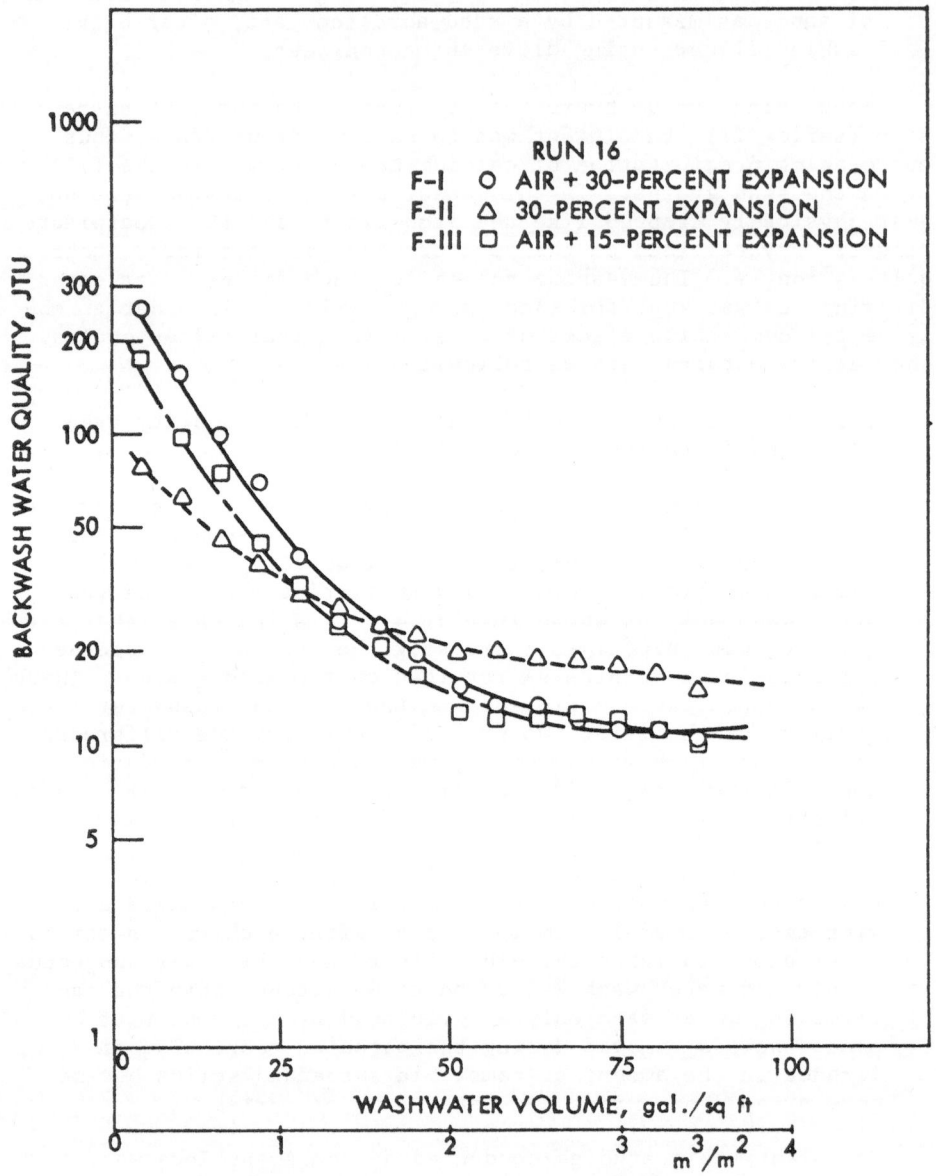

Fig. 4. Backwash water quality versus washwater volume, Run 16, series II[6].

washing the filter media cleaner with less water than water backwash alone. The results indicate higher initial washwater turbidity with air scour and lower terminal washwater quality as shown
in Fig. 5. Also, the standard abrasion test results show cleaner
filter media for the air scoured filter as shown in Table 2.

Table 1. Amount of suspended solids removed from sand grains by
the abrasion test in series II for samples of sand
taken near the surface after 75 gal. of washwater had
been used per square foot of filter area[6] (3.06 m^3/m^2)

| Series II | Suspended solids removed in mg/g of sand | | |
	Filter I air + 30-percent expansion	Filter II 30-percent expansion	Filter III air + 15-percent expansion
Run 12	2.18	5.82	4.86
Run 13	3.84	7.75	4.19
Run 14	5.98	9.81	6.22
Run 15	14.2	17.92	10.2
Run 16	16.5	24.8	10.74

THE ABRASIVE LOSS OF ANTHRACITE COAL FILTER MEDIA DUE TO AIR SCOUR

The effect of air scour on anthracite coal filter media be-
came of concern because of the possible widespread use of air
scoured filters using anthracite coal as one of the filter media.
Wide-spread use seems probable for the following reasons: (1)
the apparent benefits of air scour as an adjunct to water backwash, (2) the apparent necessity of auxiliary agitation in waste
water filtration, (3) the necessity of using dual media in waste
water filtration to achieve adequate filter run length.

The visual appearance of coal filter media when subjected to
air scour was observed in plastic filter housings. The coal appears to be in rather violent motion, especially the coal near
the top surface of the bed. The granules of the coal are in
contact with each other and rubbing against each other, unlike
in a fluidized bed where collisions and rubbing are absent. Thus
the abrasive loss of coal was of concern due to the softer nature
of coal compared with other common filter media such as silica
sand and garnet sand.

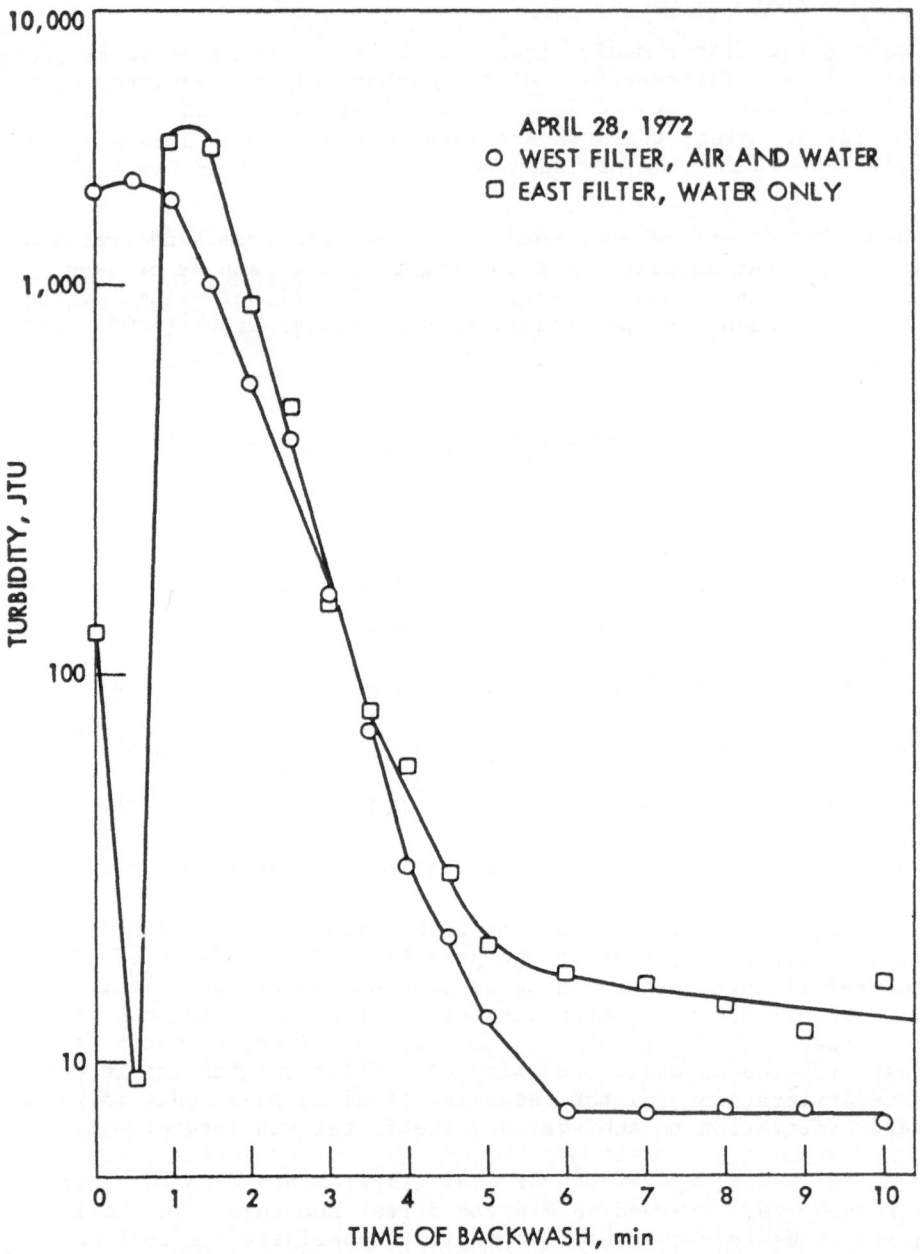

Fig. 5. Backwash water quality versus time of backwash, series III, Run 3, Ames plant (backwash rate 15 gpm/sq ft or 10.2 ℓ/sec.m^2).

An experiment was conducted to view the potential serious-
ness of the abrasive loss problem. The experiment consisted of
subjecting a filter bed of anthracite coal to essentially con-

Table 2. Amount of suspended solids removed from coal filter
grains by the abrasion test in series III for samples
of coal removed near the surface after five minutes of
wash at 15 gpm/sq ft^6 (10.2 ℓ/sec.m^2)

| Filter run series III | Suspended solids removed in mg/g of coal | |
	West filter — air scour and water backwash	East filter — water backwash only
1	1.41	2.98
2	3.14	4.44
3	1.65	6.05
4	2.24	4.64
5	0.85	5.10
6	0.68	4.84

tinuous air scour for two weeks. Two weeks of such exposure
would be comparable to about 20 years of normal filter operation
assuming 24-hour filter cycles and three minutes of air scour
per cycle. The filter was water backwashed twice a day to observe the visual loss of coal dust. Furthermore, changes in the
depth and dry weight of bed, size of media, and pressure loss
profile through the bed in both upflow and downflow were observed
to obtain evidence of changes in the filter media.

Some abrasive loss was evident by the color of the coal dust
in the water at the end of each period of air scour. The initial
backwash water was black with coal dust but cleared up in the first
two minutes of backwash.

There was no substantial loss in bed depth or bed weight as
shown in Table 3. Furthermore, there was no substantial change
in head loss either upflow or downflow indicating no substantial
change in size of the media.

A sieve analysis before and after the two-week period of air scour was conducted and the results summarized in Table 3. It is evident that the coal decreased in size slightly in this air scour exposure, about 0.02 mm in effective size, which is a negligible change.

Table 3. Changes in coal bed over a two-week period of air scour exposure equivalent to about 20 years of normal service[6].

	Initial value	After two weeks
Bed depth, in. (cm)	14.12 (35.86)	13.88 (35.26)
Bed weight dry, g	5019	4775
Total downflow headloss, ft H_2O (cm of H_2O)	0.30 (9.14)	0.29 (8.84)
Upflow rate to achieve 21 in. (53.34 cm) bed depth, gpm (ℓ/sec)	5.1 (0.32)	5.3 (0.33)
Total upflow headloss, ft H_2O (cm of H_2O)	0.35 (10.67)	0.34 (10.36)
Effective size of media (mm)	0.80	0.78
Uniformity coefficient	1.59	1.60

From this study, it is concluded that some abrasive loss of coal does occur but it is a negligible amount. The total loss of media and changes in media head loss and size were less than 3%. This is considered negligible because coal filter media suppliers usually request and are granted about 10% tolerance on the effective size of the media which they supply.

ACKNOWLEDGMENT

This work was supported by the Engineering Research Institute at Iowa State University.

REFERENCES

1. Amirtharajah, A., Optimum expansion of sand filters during
   backwash, Unpublished PhD thesis, Iowa State University
   Library, Ames, Iowa (1971).

2. Amirtharajah, A., and J. L. Cleasby, Predicting expansion of
   filters during backwashing, Journal of the American Water
   Works Association, 64: 52-59 (1972).

3. Botterill, J. S. M., Progress in fluidization, British
   Chemical Engineering 10: 26-30 (1965).

4. Botterill, J. S. M., Progress in fluidization, British
   Chemical Engineering 13: 1121-1126 (1968).

5. Camp, T. R., and P. C. Stein, Velocity gradients and internal
   work in fluid motion, Journal of the Boston Society of Civil
   Engineers 30: 219-237 (1943).

6. Cleasby, J. L., Backwash of granular filters used in waste
   water filtration, Engineering Research Institute, Iowa State
   University, Report No. 72198, E.P.A. Report, Project 17030
   DKG., Water Quality Office, E.P.A. (August 1972).

7. Coulson, J. M., and J. F. Richardson, Chemical Engineering,
   Vol. 2, 2nd ed., New York, N.Y., Pergamon Press, Inc. (1968).

8. Davidson, J. F., and D. Harrison, Fluidized Particles, New
   York, N.Y., The Syndics of the Cambridge University Press
   (1963).

9. Fair, G. M., J. C. Geyer, and D. A. Okun, Water and wastewater engineering, Vol. 2, New York, N.Y., John Wiley and
   Sons, Inc. (1968).

10. Furukawa, J., and T. Ohmae, Liquidlike properties of fluidi-
    zed systems, Industrial Engineering Chemistry 50: 821-828
    (1958).

11. Leva, Max, Fluidization, New York, N.Y., McGraw-Hill Book Co.,
    Inc. (1959).

12. Ostergaard, K., On the growth of air bubbles formed at a
    single orifice in a waterfluidized bed, Chem. Engng. Sci.
    21: 470-472 (1966).

13. Shannon, P. T., Fluid dynamics of gas fluidized batch systems,
    Microfilm copy, Unpublished PhD thesis, Illinois Institute of
    Technology, Library, Chicago, Illinois (1961).

272

14. Wilhelm, R. H., and M. Kwauk, Fluidization of solid particles, Chem. Engng. Prog. 44: 201-218 (1948).

15. Woods, C. F., Expansion and intermixing of garnet and silica sand during backwash of granular filters, Unpublished MS thesis, Iowa State University Library, Ames, Iowa (1973).

16. Zabrodsky, S. S., Hydrodynamics and heat transfer in fluidized beds, Cambridge, Massachusetts, The M.I.T. Press (1966).

17. Zenz, F. A., and D. F. Othmer, Fluidization and fluid particle systems, New York, N.Y., Rheinhold Pub. Corp. (1960).

FILTER MEDIA

A. Rushton

Department of Chemical Engineering,
The University of Manchester
Institute of Science and Technology,
Manchester, England.

1. INTRODUCTION

The principal role of a filter medium is to cause a clean separation of particulate solids from a flowing fluid with a minimum consumption of energy. Media may be broadly classified as a) those designed to recover a valuable solid product and b) those used in the clarification of a fluid, e.g. deep packed beds of sand in water clarification. In a) attempts are made to create surface deposition of the solids in recoverable form. Of course, certain media cannot be described as belonging to the a) or b) classification and create the required separation by a combination of surface deposition and particle capture in the internal interstices of the medium. The successful performance of a filter is largely dependent on the selection of a suitable filter medium. Despite the importance of the latter, little quantitative information is available which would facilitate the correct choice, and, in the present state of development, media choice usually follows experimental trials with the solid-fluid mixture. Such experiments have to be conducted with great care, since, as will be discussed below, small changes in process conditions can have an important effect on the outcome of such experiments. The task of media selection is made difficult by the multitude of media available. It is rarely clear, by inspection of the solids, which type of filter medium will prove most suitable for the separation. Exceptions here are extremely coarse solids ($> 100 \mu$m) which, in certain conditions, could be handled with perforated plate or edge-filters, and extremely fine material ($< 0.05 \mu$m), which require membrane filtration. Even in these situations, changes in the particle concentration and shape can present practical difficulties.

The inherent problems associated with media selection are illustrated quite clearly when one considers the woven fabric field. Variations in weave pattern, materials of construction, etc., produce important changes in the applicability of a particular medium to a filtration problem. The importance of correct media choice can also be illustrated graphically. In Figure 1, the yield of a constant-pressure filtration process

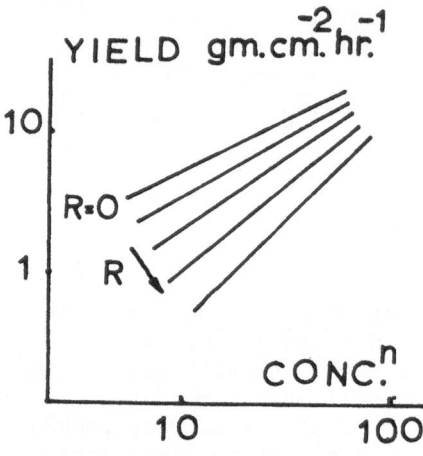

Figure 1

has been calculated at various levels of filter medium resistance, as defined in Equation 1:

$$V^2 + 2ARV/C\,\alpha_o - 2A^2\,\theta\,\Delta P^{(1-n)}/C\,\alpha_o\mu = 0 \qquad (1)$$

in which α_o is the filter cake resistance at low applied pressure and R is the resistance of the medium to fluid flow. In those cases where R may be assumed negligible in comparison with the cake resistance, simple yield equations of the form:

$$V_R = \frac{2\,\Delta P^{(1-n)}}{\mu\,\alpha_o\,C\,\theta}^{\frac{1}{2}} \qquad (2)$$

may be derived; V_R is the filtration rate per unit time per unit area. As may be seen, in Figure 1, the medium resistance has an important effect in determining the yield from such a process. It will be noted that such equations are derived by assuming that solids are surface-deposited, with no penetration into the internal pores of the medium, and with no change in R during the process. Such simplifying assumptions are rarely justified since, in practice, passage of fine particles into and through the medium cause increases in the resistance of the medium. Such processes can create conditions of media blinding, in which the permeability of the medium is reduced to zero. Internal deposition mechanisms are the principal cause of failure of application of Equations 1 and 2 to practical situations.

Some of the benefits obtained by a correct cloth or medium specification include:

a) a clean filtrate, with no loss of solids by bleeding or passage through the medium,

b) an easily discharged filter cake,

c) an economic filtration time,

d) no media deterioration by sudden or gradual blinding,

e) an adequate cloth life-time, with the first, clean performance reproduced by back-washing of the medium.

This list of requirements may be extended in practice when the construction and fitting of the medium must be considered. Thus, factors such as gasket performance, chemical and/or biological stability, strength, abrasion resistance, etc., may have to be considered in machine applications.

In view of the complexity of the situation, it will be realised that a system of classification of media action, linked to particulate properties, would be of great value in the process engineering of filtration systems.

Such a classification has been suggested(1) which is based on the rigidity of the media; the main features of this classification are presented in Table 1.

It will be observed that such a classification gives only the broadest guide to media selection.

Table 1

Media Classification

Type	Example	Minimum Trapped Particle (microns)
Edge Filters	Wire-wound tubes Scalloped washers	5-25
Metallic Sheets	Perforated plates Woven Wire	100 >5
Woven Fabrics	Woven Cloths, natural & synthetic fibres	10
Cartridges	Spools of yarns or fibre	2

It will be noted that Table 1 contains only those media upon which a severe geometric pattern is imposed. The work reported below on the permeability and filtration characteristics of media refers mainly to those media possessing geometric form or regularity, e.g. man-made woven-fibre filter cloths. The performance of random media; e.g. deep-packed beds, non-woven sheets of felt, cotton, etc.,paper, is referred to elsewhere in the text.

2. CRITERIA OF CHOICE

Three criteria by which a medium may be judged are:

1) The permeability of the clean medium.

2) The particle-stopping power of the medium.

3) The permeability of the used or deposited medium.

The permeability of the clean medium has importance in determining power requirements, e.g. fan size in gas filter stations, and in deciding the initial flowrate of fluid through the medium. As will be reported later, the initial flowrate has an effect on the cake structure of particles deposited near the medium, and, in upward filtering systems influences the size of particles deposited.

The permeability of the medium is defined by the d'Arcy equation:

$$u = B \frac{\Delta P}{\mu L} \hspace{4cm} (3)$$

in which B is the permeability, u the fluid velocity, ΔP the pressure drop over the medium, μ the fluid viscosity and L the medium depth or thickness. The question of media permeabilities is dealt with in detail below.

The particle-stopping power of the medium is, of course, of prime importance in deciding the course of a successful filtration. Much work has been reported in the literature which is aimed at the description of the medium in terms of an "equivalent pore-size", which can be related to the particle size in the deposit. Media efficiency tests are made using dilute suspensions of particles; the concentration of particles in the fluid before and after passing through the medium is measured and attention is given to proper sampling techniques and particle dispersion. The latter is of great importance since particle concentration in the fluid has a great influence in determining the probability of bridging a pore and producing a sieve-like filtration mechanism. As was pointed out above, since most of the process difficulties encountered in practice ensue when the sieve-like mechanism breaks down, specification of media pore-size is of fundamental importance.

Perhaps the most serious criterion of performance relates to the permeability of the used medium. Failure to release solids after an initial deposition also has serious economic consequences, except in those cases where disposable filter elements are intentionally used. In these cases, fluids containing extremely small amounts of solids are processed, the particles are trapped internally and no attempt is made to clean the element. An important characteristic of such elements is their solids-holding capacity.

3. PERMEABILITY OF CLEAN MEDIA

The problem of correlating the permeability of filter media with the basic dimensions of the materials composing the septum has received attention (2). Basically, the problem is to relate B, as defined in Equation 3 to variables such as fibre diameter, weave construction, etc.. Whilst the permeability of random systems is not the principal interest here, it is important to record some work reported in this area.

Perhaps the best-known equation for describing the required relationship in packed beds of particles or fibres is the Kozeny-Carman equation:

$$\frac{\mu\, u\, L}{\Delta P} = B = \frac{1}{K_o S_o^2}\; \frac{\varepsilon^3}{(1-\varepsilon)^2} \qquad\qquad (4)$$

The so-called Kozeny Constant K_o has been shown to be variable, dependent on the porosity of the deposit; rapid increases in K_o above values of ε of 0.7 have been suggested. Fibrous structures with porosity values in the range $0.7 < \varepsilon < 0.95$ have received attention and semi-empirical equations of the form:

$$B = \frac{d_f^2}{64(1-\varepsilon)^{1.5}\left(1+56(1-\varepsilon)^3\right)} \qquad\qquad (5)$$

have been reported in terms of the fibre diameter d_f and the porosity ε (3). For felted materials and air flow the equation:

$$\Delta P = k_o\, \mu\, u\, \ldots \qquad\qquad (6)$$

has been recommended, where $k = 4.29 \times 10^6 W$; W is the cloth weight in grams per square centimetre (wool, rayon and cotton).

Woven filter media can be classified into a) monofilament yarn and b) multifilament yarn. In the latter the yarn is composed of several filaments, twisted to various levels of yarn density. In the case of monofilaments, with solid yarns, where dy, the yarn diameter equals the fibre or filament diameter d_f, permeabilities are reasonably calculable by Equation 5. However, with multifilaments, $d_f < dy$, flow through the yarn must be expected, particularly in loosely-twisted yarns, and use of Equation 5 produces serious discrepancies between calculated and measured permeabilities.

Whilst an adequate correlation of the permeability of multifilament yarns is still required a useful index has been proposed which assists in a) general recognition of the type of medium being handled and b) prediction of the most probable pattern of behaviour of the medium as a filter, and when back-washed in laundering.

In multifilament cloths, fluid flow may occur through or around the permeable yarns. The degree of this flow division inter-yarn or intra-yarn has been shown to explain certain dyeing-characteristics of such cloths (4). If we define B_o as the permeability of the porous yarns, and B_1 as the permeability of the cloth if the yarns were solid, i.e. mono-filament, it may be shown that:

$$\beta = \frac{B}{B_1} = \left[1 + 1.34 \left(B_0/B_1\right)^{\frac{1}{2}}\right]^2 \text{ for } \frac{B_0}{dy^2} < 0.0017 \qquad (7)$$

where B is the overall permeability of the cloth. The β index has been shown to vary in the range $1 < \beta < 20$ within the order of accuracy of the experimental measurement necessary for the determination of B and B_0. A monofilament cloth will have a coefficient of unity since:

$$\beta = \frac{\text{Permeability of Cloth}}{\text{Permeability of Cloth if}} \qquad (8)$$
$$\text{Yarns Monofilament}$$

On the β scale, a tightly-twisted, loosely-woven multi-filament cloth will approach a monofilament cloth in character; whilst an index $\beta > 10$ indicates a tightly woven cloth in which most of the flow must be directed intra-yarn or through the porous yarns. It has been reported (5) that the β index finds use in describing the most probable course of a separation of solids and liquids. Some supporting data are reported below.

The problem of correlation of the permeability of multi-filament materials is aggravated in those media made up from natural materials such as wool, cotton, etc. In these cases, the smooth character of the so-called continuous filament (C.F.) is replaced by the hairy-random staple fibre (S.F.). Mathematical description of such systems is impossible.

In the monofilament area much more success has followed the suggestions of Pedersen (6) who adopted orifice-type formulae to correlate pressure-drop-flow information for various weave patterns.

A discharge coefficient was defined as:

$$C = \left[\rho \frac{u^2}{2\Delta P} \cdot \frac{(1-a^2)}{a^2}\right]^{\frac{1}{2}} \qquad (9)$$

where a, the effective fraction open area of the pore is:

$$a = A_0(ec)(pc) \qquad (10)$$

in which (ec) = warp yarns per centimetre
(pc) = weft yarns per centimetre
A_0 = effective area of orifice

The discharge coefficient was anticipated to be a function of the Reynolds number within the fabric:

$$C = f\ D^1\ u^1\ \rho/\mu \tag{11}$$

where $D^1 = 4\ A_0/P_r$, P_r being the orifice perimeter

and $u^1 = u/a$

The immediate problem is to calculate the α and W values for various pore configurations. In the problems solved to date it is necessary to be able to calculate α and W for plain and twill pores, Figure 2.

For a plain pore:

$$a = \frac{\emptyset}{2}\ \ln\ \frac{1+\sqrt{(1+\emptyset^2)}}{\emptyset} + \frac{\sqrt{(1+\emptyset^2)}}{2\emptyset} - (ec)d_2-(ec)(pc)d_1^2$$

$$\left\{ \sqrt{\left[\left(\frac{(1/ec)-d_2}{d_1} \right)^2 + 1 \right]} \right\} \tag{12}$$

$$P_r = 2\sqrt{\left[\left(\frac{1}{ec} - d_2 \right)^2 + d_1^2 \right]}\ \xi\left[\left(1+ \frac{d_1^2}{(\frac{1}{ec}-d_2)^2} \right)^{-\frac{1}{2}},\ \ \pi/2 \right]$$

$$+ \frac{2}{pc}\sqrt{\left[1 + (d_1+d_2)^2 pc^2 \right]} \tag{13}$$

$$a = \frac{\emptyset}{2}\ \ln\ \frac{1+\sqrt{(1+\emptyset^2)}}{\emptyset} + \frac{\sqrt{(1+\emptyset^2)}}{2\emptyset} -(ec)d_2-\frac{(ec)(pc)}{2}$$

$$\left\{ d_1^2\sqrt{\left[\frac{\frac{1}{ec} - d_2}{d_1}^2 + 1 \right]} + d_1(\frac{1}{ec} - d_2) \right\} \tag{14}$$

$$P_r = \sqrt{\left[(\frac{1}{ec} - d_2)^2 + d_1^2 \right]}\ \xi\left[\left(1 + \frac{d_1^2}{(\frac{1}{ec} -d_2)^2} \right)^{-\frac{1}{2}},\ \ \pi/2 \right]$$

$$+ (\frac{1}{ec} - d_2) + \frac{1}{pc} + \frac{1}{pc}\sqrt{\left[1 + (d_1+d_2)^2\ pc^2 \right]} \tag{15}$$

where $\emptyset = \left[(ec)\ (d_1+d_2) \right]^{-1}$ and $\xi\left(k,\ \pi/2 \right)$ is the elliptical integral of the second kind with variable k.

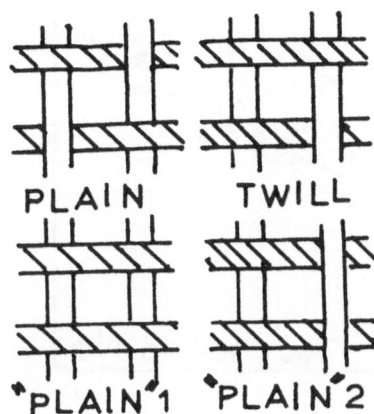

Figure 2

Figure 2 depicts the four warp yarn configurations which are possible in a single layer monofilament fabric. Pedersen tested the analysis by comparison with air permeability data on plain and 2/2 twill fabrics; a successful correlation was obtained, producing better results than those based on the 'projected pore-size'.

Recently (7) the more complicated 2/1 twills and 5/1 sateens have been analysed and flow data extended to water flow. The results are depicted in Figure 3a; use of the simple projected diameter approach (pore viewed from above) gives the results presented in Figure 3b. The flow data in the range 1<Re<10 has been shown to be represented by:

$$C = 0.17 (Re)^{0.41} \qquad (16)$$

for water flow, in plain and twill fabrics, with a maximum error of $\pm$ 18%.

In order to correlate the 2/1 twill and 5/1 sateen data it is necessary to define a 'flow-cell', Figure 4, which is repeated in the pattern of the cloth. In the case depicted (2/1 twill) the cell is made up of 6 twill pores and 3 plain pores; a weighting procedure is proposed so that:

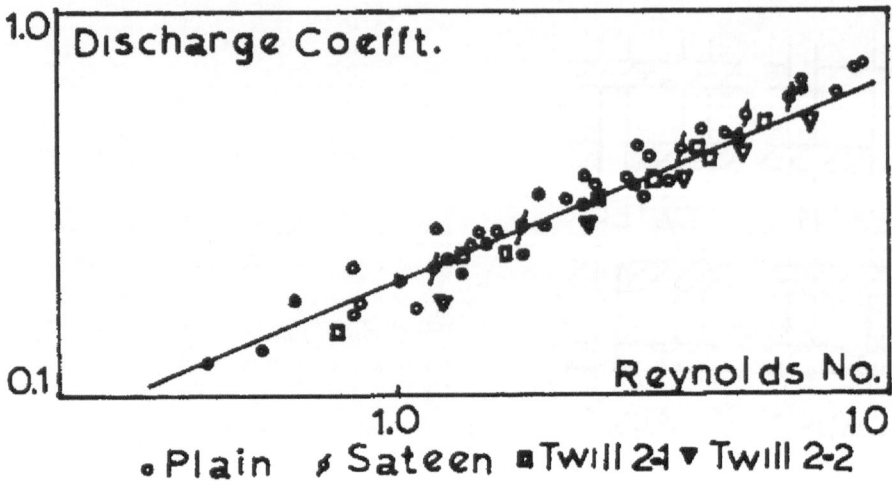

Figure 3a

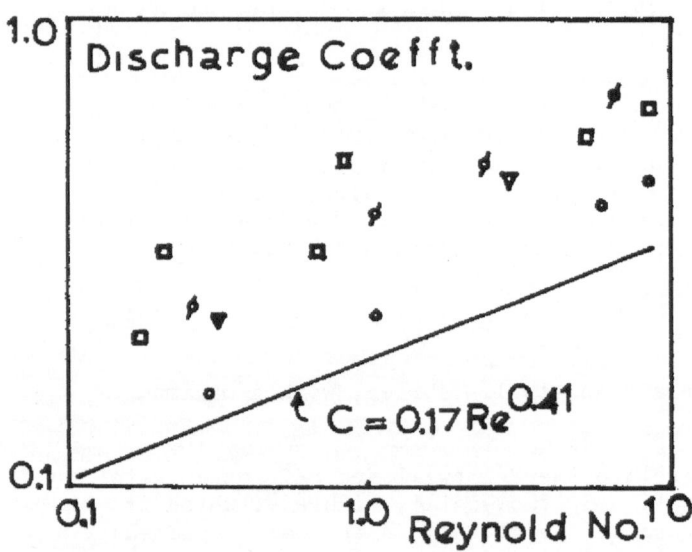

Figure 3b

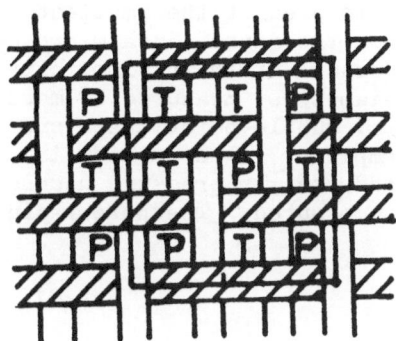

Figure 4

twill and plain pores respectively. The orifice perimeter
calculation is treated similarly.

The empirical approach described above may be compared
with the analytical techniques described by Hutson (8). Using
an approximate solution to the Navier-Stokes equations for
parallel cyclinders it was shown that

$$B = \frac{\mu \, V \, L}{\Delta P} = \frac{d_f^2(1-x)}{x \, f(x)} \qquad (18)$$

where $x = d_f/h$; d_f is the cylinder diameter and h the
space between cylinders. The function, $f(r)$, for $r<0.6$ may be
obtained from the expression:

$$\frac{8\pi}{f(x)} = 3.31571 - 2 \ln_e 10x + 1.6449(x^2) - 0.67644(x^4) \qquad (19)$$

These equations give good agreement between theory and
practice for plain weave wire gauzes and nylon cloths.

4. PARTICLE RETENTION AND PORE-SIZE MEASUREMENTS

The size and shape of the pore in the medium will determine
the feasibility of a complete separation by sieving, particularly
with media of the edge, perforated plate, simple wire or monofilament type filters. Where random fibres, scintered or porous

elements, staple fibre cloths are used, the pore size of the medium has less significance or use in predicting media behaviour. In simple woven cloth; the projected 'square' opening is directly calculable from mesh counts and the diameter of wire, and such data are used to predict the smallest spherical particle which can be retained on the mesh. Such microscopic count methods are attractive, because of their simplicity and have been compared with 'pore-diameters' measured by more complicated techniques such as a) bubble-point tests and b) permeability tests. In a) a sample of filter medium is submerged in a wetting liquid and the air pressure necessary to force air through the fabric is measured. The pore radius is calculated from:

$$r_{bp} = 2 \ \tau / \ \Delta P \qquad\qquad (20)$$

where r_{bp} is the bubble-point radius, τ is the surface tension of the fluid and ΔP the applied pressure

The pore-radius may also be inferred from permeability tests and the use of Equation 21:

$$r = \left[\frac{K_o B}{t} \right]^{\frac{1}{2}} \qquad\qquad (21)$$

Careful experimental determinations of r, r_{bp} and r_c (pore radius by microscopic count) have established the following simple relations for woven wire or monofilament cloths:

$$r_c = 1.26 \ r \qquad\qquad (22)$$

$$r_{bp} = 1.58 \ r \qquad\qquad (23)$$

No such relationships are available for multifilament cloths since the permeability is not accounted for by inspection or bubble-test. In the latter, the larger inter-yarn pores are measured. Calculations on the multifilament yarns show that, in such cases, the pore-size within the fibre r_f is generally smaller than the pore size between the yarns and $r_f < r < r_y$.

In liquid filtration trials using media of random pore-structure it is usual to report three particle sizes: a) where 100% capture is attained and b) two other, arbitrary levels of retention, e.g., 90% and 10% capture. Again it is usual to use a test mixture which is typical of the industrial problem at hand. A wide varity of 'standard' test powders have been proposed. In the gaseous field, both solid and liquid tests are used.

Generally the test particles are much smaller than those
employed in liquid testing, e.g., 0.03-5 μm (sodium chloride);
0.03 - 1.2 μm (methylene blue). With liquids, e.g. dioctyl
phthalate (D.O.P), closer control on the particle size spread
is possible and tests using droplets, all at 0.3 μm, are
feasible. The interpretation of such trials are made difficult
by the small size of the material passing; the filtration
mechanisms involved: direct interception, gravitational settling,
inertial impaction, diffusion and electrostatic attraction have
various degress of importance as the particle size decreases.
This situation generally leads to a particular particle size
having maximum penetration through such filters. Of fundamental
importance in this field is the generation of electrostatic
charges to fabrics during the passage of air; the charging
characteristics of filter materials have been arranged in a
so-called triboelectric series: wool, glass, polyamide, cotton,
polypropylene, polyethylene. In the latter the highly positively
charged wool compares with the negatively charged polyethylene.
Such charging characteristics for particles are less well
defined, although some progress has been made in a broad
classification.

As mentioned above, the efficacy of such tests in aiding
filter media selection is often criticised since changes in
process conditions can often bring about plant failure,
undetected at the testing stage. Again, many separations would
proceed more smoothly if surface deposition conditions could be
produced during media life.

Experiments (9) have shown that particle size, pore-size
and particle concentrations are important in determining the
onset of pore-bridging and, therefore, surface depositions.
The spread in particle size and skewness of the size distribution
curve are also factors of importance, as are the pore and
particle shape and cloth type.

In the work reported below the concentration of particles
of mean size $\bar{d}$, required to bridge monofilament pores of size d,
are reported. The pore sizes d are inferred from permeability
measurements. Relationships of the type:

$$d = K_1 \bar{d} \; c^m \qquad\qquad 0.002 < c < 2 \quad \% \; w/w \qquad\qquad (24)$$

have been derived for various materials of industrial
interest; values of k_1 and m are reported in Table 2.
Some data are included on straight capillary tubes.

Table 2

Material	Pore Type	K_1	m	C(%w/w)
Kieselguhr	Capillary	1.67	0.29	2.22
Dicalite 438	Capillary	2.02	0.31	2.25
50%CaCO$_3$/Kieselguhr	Capillary	1.44	0.23	2.12
Kieselguhr	Plain Weave	20.8	0.26	0.002-0.2
50%CaCO$_3$/Kieselguhr	Plain Weave	32.7	0.38	0.005-0.06
Magnesium Carbonate	Plain Weave	13.0	0.35	0.01-0.2
Dicalite 418	Plain Weave	16.0	0.29	0.007-0.2
Dicalite 438	Plain Weave	702.0	0.93	0.003-0.005
Calcium Carbonate	Plain Weave	10.2	1.04	0.3-1.9

A typical bridging curve is shown in Figure 5 where the

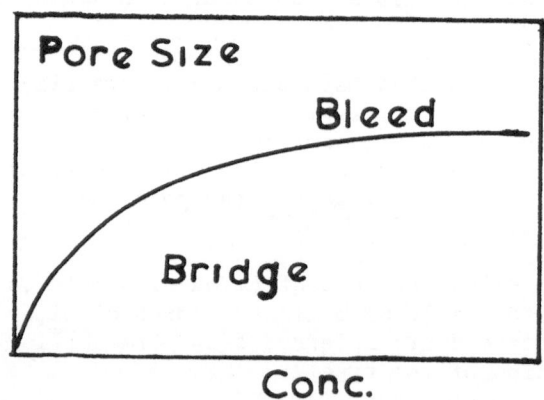

Figure 5

clear zone between pore bridging and particle penetration is
seen; a further point is that for a particular particle size, a
maximum pore size exists across which bridge formation is
impossible and a complete loss of solids occurs.

The question of weave pattern on pore bridging has been

examined (9). The dimensions of 'bridging pores', which have
four boundary yarns in the plane first contacted by the impinging
solids, was determined in the various 'unit cells' referred to
above. At equivalent permeability, the pore sizes of each cloth
will differ. For example, the pore sizes in 5-1 sateen will be
much smaller than those in a 1-1 plain cloth of equal permeability,
and therefore will have a better bridging capability.

5. RESISTANCE OF DEPOSITED MEDIA

The resistance of a medium after passage of a certain
volume of filtrate will depend on the mode of deposition of the
particles. In tests associated with the separation of particles
from fluid, experimental measurements of the amount of filtrate
collected at various times are recorded, for pre-determined values
of fluid pressure, particle concentration, etc. The filtrate
volume-time relationship is then examined to ascertain the
principal modes of deposition. If a sieve-like action is obtained,
with no change in medium resistance, elementary theory suggests
that graphical plots of $(\theta/v, V)$ should produce linear plots;
the slope of the latter is determined by the resistance of the
filter cake and the intercept on the θ/v axis is determined by
the medium resistance. Such 'cake filtration' conditions may be
described by the use of p = o in the expression:

$$\frac{d^2\theta}{dV^2} = K_2 \left(\frac{d\theta}{dV}\right)^P \tag{25}$$

Non-linear $(\theta/v, V)$ relationship have been attributed to
other modes of deposition than cake filtration. For example,
if the pores of the medium are blocked completely by the arrival
of solid particles, the 'blocking law', p = 2 is obtained.
Other experimental results indicate a gradual blinding of the
medium, or 'standard law', p = 1.5; the presence of a fourth
'law', p = 1.0 has been demonstrated practically, although has
no physical interpretation.

It is of interest to record the views of Gonsalves (10) who,
commencing from a different physical model than a gradually
blocking pore, derived the p = 1.5 relationship. This demonstrates
that agreement between experiment and the predictions of a certain
model for the process is no guarantee that particles are, in fact,
being separated in the manner assumed in the model. Such models
must be viewed merely as convenient for discussion purposes.

The importance of the concentration of the prefilt in
determining the mode of deposition has been thoroughly investigated
by Heertjes and is discussed elsewhere in this text. In most

practical cases all modes of deposition will be taking place
simultaneously, particularly during the early life of the filter
cake when the medium is actually separating the two phases.
After these initial layers are deposited subsequent cake growth
ensues upon the first layers of particles and it is the resistance
of the clean medium plus the initial deposits, which are often
not removed during other parts of the filtration cycle, which
determines the course of subsequent filtrations. Quite often,
in choosing a filter medium, certain poor characteristics of the
medium are accepted, if other more serious factors can be
alleviated. Thus, it may be necessary to choose a medium which
bleeds particles initially, to produce easily discharged cakes;
the bleeds may be recovered by recycling the liquors.

Measurements (5) have been reported of the combined resistance of single layer deposits of particles on monofilament cloths,
using a wide range of particles sizes. In general, small increases
in the combined resistance R_T were obtained for large particles
where $d > d_p + d_f$; where d is the particle size, d_p the pore size and
d_f the fibre size. In this condition, the large particle
interferes with the siting of further particles in adjacent pores,
resulting in a low increase in flow resistance. As $d \rightarrow d_p + d_f$,
a rise in resistance is recorded to a maximum at $d \approx d_p + d_f$;
thereafter, further decreases in particle size results in bleeding
of particles. Maximum increases in the ratio R_T/R, deposited over
used resistances of the order of 6 have been recorded. This type
of result is often obtained in $(\theta/V, V)$ plots of a linear character
which suggest 'cake filtration' conditions; the inferred cloth
resistance, taken from the intercept on the t/v axis are often
several times larger than clean media resistances.

An equation of the form:

$$R_T = \psi R \qquad\qquad\qquad (26)$$

has been reported for deposits of glass spheres and sand on
monofilament cloths, where

$$\psi = 1 + K_3/(d/d_p + d_f)^q \ , \ 0.6 < d/d_p + d_f < 2.0 \qquad\qquad (27)$$

and

	K_3	q
glass	2.25	1.65
sand	0.07	2.94

As discussed elsewhere in the text, the medium resistance,
by deciding the initial filtration velocity of the deposits
can affect the value of cake resistance. Relationships of the
form:

$$\alpha = \text{Const exp} \ -\left\{\frac{K_4}{\overline{\mu}R}\right\}^{K_5} \tag{28}$$

are suggested from such work. Evidence has been collected (13) which demonstrates that (α,R) relations of this form are found only where cake filtration conditions obtain throughout the deposition. The presence of particle bleeding or partial pore plugging produces a random relationship between α and R; where cake conditions apply a low R value is associated with low α values.

Deposition studies using multifilament cloths of various weave patterns to produce the $R_T = \psi R$ type of relationship gave results for ψ which were not readily linked with cloth structure. Generally, the ψ values recorded were lower than those of mono-filament media; continuous filament cloths tended to produce higher ψ values than staple fibre systems, other factors being the same.

An interesting fact which emerged from such work using small particles where $d \rightarrow d_f$ was that the cloth structure has an effect in determining the mode of particle deposition. Apparently the high β factor cloths produce filtration conditions describable by p = 1.0 in Equation 25 whilst low β cloths, which should demonstrate monofilament characteristics tended to produce the result p = 1.5. Some supporting data are presented in Table 3 .

Table 3

Particle Concentration 0.001 gm/l
Mean Particle Size   0.7 μm
Largest Particle   2.0 μm

Cloth	β	d_y	d_p (microns)	d_f	Mode P
Polyester D	1.34	10	7	0.8	1.5
Polyamide A	2.00	21	17	4.2	1.5
Polyester E	2.66	20	15	4.4	1.5
Polypropylene	16.0		7.8		1.0
Polyamide B	16.1		5.0		1.0
Polyester B	>20		6.0		1.0

Media Blinding

Studies of the onset of media blinding have been reported for several multifilament cloths of various weave patterns. Prior to such studies, the data on the progressive change in media resistances with alternate filtration and back-washing were extremely scanty. The situations examined were free of problems associated with the so-called 'filter effect', in which a gradual increase in medium resistance is observed by the passage of superficially clean water. This effect has been attributed to the presence of micro-organisms and algae in the process fluid and may be eliminated by careful pre-filtering and the use of a small amount of oxidising agent. Again, the substances studied did not tend to adhere to the media by virtue of their stickiness, chemical adhesion, etc. Filtration studies were made with dilute suspensions of calcium and magnesium carbonate on multifilament media of various configuration. The results indicated that low β factor cloths tended to demonstrate monofilament behaviour with large bleed percentages; the extent of the bleeding and the overall filtration rate was largely determined by the concentration of the prefilt. In those cases where the particle size was of the same order of magnitude as the pore size in the yarns, particle penetration into the yarns occurred and the removal of such deposits was made difficult by the fact that, on the reversal of flow in backwashing, large proportions of wash water channelled through the inter-yarn pores. The overall effect was a gradual and continuous accumulation of particles in the yarn, the medium resistance increasing in regular steps. This behaviour contrasted with the high β factor cloths where, in those cases where $d \rightarrow d_f$, particle penetration in the yarns was removed by the more efficient back-washing. In general, however, the permeability of the higher β cloths was such to produce lower overall filtration cycles. Thus the blinded low β factor cloths, having approached an essentially monofilament character could still be viewed as the more success-ful application, depending on the process requirements.

Filter Machine Effects

Filtration results, obtained in ideal laboratory conditions, are often found to be of questionable application in practice because of the influence of the filter machine or mode of operation. For example, the permeability of filter media in conditions of centrifugation (10) is generally lower than that measured in normal trials. Table 4 contains measurements of cloth resistance made at various rotational speeds; it is seen that quite serious increases in flow resistance occur due to yarn or fibre compression effects during rotation.

Table 4

Cloth	Type	β	0	Rx10^{-6}@ various r.p.m.			
				1100	1300	1500	1700
Polyester A	Plain; 100%CF	>20	54.9	395	605	678	743
Polypropylene C	Plain; 100%CF	>20	41.2	356	454	511	538
Polypropylene E	Plain; 100%SF	3.5	15.9	96	100	106	116
Polyamide D	Twill; 88%SF	>20	16.6	99	119	131	141

Other published data on centrifugation reports increased media resistances five times those determined in flow cells. Apparently, continuous filament cloths give higher resistances than staple fibre cloths at the same speed; most cloths demonstrate an approach to a constant maximum resistance at 2000 r.p.m.

The effect of medium resistance in upward filtering machines such as the rotary vacuum filter has also received attention. It will be realised that in many cases, the overall resistance of the medium is augmented by the need to pre-coat the medium. The initial filtration velocity of the system will thus be determined by the available vacuum, the level of submergence and the combined cloth-precoat resistance. In many cases, the drag forces generated by the upward flowing fluid are insufficient to support the mass of growing cake against the simultaneous stripping action of the agitator, used to prevent sedimentation of the particles, and the gravity forces acting on the cake. In certain circumstances, cake growth ceases at a certain point on the drum surface; the passage of liquid is not prevented, however, and many so-called filtrations are, in fact, combined filter-thickener processes. The effect of medium resistance on the course of upward filtering systems (12) is shown in Fig. 6; results obtained in downward deposition are also shown. In laboratory leaf testing where agitation conditions are not sufficient to provide a complete supply of particles to the filtering zone, sedimentation effects will produce the same action as the agitator in the rotary filter. In contrast, a buchner flask, involving downward deposition will produce filter cakes of constant average resistance. The loss of solids in leaf tests tends to produce non-linear $(\theta/V, V)$ graphs and make the calculation of α difficult. In general, where experimental work is aimed at a correlation of filtration systems in terms of medium resistance R and cake resistance α, care should be taken to ensue that the level of R chosen is not producing abnormalities in the filtration process.

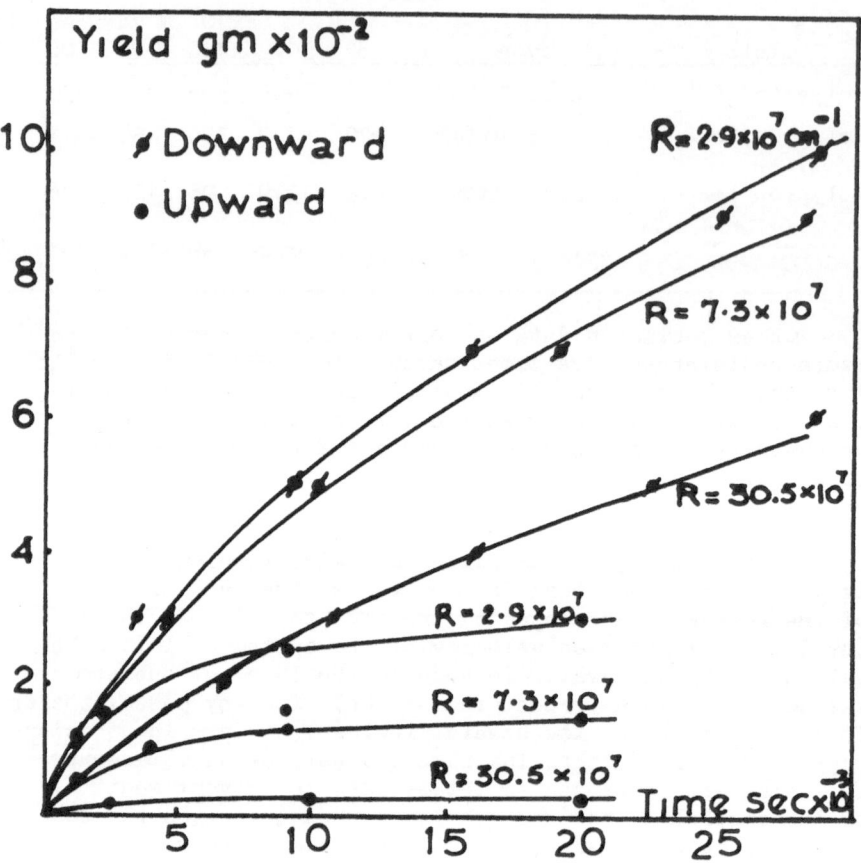

Figure 6. Effect of R on Yield

NOMENCLATURE

A	= filter area	cm^2
A_o	= effective orifice area	cm^2
a	= effective fraction open area of pore	–
a_T, a_p	= effective fraction open area of plain and twill pore respectively	–
B	= cloth permeability	cm^2
B_o	= yarn permeability	cm^2
B_1	= cloth permeability, solid yarns	cm^2
C	= solids concentration	$gm\ cm^{-3}$
D'	= diameter term in equation 11	cm
$\bar{d}$	= mean particle size	cm
d	= permeability pore diameter	cm
d_f	= fibre diameter	cm
d_y	= yarn diameter	cm
d_1	= weft yarn diameter, monofilament cloth	cm
d_2	= warp yarn diameter, monofilament cloth	cm
h	= distance between fibre	cm
k_o	= constant, equation 6	–
k	= variable, equation 15	–
K_o	= Kozeny constant, equation 4	–
K_1	= constant, equation 24	–
K_2	= constant, equation 25	–
K_3	= constant, equation 27	–
K_4	= constant, equation 28	–
K_5	= constant, equation 28	–
L	= bed depth or medium thickness	cm
m	= exponent equation 24	–
n	= compressibility index, equation 1	–
ΔP	= pressure drop over system	$dynes\ cm^{-2}$
P_r	= wetted perimeter of pore	cm
p	= exponent, equation 25	–
R, R_T	= clean and deposited medium resistance	cm^{-1}
r, r_{bp}, r_c	= permeability, bubble-point and microscopic count radius of pore	cm
S	= specific surface	cm^{-1}

u	= fluid velocity	cm.sec^{-1}
u'	= effective fluid velocity in pore	cm.sec^{-1}
V	= volume of filtrate	cm^3
V_R	= volumetric rate per unit area	cm.sec^{-1}
W	= cloth weight	gm.cm^{-2}
α_o	= filter cake resistance	cm.gm^{-1}
β	= cloth index, equation 8	-
ϵ	= porosity	-
θ	= time	sec.
μ	= fluid viscosity	gm.cm^{-1}.sec^{-1}
ρ	= fluid density	gm.cm^{-3}
τ	= surface tension	dynes.cm^{-1}
ψ	= ratio of resistance of used to clean media, equation 26	
ec	= warp yarn per unit length	cm^{-1}
pc	= weft yarn per unit length	cm^{-1}

REFERENCES

1. D.B. Purchas, Industrial Filtration of Liquids,
   Leonard Hill, London, 1971.
2. Rushton, A. & Green, D.J. Filtration and Separation
   Nov/Dec. 1968.
3. Davies, C.N., Proc.Instn.Mech.Engrs. 1952, 1B, 185.
4. McGregor, R. J.Soc.Dyers.Colour 1965, 81, 429.
5. Rushton, A. Chem.Engr.London, No 237, April 1970, p88.
6. Pederson, G.C. 64th AIChE Meeting, New Orleans, March 1969.
7. Rushton, A. & Griffiths, P.V.R. Trans.Inst.Chem.Eng.
   Vol 49, 1971, 49.
8. Hutson, V.C.L. Chem.Engr. London 1969, No 232, p362.
9. Rushton, A. & Rushton, Alan Filtration and Separation,
   May/June 1972.
10. Gonsalves, H., Rec.trav.chim. 69, 873, (1950).
11. Rushton, A. & Spear, M. Filtration and Separation,
    May/June 1970.
12. Rushton, A. & Rushton, Alan, Filtration and Separation
    May/June 1973.
13. Rushton, A. & Griffiths, P.V.R., Filtration and Separation
    Jan/Feb 1972.

FORMATION OF FILTER CAKES AND PRECOATS

P.M. Heertjes

Professor in Chemical Engineering,
University of Technology Delft

Filtration is the operation of separating - more or less completely - a heterogeneous mixture of a fluid and particles of a solid
by means of a filter medium which permits the passage of the
fluid but retains the particles.

From a physical point of view filtration is basically a
process of flow. It consists of flow of the slurry up to the
filter medium and flow of the fluid through the cake and the
medium.

The second important physical phenomenon in the filtration
process is that the solid particles are deposited suddenly and
discontinuously over or in the filter medium.

The object of filtration theory - including that of the
formation of a cake - is to investigate the interrelation between
the properties of the suspension to be filtered and of the filter
medium on the one hand, and the operational conditions on the
other hand.

Fluid dynamics have been extensively studied for porous masses. Both cake and filter medium are such masses. How much of
this knowledge can be used to evaluate filtration problems will
depend on how much can be predicted from the properties of the
slurry, the filtration conditions and the properties of the
filter medium during the building up of a cake.

Already a superficial inspection of the situation will
enable one to draw some general conclusions in this respect.

By nature the filter medium is inhomogeneous. In general
the openings (pores) are not uniform in size, have erratic forms
and are unevenly distributed over the surface. It has to be remembered that the criterion for this, the distance between the
openings, generally is of the same order of magnitude as the size
of the pores - the size of the particles is normally smaller than
the size of the pores - and also that flow through the medium
takes place through the pores only. Therefore, over the surface
of the filter, the local flow-rate of fluid will show large
differences. This must signify that a cake formed on the filter,
perpendicular to the main direction of flow, will be inhomogeneous. Also it can be said that a slurry, however well stirred
it may be, will never be homogeneous. Furthermore, in most
filtration apparatus, the velocity profile will be most irregular
because, inter alia, of large wall effects. All this adds to the
inhomogeneity of the first layers of the cake.

Because the number of passages in a cake is very large
compared with the limited number of openings in the filter medium,
the basic structure of a cake will be determined by its first
layers. It has to be noted that, if in the first layers a cake
contains less dense parts, by further increase of the cake thickness, because of the local higher flow at these parts again a less
dense cake will be formed. The reason for this will be given in
one of the next paragraphs. Therefore, the whole cake will be inhomogeneous perpendicular to the direction of flow. The initial
micro-inhomogeneity will easily lead to macro-inhomogeneity by
compression and consolidation. Ample proof exists about this inhomogeneity. A few examples will now be given. In all these
examples great care had been taken to form even cakes, by a proper
choice of the hydrodynamic conditions. As a fluid the discussion
will be based on systems with a liquid (mainly water) as a fluid.

If cakes from different substances, always kept under water,
varying in thickness from 2 - 30 cm and of quite different origin
as far as the treatment to which they had been subjected is concerned (run through with water for long and short periods of
time, at low and at high pressures) are percolated with a pot-
assium permanganate solution its flow pattern can be studied by
carefully peeling off the cake layer by layer parallel to the
filter and comparing the coloured and not coloured patches.

It was found in all cases that only part of the cake was
coloured, showing marked preference regions of flow of a rather
erratic nature. In most cases a wall effect was observed. Also
a channel in the centre. No essential difference could be found
between cakes formed by sedimentation or by filtration nor
between cakes consolidated at low or high pressures, although the
cakes brought in by filtration showed a somewhat more regular
flow pattern.

Also, for a cake of calcium carbonate (2 - 3 cm thick)
percolated with water, by interruption of the flow for a few
hours, at the end of that period a star shaped crack appears
in the centre. If flow is then resumed again, the crack slowly
fills up and either a new crack appears about 1 cm from the wall,
or - diminishing from the top of the cake to the filter - the
cake detached itself from the wall.

When the flow is again interrupted for some time, the
original crack reappears, the crack near the wall, if present,
disappears. This could be repeated several times.

No appreciable difference in porosity measured from 1 cm^3
samples could be found between places near the wall and in the
centre of the cake.

Inhomogeneity can be suppressed by the use of filters of a
regular form and by constant hydrodynamic conditions over the
whole filter surface.

Besides its inherent inhomogeneity it has also to be
recognized that a cake is formed under streaming conditions with
the result that a cake will have a density far from that of the
particles in their state of densest packing. The structure of
the cake is stabilized by flow. Every cake formed by filtration
therefore is in a state of instability. Therefore every change in
the magnitude of the stabilizing forces will cause a change in
the packing of the cake and thus of its resistance against flow.
This holds mutatis mutandis also for disturbances alien to the
process of filtration proper.

A conclusion following from the last consideration is then
that no direct measurement of characteristic parameters of a cake,
such as porosity and porosity distribution, is possible without
disturbing the cake. Thus information under filtration conditions
has to be gained by indirect means. Any information obtained by
other means such as in the compressibility cell, has to be handled
with great caution.

Finally it follows from the type of operation that the cake
and the filter medium influence each other.

The superficial inspection given indicates amongst other
things that the behaviour of the system studied, has to be derived
from the change of the parameters of simple flow equations.

MATHEMATICAL DESCRIPTION OF THE FILTRATION PROCESS

Because relatively large surfaces exist in the cake and in the filter, for isothermic flow by far the main loss of pressure in the liquid will be caused by friction. Because of the complexity of the hydrodynamic situation, the simple Fanning equation for flow through porous masses will be used.

If V_f is the volume of the filtrate, ΔP the pressure difference in the liquid and θ the time, the Fanning equation may be written in differential form as :

$$W = \frac{1}{A} \frac{dV_f}{d\theta} = \frac{1}{\eta^\gamma} \frac{d(\Delta P)^\delta}{dR} \qquad (1)$$

which defines the resistance R of the filter plus cake. For viscous flow, $\gamma = \delta = 1$; for turbulent flow $\gamma = 0.11$, $\delta = 0.55$. If the filter medium and the cake can be distinguished separately, the resistance of the cake R_c, the pressure loss over the cake $(\Delta P)_c$, and the corresponding values for the filter medium R_f and $(\Delta P)_f$ for viscous flow are related by the equations :

$$R_c + R_f = R \qquad (2)$$

$$(\Delta P)_c + (\Delta P)_f = \Delta P \qquad (3)$$

and :

$$\frac{(\Delta P)_c}{R_c} = \frac{(\Delta P)_f}{R_f} = \frac{\Delta P}{R} \qquad (4)$$

It has to be recognized from the beginning that R_f need not be a constant and that W can depend on the position in the system and that it decreases with time because R and R_c increase. The value of R depends on the amount of material in the cake. If a specific resistance, r, for a cake-element containing unit weight of particles per unit cross-section of filter is defined, it follows that $dR_c = rd(g)$. The symbol g represents the weight of particles in the cake per unit of filter surface.
Recognizing that r need not be constant in the direction of flow, even for homogeneous cakes, we can define a mean specific resistance $(\bar{r})$ in the usual manner :

$$R_c = \int_o^g r \, (dg) = \bar{r}g. \tag{5}$$

To calculate g one must realize that the cake and the filter medium contain liquid. Therefore at time θ, when V_i of fluid has passed the slurry-cake interface, a volume

$V_i(1 + \frac{c}{\rho_s})$ of slurry has been handled (c = concentration of particles of density ρ_s in weight per unit volume of liquid).

If it is assumed that the filter contains a negligible amount of particles, $g = cV_i/A$.

The volume V_c which passes the interface between the cake filter medium is smaller than V_i because an amount V_r is retained in the cake. Neglecting the amount of liquid in the filter medium, which is generally permissible, the following set of equations is valid (V_f = the amount of filtrate).

$$V_c = V_f \tag{6}$$

$$V_i = V_f + V_R \tag{7}$$

$$V_R = g \, \frac{\bar{\varepsilon}}{(1 - \bar{\varepsilon})\rho_s} \, A \tag{8}$$

($\bar{\varepsilon}$ = mean porosity of the cake), therefore :

$$g = cV_f/A \left| 1 - \frac{\bar{\varepsilon}.c}{(1 - \bar{\varepsilon}) \, \rho_s} \right| \tag{9}$$

The porosity of the slurry

$$\varepsilon_s = \rho_s/(\rho_s + c) \tag{10}$$

Thus :

$$g = cV_f/A \left| 1 - \frac{(1 - \varepsilon_s)}{\varepsilon_s} \frac{\bar{\varepsilon}}{(1 - \bar{\varepsilon})} \right| A \qquad (11)$$

For very dilute suspensions it follows that $V_i \simeq V_f$. For the limiting case of slurries where $\varepsilon_s = \bar{\varepsilon}$, $V_f = 0$.

We will discuss now what is the influence of the operative conditions on R and r and to what conclusions it leads for the formation of the cake.

THE INFLUENCE OF THE CONCENTRATION OF THE SLURRY.

If a very dilute suspension is filtered, as in clarification processes, each particle moves separately from the others and will follow the streamlines of flow directed towards the pores in the filter. The result will be that a particle will either enter a pore or will cover the pore opening, depending on the ratio of pore diameter to particle diameter. This will result in blocking of the pore and finally of the filter.

If the concentration c is increased more particles will arrive near the pores at the same time and blocking will decrease. Finally, a situation will be reached, in which the particles hinder one another because they attempt to enter the pore together. The result will be that a bridge of particles is formed over the pore and the resistance of the filter medium proper will remain unchanged. Thus a cake is formed. This cake will be in a rather dense packing. Further increase of the concentration will cause a decrease in packing density of the cake, and therefore of its specific resistance which will reach a constant minimum value at a high concentration.

The limiting cases for blocking filtration therefore are complete blinding of the filter medium and a cake filtration.

Let us consider a system consisting of a horizontal filter with a large number N of circular pores per unit of surface by means of which a suspension of concentration c is filtered, which contains spherical particles of weight a and of a diameter larger than the diameter of the pores and smaller than the distance between the centres of two pores. The concentration be

such that in the beginning each particle can follow the streamline through the pores.

After a certain time θ a volume V has been filtered off.
This amount V will be considered to have flown through in S
steps, each step filtering an equal amount V'. Therefore :

$$V'S = V$$

The number of steps has been chosen in such a way that the
number of particles it contains can reach the open part of the
filter in the time interval to filter off a volume V'. S will
increase with the concentration.

If the filter has a surface A, after the first step the
number of pores completely blocked will be V'c/a and therefore
the blocked surface will be V'c/aN, if it is assumed that each
particle reaching an open part of the filter blocks one pore
completely.

The surface of the filter A' still open after filtering off
V' therefore will be :

$$A' = A - \frac{V'c}{aN} = A \left(1 - \frac{g}{SaN}\right) \tag{12}$$

In the second step again a volume V' will be filtered. Because
part of the pores are blocked, now not all the particles will
follow exactly the streamlines of the fluid. Let us assume as a
limiting case that the path of the particles over the whole
length of the container over the filter does deviate less from
the vertical than the distance between two pores, whereas the
liquid can move freely. In that case the concentration c' of the
second portion to be filtered will be less than c and is equal
to :

$$c' = c \frac{A'}{A}$$

This phenomenon will be called the dilution effect. The still
open surface A" after the second portion has been filtered will
be :

304

$$A'' = A' - \frac{V'c'}{aN} = A' - \frac{V'c}{aNA} A' = A' (1 - \frac{g}{SaN}) \qquad (13)$$

Combination of (12) and (13) gives :

$$A'' = A (1 - \frac{g}{SaN})^2 \qquad (14)$$

Using the same reasoning in the following steps, after S steps the free surface A^S can be expressed by :

$$A^S = A (1 - \frac{g}{SaN})^S \qquad (15)$$

Considering that the open filter by definition has a resistance R_f :

$$\frac{A}{R} = \frac{A^S}{R_f} \qquad (16)$$

and therefore :

$$R = R_f (\frac{SaN}{SaN - g})^S \qquad (17)$$

The same type of equation can be derived for other manners of blocking, e.g. in the pores.

Taking into account that even in the case of complete blocking of the pores, the dilution effect need not always to take place, due to the fact that the particles will always follow the streamlines of flow to a certain extent; that unavoidable vibrations of the apparatus influence the movement of the particles and that the slurry will not always be homogeneous, the cases investigated can be represented by one general equation of the form :

$$\frac{R_f}{R} = (1 - \frac{g}{OaN})^S \qquad (18)$$

or

$$R = R_f \left(\frac{OaN}{OaN - g}\right)^{S} \tag{19}$$

The differential specific resistance r per unit of weight of solid substance per unit surface of the filter defined by : dR = r d(g) is :

$$r = \frac{R_f}{aN} \cdot \frac{S}{0} \cdot \left(\frac{OaN}{OaN - g}\right)^{S+1} \tag{20}$$

The limiting value for g = 0 is $(r_{lim})_{g \to 0} = \frac{R_f}{aN} \frac{S}{0} \simeq \frac{R_f}{aN}$ \hfill (21)

In these equations 0 and S (increasing with the concentration) can range from unity to any value greater than 1. In general, even with one particle blocking one pore 0 $\neq$ S, the limiting case being that 0 = S = 1, absolute blocking of a pore by one particle.

For S = 0 = ∞ equations (20) and (21) transform to :

$$R = R_f \exp \left(\frac{g}{aN}\right) \tag{22}$$

and

$$r = \frac{R_f}{aN} \exp \frac{g}{aN} \tag{23}$$

There is no sound physical background for these equations, but they can be used as an approximation of the two foregoing equations for large values of S and 0 and these are easier to handle.

It seems worth while to try to indicate the factors governing the magnitude of the parameters S and 0. Excluding the vibrations of the apparatus for which it is nigh to impossible to predict anything in this respect, there must, amongst other things, be an influence of the concentration of the slurry, of density differences between particle and liquid, of the size-spectrum of the particles and of the surface tension between particle and fluid. For the parameter 0 there is in general also an influence of the form of the pores and of the particles. As far as the concentration is concerned it may be realized that each particle will be surrounded by a volume of liquid of $\underline{a}$. Therefore the distance between the particles will be directly proportional to $(\frac{a}{c})^{1/3}$. The slurry can be visualized to be built up of layers containing $(\frac{c}{a})^{1/3}$. The path each particle has to travel from the layer nearest to the filter is : $(\frac{a}{c})^{1/3}$. The larger the concentration, the smaller therefore will be the length of this path. The smaller this length, the less the chance will be that a particle comes into one of the streamlines directed towards the open pores and the bigger will be S and O. In connection with the foregoing the ratio between the amount of particles in one layer and the number of pores in the filter N will also be important. In the domain below unity, the larger this ratio the greater will be the chance that a particle will enter a pore, therefore the smaller will be S and O. An increase in concentration results in a shorter path the particle has to travel, therefore in a larger S and O.

It however also signifies a bigger number of particles per layer per pore which will tend to decrease S and O. Which of the two factors will prevail cannot be predicted, although, because the number of particles per pore is very small in blocking filtration, the path influence seems to be the largest.

A small number of pores will give rather curved streamlines near the pores which causes through the inertia that the path of the particles will show a greater tendency to deviate from the streamlines, increasing S and O. This will also depend on the size of the particles, on the density difference and on the filtration velocity.

Increase of the density difference between liquid and particle will tend to increase S and O.

By increasing the concentration, accompanied by an increase of S and O, the influence of g on the value of r will decrease. Before very large values of S and O are reached, at a transition concentration c_t, blocking filtration changes to cake filtration. An example is given in Fig. 1.

The cake formed at concentration c_t will have a resistance r_t which will be about equal to $(r_{lim})_{g \to 0}$. Therefore :

$$r_t = \frac{R_f}{aN} \qquad (24)$$

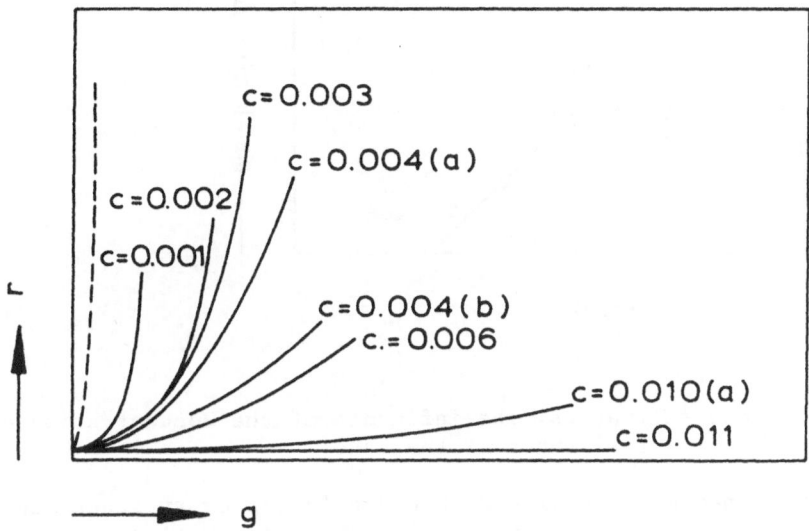

Dotted line : complete blocking, blinding of the filter medium
Horizontal line : first cake filtration

Fig. 1. The specific resistance r plotted as a function of the
amount g on the filter for different concentrations c of the
slurry filtered.

Now the packing density of a cake will depend on the number
of particles arriving at the same time interval at a pore.
Therefore r will depend on c/aN. The type of relation between
the two will take the form of a probability distribution. Thus,
including limits :

$$\frac{r - r_\infty}{r_t - r_\infty} = \exp - (m \frac{c - ct}{aN})$$ (25)

In this r_∞ is the resistance at an hypothetical infinite
concentration and m will depend on the particle size and form.
Also, r_∞ will depend indirectly on the filter properties, that
is on R_f and on N.

The form of the equation given (see Fig. 2) indicates that
r falls off very rapidly with the concentration.

308

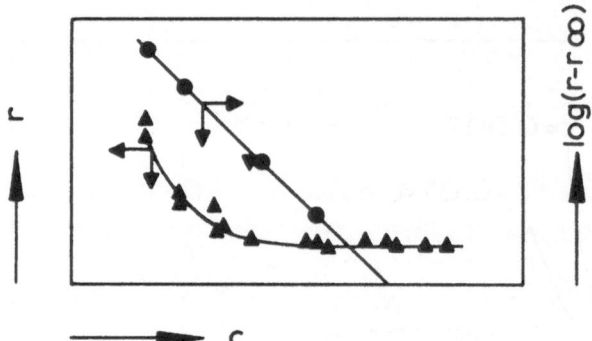

Fig. 2. Cake filtration. The influence of the concentration c on the specific resistance of the cake.

The phenomena described should hold for every substance but the concentration range over which they occur will depend on the system. The phenomena are just as important for clarification and for deep-bed filtration in which only sieving action occurs.

It should be noted that even if a slurry of a concentration for which cake filtration occurs is filtered, the phenomena just discussed can be met. If the filter has e.g. been cleaned with water before use, there is a dilution effect which can cause an extra increase in resistance, both by possible clogging of the filter or by increasing r for a short period.

THE INFLUENCE OF W_0

The same type of effect as observed by changing the concentration of the slurry will occur on changing the initial filtration velocity W_0, at constant concentration. If the cake is formed at a low velocity of flow, the particles will follow the streamlines easily and the result will be a dense cake. If the velocity is high, the packing density will decrease. The limiting cases will be that of a cake formed at zero velocity of flow in which the cake resistance should approach the resistance of a most closely packed cake, r_a, and that of a cake with the same properties as the slurry. Some results to substantiate the above are given in Fig. 3.

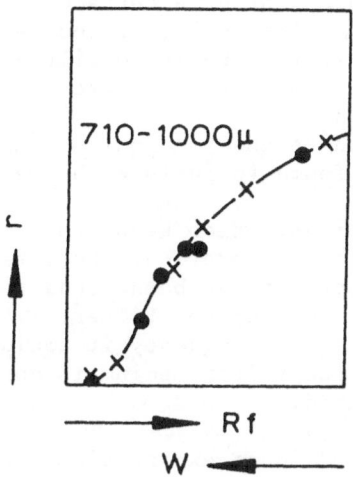

Fig. 3. The influence of the initial filtration velocity, W_0, on the specific resistance of a cake of polystyrene beads of 710 - 1000 μ

Again an exponential function appears to correlate the results quite satisfactorily :

$$r = r_a \exp - (\frac{W_0}{f})^d \qquad (26)$$

The values found for r_a by extrapolation are very close to the resistances of layers of particles consolidated to a very dense packing by prolonged flow and vibrations.

THE INFLUENCE OF VIBRATIONS

As said the filter cake has an unstable structure, stabilized by some force related to the flow velocity. In general an extremely loose packing exists and disturbances will easily cause changes in this structure.

A cake can therefore be consolidated if the flow velocity becomes too small. This can be quite sudden, as in the so-called retarded-packing compressibility.

The phenomenon is that as soon as a cake has reached a certain thickness and the velocity drops below a certain critical value, the cake starts to consolidate. This will, of course, occur where the situation is the least favourable to filtration, that is, near the filter medium where the cake pressure is largest. It results in an increase in the slope of the R-V lines from this point on. The situation can be analysed quite simply and the appropriate equation can be found to include this effect.

Also the existing balance of forces, which keeps the cake in a certain structure, can be disturbed by vibration. The effects of vibrations are difficult to predict. It has been found that vigorous vibrations can loosen the cake from the filter. With vibration of a not too large aplitude and frequency it could be shown that whilst the resistance of the filter medium is not effected, the cake becomes consolidated.

The relationship can be expressed in an exponential form :

$$\frac{r_F - r}{r} = i \ exp - (\frac{h}{F})^d \qquad\qquad (27)$$

In this equation r_F is the resistance of a cake vibrated with a frequency F, and i, h and d are parameters which depend on the system and on the amplitude of the vibration.

It has been found that the resistance of cakes formed under vibration show the same type of dependency on the initial filtration velocity as have non-vibrated cakes.

Cakes of high resistance show a much larger relative increase of resistance on vibration than do cakes with a low resistance. The stabilizing forces in the first case are much smaller than in the second case, so that the balance of forces is more easily disturbed. Or, in other words, loosely packed cakes of one system are more stable than densely packed cakes.

Vibration of cakes produces phenomena comparable to those observed in flow consolidation with respect to the occurrence of unrest; these phenomena are marked by alternate decreases in the flow rate.

Phenomena of the kind mentioned can most clearly be observed if a cake is initially formed by filtration without vibration, if in a second period of filtration, vibrations are applied; and thereafter in a third period the filtration is continued in the normal way. Two examples of such experiments are presented in Fig. 4, in which the R-V relationships for such filtrations

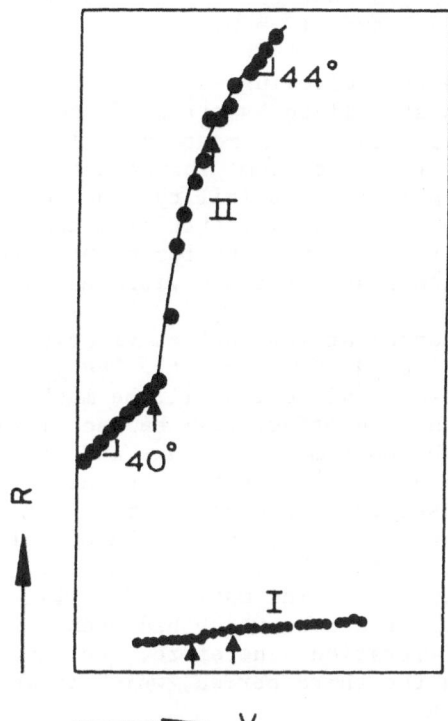

Fig. 4. The R-V relationships for cakes formed without
vibrations, with vibrations (arrows) and again without
vibrations

carried out under constant pressure are given. The period of
vibration occurs between the two arrows.

The cake of example I had been formed at a high initial
velocity of flow so its specific resistance is small, therefore R
increases slowly with V. During vibration the specific resistance
increases somewhat; thereafter the specific resistance of the new
cake is about the same or somewhat smaller than in the first
period.

The cake of example II had been formed with a much lower
initial velocity. Here the specific resistance is larger and the
effect of vibration is more pronounced. After a period of unrest
with continued normal filtration the new cake formed has a

resistance larger than the cake formed in the first period.

The limiting case, not shown, is that the R-V line on extra-polation passes through R_f on the R axis (V = 0).

The explanation seems to be the following. A cake formed under a high velocity of flow is stabilized by strong forces, so that consolidation is small and starts at the most unstable place in such a cake which is near the filter medium. The effect is comparable with the retarded packing compressibility. The top layers of the cake will not be compressed by vibrations; there-fore on stopping the vibrations the cake will be formed with the same density and the same resistance as for a non-vibrated cake.

On the other hand, a cake formed at low initial velocity of flow will have a high resistance but a low stability. The layer of cake which will be consolidated on vibration will be much thicker than in the first case and the effect will be much larger. Moreover, the effect starts at the bottom of the cake and grows upward. The relative effect will decrease with time because the top layers are less dense. So instead of remaining straight the R-V curve will be concave downward. Finally, the top layer will be affected. The new cake formed in the third period will, as al-ways, have the density of the top layer. The maximum density possible is that of the top layer of a cake which had been vibrated from the beginning of filtration. Therefore, the limit-ing case is that the R-V line in the third period, which again is a straight line, passes through R_f.

Technically, vibrations are important because they often occur in practice, sometimes in the form of a pulsating flow, which has the same effect.

PRECOATS

Partial blocking of the filter medium can be very helpful in ensuring clarity of the filtrate. It is a regularly occurring phenomenon in sludge filtration. However, especially with slimy substances, blocking can be too cumbersome. The rate of filtration can be reduced too rapidly and the filter media are difficult to clean. In such cases filter aid is added or a precoat is applied on the filter medium.

The use of the precoat allows a great increase in the number of possible flow paths as compared to those in the filter and it also prevents the slimy substance reaching this filter. By the addition of filter aid the concentration in the prefilter is in-creased considerably, with all the advantages mentioned before.

The slimy substance will adhere to the particles of the filter aid.

A good filter aid, such as diatomite, perlite, cellulose, asbestos fibres, non-activated carbons, and the like, must form a cake of high porosity (0.85 - 0.90). It must have a low surface area and must have a good particle size distribution with non-uniform sizing. High porosity provides room for the filterable solids to be deposited and yet for the aid to still retain a high percentage of channels open for flow. Too many fine particles would cut down flow, too many large particles would give poor clarity. This necessitates a good particle size distribution. The choice of a proper filter still asks for experience, supported by small scale experiments. However, the theory as given before can also be applied to the filtration of suspensions with filter aids or for the formation of precoats.

The form of the particles in filter aids, however, results in a decrease of the influence of the factors mentioned. To give an example, the influence of the initial filtration velocity as given is present, but it is not very large. Moreover, the particles are of such a shape that consolidation of the cake as such is not very pronounced. If, however, slimy substances are present, the particles can move with respect to each other more easily and consolidation increases. Also the cake as such, by nature of the filter aid taken, already has a high porosity, which gives raise to the formation of a cake of high porosity as wanted. On the other hand, this reduces the influence of the factors mentioned.

CONCLUSION

It follows from the material presented that some insight in the factors governing the formation of cakes and precoats has been obtained. On the other hand it is quite clear that the amount of knowledge in this respect is still meagre. Factors such as particle size distribution, form factors and the like still ask for extensive studies.

314

LIST OF SYMBOLS

A   = surface of filter
a   = weight of one particle
c   = concentration of slurry per unit of volume
d   = exponent
f   = parameter
F   = frequency of vibration
g   = weight of dry cake per unit of surface
h   = parameter
i   = parameter
m   = parameter
N   = number of pores in filter medium per unit of surface
O   = parameter
P   = liquid pressure
R   = overall resistance
R_f = resistance of filter
R_c = resistance of cake
r   = specific resistance of a cake of unit weight per unit
        of surface
S   = exponent
V   = volume of liquid
W   = flow velocity in system
γ, δ = exponents
ε = porosity
η = dynamic viscosity of liquid
θ = time
ρ_s = density of solid substance

   The above quantities may be expressed in any set of
consistent units in which force and mass are not defined
independently.

REFERENCE

Heertjes, P.M., Filtration, *Trans.Instn.Chem.Engrs.*, 42, T266,
1964.
More literature references are to be found in this paper.

COMPRESSIBLE CAKE FILTRATION

Frank M. Tiller
M. D. Anderson Professor of Chemical Engineering
University of Houston
Houston, Texas   77004   U.S.A.

1.   FLOW THROUGH INCOMPRESSIBLE POROUS MEDIA

1.1   Introduction

Basic laws governing the flow of liquids through uniform, incompressible beds serve as a basis in developing formulas for more complex, non-uniform, compressible cakes. Substantial quantities of data are available for the flow of air and water through widely different kinds of solids. Those data serve as a basis for various types of mathematical formulation.

1.2   Darcy's Law[4]

In 1855-56, Darcy carried on a series of experiments involving the flow of water through sand. He found the rate to be proportional to the head, thus indicating that the flow regime was laminar. Although Darcy did not include viscosity in his original formula, it is customary to write his equation in the form

$$\frac{dp_L}{dx} = \frac{\mu}{K} q \tag{1.1}$$

where p_L is the hydraulic pressure, x the distance through the cake, K the permeability, and q the superficial velocity expressed as flow volume/(unit cross-sectional area)(time). In fixed beds, it is assumed that the solids are stationary and

that q is constant. In filtration, it is customary to write
Equation (1.1) in the form

$$\frac{dp_L}{dw_x} = \mu \, \alpha \, q \qquad (1.2)$$

where w_x is the mass of dry cake per unit area deposited in
distance x from the medium which serves as support. The flow
rate may not be constant throughout the cake. The differential
dw_x can be written as

$$dw_x = \rho_s (1 - \varepsilon) dx \qquad (1.3)$$

Substituting (1.3) in (1.2) yields

$$\frac{dp_L}{dx} = \mu \rho_s (1 - \varepsilon) \alpha q \qquad (1.4)$$

The permeability can be related to the filtration resistance
by comparing Equations (1.1) and (1.4)

$$K = 1/\rho_s (1 - \varepsilon) \alpha \qquad (1.5)$$

The dimensions of K are L^2, and the dimensions of α are L/M.
Generally α varies less than K with pressure.

1.3 Moving Solids

If the solid particles are moving, Darcy's law should be
modified to include the velocity of the fluid relative to the
particles. While for most filtrations, the velocity of the solid
particles is not large, Shirato et al[14] showed that in the filtration of thick slurries in short time intervals, it is necessary to include the velocity of the solids. The average velocity u of the liquid is related to the superficial value q

$$u = q/\varepsilon \qquad (1.6)$$

The actual liquid velocity will vary as it passes through the
interstices of the porous cake, and Equation (1.6) represents
the average value.

In a filter cake under pressure, the cake will be continual-
ly compressed. As a result, the solids move toward the supporting medium. The velocity of the solids may be an appreciable
fraction of the liquid velocity where thick cakes are built up
in short periods as in rotary vacuum filtration. If r represents

the superficial volumetric flow rate of the solids, then

$$u_s = r/(1 - \varepsilon) \tag{1.7}$$

represents the velocity of the solid particles. The velocity of the liquid relative to the solid is

$$u - u_s = \frac{q}{\varepsilon} - \frac{r}{1-\varepsilon} = \frac{1}{\varepsilon}(q-er) \tag{1.8}$$

where $e = \varepsilon/(1 - \varepsilon)$ is the local void ratio. To preserve the form of Equation (1.2), it is written

$$\frac{dp_L}{dw_x} = \mu\alpha\varepsilon \; (u - u_s) = \mu\alpha \; (q - er) \tag{1.9}$$

1.4 Turbulent vs. Viscous Flow

Many authors (6, 13, 18, 20) have investigated flow through porous media. In 1927, Kozeny[12] remarked on the mass of observation which were available and upon the need for a simple formula to correlate existing experimental data. While a substantial number of investigators have made valuable contributions to the literature, the simple formula sought by Kozeny is still lacking. Experimentation is still an essential part of the calculation of permeability or filtration resistance, particularly when compressible cakes are involved.

A large quantity of experimental data are available for the flow of water through sand, clay, various soils, and various tower packings. Many modified Reynolds number-friction factor plots have been published. While the numerical magnitudes of both the friction factor and the Reynolds' numbers are different from the values occurring in pipe flow, there is a marked similarity of the nature of the curves. Viscous flow prevails over the greater portion of the data. There is a gradual transition from viscous to turbulent flow which would indicate that there is probably a combination of viscous and turbulent flow occurring in addition to contraction and enlargement losses. Thus it would not be expected that the sharp break from laminar to turbulent flow observed with pipes would occur in filtration. In most filtrations, it has been found that the flow is viscous; and, consequently, modifications of Darcy's law may be employed in the development of fundamental equations.

In filtration of moderately resistant materials, flow rates of $2.5 m^3/(m^2)$ (hr.) would be considered satisfactory. Translating this rate to a superficial velocity leads to a value of

0.07 cm./sec. Even with 30 percent voids, the average velocity in the interstices only reaches about 0.2 cm./sec. With such low velocities, it is not surprising that the flow is usually laminar.

In beds of large particles, the flow rate may become turbulent. It is customary to write an equation for the pressure drop in the form

$$\Delta p / L = C_1 u + C_2 u^2 \qquad (1.10)$$

where C_1 and C_2 have usually been considered constants for a given medium. Fahien and Schriver[2] developed an equation in which their constants equivalent to C_1 and C_2 were functions of the Reynolds number.

1.5 Modified Friction Factor and Reynolds' Number

Various modifications of the Reynolds number and friction factor have been utilized in the literature. In fluid mechanics, the frictional pressure drop Δp is given by

$$\Delta p = f \rho \frac{q^2}{2} \frac{L}{D} \qquad (1.11)$$

The Reynolds number is defined by

$$N_R = D q \rho / \mu \qquad (1.12)$$

The first change in Equations (1.11) and (1.12) has been the substitution of the particle diameter D_p for D. However, it should be recognized that except for idealized beds, there are generally present a large variation of irregularly shaped particles which cannot be characterized by a single parameter such as D_p.

Blake[1] proposed the following modifications of f and N_R:

f replaced by $2f\epsilon^3 / (1 - \epsilon)$

N_R replaced by $N_R / (1 - \epsilon)$

The Blake proposal pre-dates the kind of modification which has been based upon the Kozeny[12] equation. Ingmanson et al[11] in dealing with flow through wire screens introduced the following

f replaced by $p_f \epsilon^3 / (1 - \epsilon) S_o \rho q^2$

N_R replaced by $\rho q / \mu (1 - \epsilon) S_o$

The factors $\varepsilon^3/(1 - \varepsilon)$ and $1/(1 - \varepsilon)$ are retained, but the specific surface whose dimensions are reciprocal feet replaces D_p. For a shpere, $S_o = 6/D_p$.

Ward[20] replaced the diameter D_p by the reciprocal of the square root of the permeability as defined by the Darcy equation in which

$$K = \mu q L / p_f \qquad\qquad (1.13)$$

Then the modified expressions are given by

$$f \text{ replaced by } p_f K^{1/2}/\rho L q^2$$

$$N_R \text{ replaced by } q\rho/K^{1/2}\mu$$

Some authors have been careless about defining friction factors for flow in porous solids. It is essential to carefully analyze the formulae and units.

One useful method for differentiation among authors is to observe the value of f in the fully turbulent range and the value of N_R in the transitional region. The following comparison of a few authors data illustrates the necessity for care:

	N_R(transitional)	f(turbulent)
Ergun[6]	100	1.75
Martin, McCabe and Monrad[13]	100	2-20
Ingmanson, Han, Wilder, and Myers[11]	100	2.5
Fahien[7]	1	1.0
Ward[20]	1	0.55
Han and Ingmanson[8]	10	0.12

1.6 Correlation of Experimental Data

While progress has been made on correlating experimental data, there is no single curve on a friction factor plot which represents all solids. The present situation is not unlike that prevalent in the early days when friction factor plots showed one curve for smooth and one curve for rough pipe. Some investigators have shown a series of parallel curves on friction factor plots for porous media, but most have attempted to reduce all of the data to a single curve. Such a procedure is particularly tempting in the viscous range. However, the complexity and changing nature of small particles makes it doubtful that permeabilities can be calculated theoretically for real cases in the near future.

Ergun's[6] plot in Figure 1.1 is typical of what is encountered in the literature. While the correlation appears fairly good, other authors[15,20] have indicated that substantial errors may accrue from utilizing Figure 1.1 as a predictive tool. Similar graphs have been presented by other investigators.

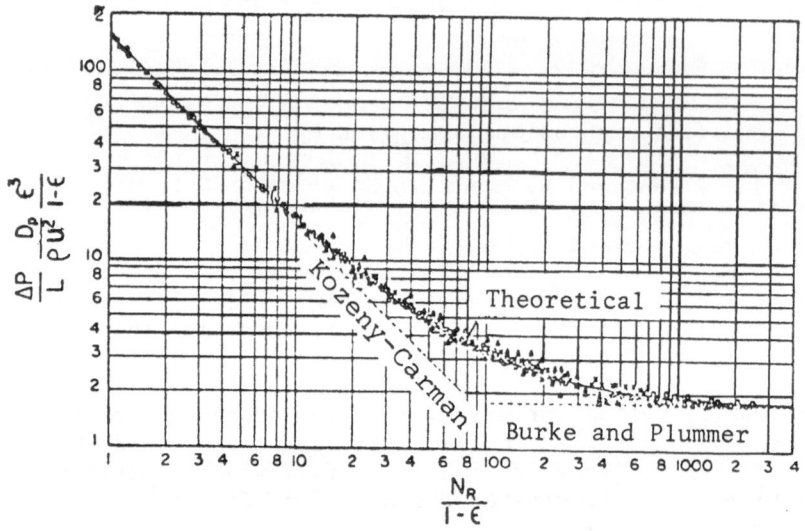

Ergun's Correlation
Figure 1.1

A range of values for D_p could be used in the Ergun correlation. From various types of measurements, one would obtain different "equivalent" sphere diameters such as:

Equivalent Stokes settling velocity
Equivalent surface area
Equivalent projected area
Equivalent perimeter
Equivalent volume

With a wide variety of "equivalent" diameters, it is apparent that no single correlation can result. Using a larger diameter for the curves given in Figure 1.1 would move them up and to the right.

Ergun[6] and Suskind and Becker[18] correlated their data for flow through porous solids in the form

$$f^* = c_1 + c_2/N_R^*$$

(1.14)

where f* and N_R* are modified values. In terms of the basic
variables, (1.14) can be written as

$$\frac{\Delta p}{L} = c_1 \mu \frac{(1 - \varepsilon)^2}{\varepsilon^3} \frac{q}{D_p^2} + c_2 \frac{1 - \varepsilon}{\varepsilon^3} \rho \frac{q^2}{D_p} \qquad (1.15)$$

Ergun gives the values of c_1 = 150 and c_2 = 1.75, while Susskind
and Becker report values of c_1 = 166.8 and C_2 = 0.84.
Stepochkin[17] used the same basic form as shown in (1.14) modified as follows:

$$\frac{\Delta p}{L} = c_3 \left[\mu \frac{(1 - \varepsilon)^2}{\varepsilon^3} \frac{q}{D_p^2} + \frac{1}{120} \rho \frac{(1 - \varepsilon)}{\varepsilon^3} \frac{q^2}{D_p} \right] \qquad (1.16)$$

where c_3 is a constant related to the material. An accuracy of
$\pm$ 10% was claimed for Equation 1.16. A variation of approximately six-fold was reported for the value of C_3.

There is some doubt about the form of (1.14) in which viscosity is completely absent from the inertial term. In pipe
flow; the viscosity enters to the 0.1 to 0.3 power in its effect on pressure loss. It might be expected that a similar
effect would be present in porous media. Fahien and Schriver[8]
proposed an equation like (1.16) in which the constants are
functions of modified Reynolds numbers.

1.7 Kozeny Equation[12]

Many attempts have been made to develop analytical expressions for the permeability or filtration resistance. The Kozeny
equation is perhaps the best known of many attempts. It is an
improvement over Darcy's simple formula and might be likened
to the Van der Waals equation of state. There are times when
it works fairly well as with large particles and low porosities.
It is excellent for demonstration purposes in that it illustrates the important effect of porosity and particle size. However, no one would use it seriously in cake filtration for the
prediction of resistances. Based upon the Poiseuille equation
Kozeny developed a widely used expression for the flow of fluids
through porous beds. The constants of the Poiseuille equation
for viscous flow through a circular pipe will not be the same
as those for other cross-sections. However, it might be expected that the variables such as viscosity, velocity, and hydraulic radius for a non-circular channel would enter a flow
equation in approximately the same manner as for a circular section. The Poiseuille equation for flow through a circular pipe
can be written as

$$\frac{dp_L}{dx} = \frac{32\mu u}{D^2} = \frac{k\mu}{R_H^2} u \tag{1.17}$$

where the hydraulic radius and the constant 32 have been replaced respectively by R_H and k. The hydraulic radius which
is D/4 for circular conduits can be related to the parameters
involved in porous beds by the following stratagem.

$$R_H = \left(\frac{\text{flow area}}{\text{wetted perimeter}}\right)\left(\frac{\text{length of path}}{\text{length of path}}\right) \tag{1.18}$$

$$= \frac{\text{void volume}}{\text{surface of solids}}$$

Since the value $\varepsilon/(1 - \varepsilon)$ represents the ratio of the void
volume to the volume of the solids, the void volume is given by

$$\text{void volume} = \frac{\varepsilon}{1 - \varepsilon} \text{ (volume of solids)} \tag{1.19}$$

Substituting (1.19) in (1.18) yields

$$R_H = \frac{\varepsilon(\text{volume of solids})}{(1 - \varepsilon)(\text{surface area of solids})} \tag{1.20}$$

The surface area of the solids per cubic per cubic foot of
solids (exclusive of voids) is termed the specific surface, S_o
which can be substituted in Equation (1.20) to yield

$$R_H = \frac{\varepsilon}{1 - \varepsilon}\frac{1}{S_o} \tag{1.21}$$

The true average velocity u in the interstices of the solids is
given by q/ε. Making substitutions in the modified Poiseuille
equation leads to

$$\frac{dp_L}{dx} = \frac{k}{g}\mu\frac{(1 - \varepsilon)^2 S_o^2}{\varepsilon^3} q \tag{1.22}$$

Comparing Equation (1.22) with (1.4) and (1.5) leads to a
relation between the permeability K, the filtration resistance
α, and Kozeny's contant k, thus

$$\alpha = k\frac{(1 - \varepsilon)}{\rho_s \varepsilon^3} S_o^2 \tag{1.23}$$

$$K = \frac{1}{k} \frac{\varepsilon^3}{(1 - \varepsilon)^2} \frac{1}{S_o^2} \qquad (1.24)$$

Kozeny's constant has been frequently said to have the value of 5.0. The mere fact that an anistropic solid has the same value of ε and S_o but quite different directional permeabilities is sufficient to cast doubt on the constancy of k. Happel and Brenner[9] present theoretical values for k as derived from equations representing flow through assemblages of cylinders and spheres based on the work of Sparrow and Loeffler[16] and Emersleben[5]. In Figure 1.2 the variation of k with porosity is shown. A curve representing the following equation as proposed by Ingmanson and Andrews is shown as a dotted curve on Figure 1.2.

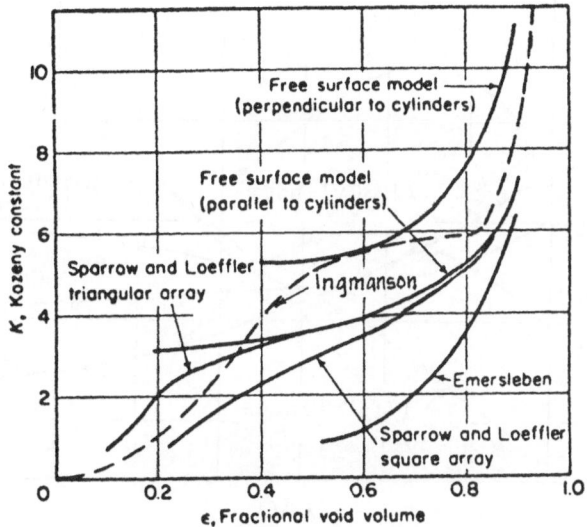

Kozeny's Constant vs. Porosity
Figure 1.2

$$k = 3.5 \frac{\varepsilon^3}{(1 - \varepsilon)^{0.5}} [1 + 57(1 - \varepsilon)^3] \qquad (1.25)$$

The values of k as illustrated in Figure 1.2 do not follow the trends normally encountered with compressible cakes. From measurements of filtration resistance and porosity made in compression-permeability cells, it is possible to calculate kS_o^2

in Equation (1.23). It is not possible to obtain k unless independent measurements are made of S_o. Tiller[19] showed that kS_o^2 increased as porosity <u>decreased</u>. Either k or S_o would have to increase in order to account for the phenomena discussed by Tiller. The apparent conflict between the values in Figure 1.2 and observed behavior of kS_o^2 remain to be resolved.

In Figure 2.3, values of k as reported by Coulson[3] are shown. The dotted line corresponds to values based upon Equation (1.25). It is apparent that there is only a weak association of k with the value 5.0. In the porosity range between 0.35 and 0.45 corresponding to values encountered in incompressible beds with large particles in the range of 500 microns, values of k might be expected to vary from 4-6. While this might appear to be a large variation, it is no greater than variations in friction factors for rough pipe.

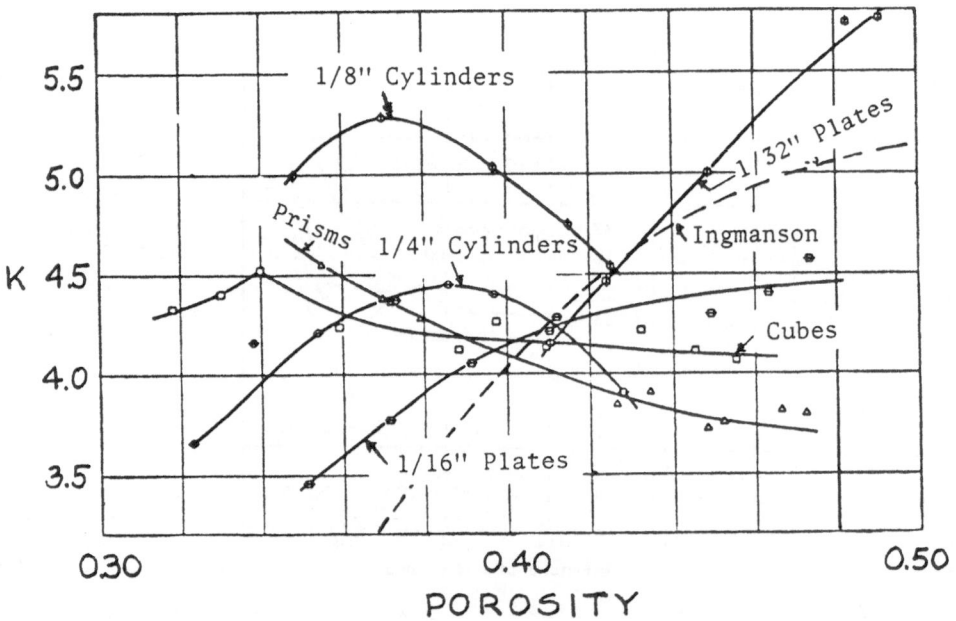

Kozeny's Constant vs. Porosity
Figure 1.3

Snyder and Stewart[15] solved numerically the Navier-Stokes equation for flow through dense cubic and simple cubic arrangements of uniform spheres. They employed Galerkin's method of trial solutions and produced velocity and pressure profiles. Numerical values for dense cubic packing obtained by Snyder and

Stewart were found to fit Equation 1.14 with C_1 = 150 and C_2 = 0 as the turbulent range was not included. Thus their value for C_1 coincided exactly with that found by Ergun[6].

References

1. Blake, F. E., The Resistance of Packing of Fluid Flow, _Trans. Am. Inst. Chem. Engrs._, 14, 415, 1922.

2. Carman, P. C., Fluid Flow Through Granular Beds, _Trans. Inst. Chem. Engr._ (London), _15_, 150, 1937.

3. Coulson, J. M. "The Flow of Fluids Through Granular Beds Effect of Particle Shape and Voids in Streamline Flow", _ibid._, _27_, 237, 1949.

4. Darcy, Henry, _Les fontaines publiques de la ville de Dijon_, see M. King Hubbert, Publication 104, Shell Development Co., Houston, Texas or Jour. Pet. Tech., 8, 222, Oct., 1956.

5. Emersleben, O., _Physik. Z._, 26, 601, 1925.

6. Ergun, Sabri, Fluid Flow Through Packed Columns, _Chem. Eng. Progr_. 48, 89, 1952.

7. Fahien, R. W. and C. B. Schriver, "_The Effect of Porosity and Transition Flow on Pressure Drop in Packed Beds_," Presented at Denver meeting of AIChE (1962).

8. Han, S. T., and W. L. Ingmanson, A Simplified Theory of Filtration, _TAPPI_, 50, No. 4, 176, 1967.

9. Happel, John and Howard Brenner, "_Low Reynolds Number Hydrodynamics_," p 393-400, Prentice-Hall, Inc., Englewood Cliffs, N. J., 1965.

10. Ingmanson, W. L., and B. D. Andrews, High Velocity Water Through Fiber Mats, _TAPPI_, 46 (No. 3), 150, 1963.

11. Ingmanson, W. L., S. T. Han, H. D. Wilder, and W. T. Myers, Jr., Resistance of Wire Screens to Flow of Water, _TAPPI_, 44, No. 1, 47, 1961.

12. Kozeny, J., Soil Permeability _Sitzer-Akad. Wiss. Wien, Math Naturw. Klasse_, 136, 11a, 271 1927 . English translation available from Frank Tiller, University of Houston.

13. Martin, J. J., W. L. McCabe, and C. C. Monrad, Pressure

Drop Through Stacked Spheres Effect of Orientation, Chem. Eng. Progr., 47, 91, 1951.

14. Shirato, Mompei, M. Sambuichi, H. Kato, and T. Aragaki, Internal Flow Mechanisms in Filter Cakes, AIChE J. 15, 405, 1969.

15. Snyder, L. J., and W. E. Stewart, Velocity and Pressure Profiles for Newtonian Creeping Flow in Regular Packed Beds of Spheres ibid, 12, 167, 1966.

16. Sparrow, E. M., and A. L. Loeffler, Jr., Longitudinal Laminar Flow Between Cylinders Arranged in Regular Array ibid, 5, 325, 1959.

17. Stepochkin, B. F., "A two-term Equation for the Resistance of Porous Media", Int. Chem. Engr., 3, 64, 1963.

18. Susskind, Herbert and Walter Becker, "Pressure Drop in Geometrically Ordered Packed Beds of Spheres", AIChE Jour., 13, 1155 1967 .

19. Tiller, F. M., The Role of Porosity in Filtration, II, Analytical Formulas for Constant Rate Filtration, Chem. Eng. Progr. 51, 282, 1955.

20. Ward, J. C., Turbulent Flow in Porous Media, Proc. Am. Soc. Civ. Engr., HY5, 90, 4019 1964 .

2. POROSITY AND FILTRATION RESISTANCE

2.1 Cake Filtration

Filter operations can be divided into two broad categories of cake and depth filtration. In cake filtration, the slurry particles are stopped at the surface of a supporting porous medium while the fluid passes through. In depth filtration, the particles are captured in the interstices of the solid, and no cake is formed. In many processes, a stage of depth filtration may precede the formation of a cake. The first particles may enter the medium, and there may be a lag before a cake begins to form. Smaller particles will pass into the medium while larger particles will bridge the openings and start the build-up of a surface layer.

In general, depth filtration is used for taking out small quantities of contaminants such as in the filtration of municipal water. Cake filtration is primarily employed for more concen-

trated slurries. Both processes may be used together. The
filtrate from a cake filtration may contain small particles which
have passed through the medium and must be removed in a polishing
step.

2.2 Approach to Filter Theory

In dealing with the mathematics of filter operations, it is
necessary to divide the analysis in two parts. In the first
part, the internal flow mechanism within the cake is considered.
In the second part, the external conditions imposed upon the
filter cake by the pumping mechanism must be harmonized with
results arising from the internal portion of the cake. In treating the internal flow within the filter solids, the distribution
of hydraulic pressure, p_L, porosity, ε, and internal flow rate
q are determined as functions of the distance, x, through the
cake. The external pumping mechanism and supporting medium
essentially provide boundary conditions which must be satisfied
at the extremities of the cake. The pump characteristics relate the rate of flow at the exit of the cake to the applied
pressure. The resistance of the medium determines the pressure
at the outlet side of the cake.

Ultimately it is desired to have filtrate volume, v, and
applied pressure, p, as functions of time, t. In simplified
situations, it is possible to obtain analytical relations p, v,
and t.

2.3 Porosity Variation

When suspended solids are deposited during cake filtration,
liquid flows through the interstices of the compressible bed
in the direction of decreasing hydraulic pressure gradient. The
solids are retained by a screen, cloth, porous metal, or other
solid bed known as the septum or filter medium. The solids
forming the cake are compact and dry at the medium whereas the
surface layer is in a wet and soupy condition. The porosity
is a minimum at the point of contact between the cake and medium
where x = 0, Figure 2.1, and a maximum at the surface (x = L)
where the liquid enters. The drag on each particle is com-
municated to the next particle; and consequently, the net solid
compressive pressure increases as the medium is approached,
thereby accounting for the decreasing porosity.

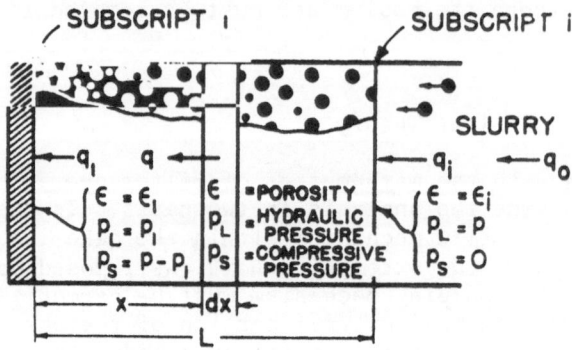

Schematic Diagram of Cake
Figure 2.1

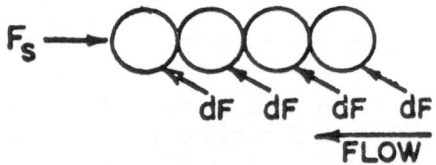

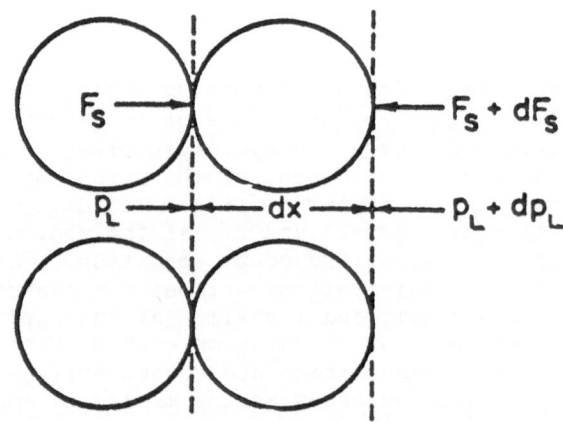

Frictional Drag
Figure 2.2

The Navier-Stokes equations should be used to obtain equations relating friction to flow rates and bed geometry. For small particles, the situation is complicated by colloidal effects at the particle-fluid interface. It is doubtful that the usual boundary conditions involving zero velocity at the walls and constant viscosity would be sufficient for solution of a real problem. Continued rearrangement of the particles and the extreme sensitivity of porosity to changes in pressure make it difficult to define much less solve a problem involving flow through compressible, porous media.

When fluid flows around a particle, the integral of the normal component of the force leads to the form drag while integrations of the tangential component yields the frictional drag. When the integration are carried out over the entire particle surface, there must only be point contact between particles. Up to the present time, there has been no exact analysis made of the flow process involved in compressible beds. In the simplified approach used herein, it is assumed that the particles are in point contact (Figure 2.2) and that the liquid completely bathes each particle and communicates the liquid pressure uniformly in a direction along a plane perpendicular to the direction of flow. Under this assumption, the hydraulic pressure p_L is effect over the entire cross section of the cake as the area of contact is negligible. The net force on the total mass within the differential distance, dx, is given by

$$\text{Net force} = F_s + dF_s + A(p_L + dp_L) - F_s - A dp_L \qquad (2.1)$$

This force equals the product of the mass within dx and the acceleration. The differential mass includes both the mass of liquid $\varepsilon \rho A\ dx$ and the mass of solid $(1 - \varepsilon)\rho_s A\ dx$. Although the solid actually moves in the cake toward the medium, the acceleration is negligible. The average velocity of the liquid is generally less than 0.001 m/sec., and the acceleration is so small as to be negligible. Consequently Equation 2.1, can be written as

$$dF_s + A\ dp_L = 0 \qquad (2.2)$$

A pseudo-solid compressive pressure is defined as $p_s = F_s/A$ yielding

$$dp_s + dp_L = 0 \qquad (2.3)$$

It should be noted that F_s is the total frictional drag at an arbitrary cross-section and that it is transmitted through the point contacts of the particles. Thus, the area A does not correspond to the surface over which f_s acts. The general notion of "point of contact" is not well-understood. Where particles enter into loose combination with the surrounding fluid, it becomes difficult to define what is "hard solid". The "soft solid" surrounding the hard core has different properties and is subject to easy degredation by shear.

In actual cakes, there is a small area of contact A_c between particles, and solid pressure could be defined as F_s/A_c. This definition serves no useful purpose excepting as it relates to deformation of the particles. However, for area rather than point contact, Equation 2.3 would have to be modified to the form

$$dF_s + (A-A_c)dp_L = 0 \tag{2.4}$$

On integration, this equation yields

$$p_s + \left(1 - \frac{A_c}{A}\right) p_L = p \tag{2.5}$$

When $A_c = 0$, this reduces to

$$F_s/A + p_L = p_s + p_L = p \tag{2.6}$$

As p_s cannot be directly measured directly, it is convenient to define the solid compression pressure as $p - p_L$. At the present stage of filtration theory, it appears unnecessary to take the area factor into account.

The "derivation" which has just been presented suffers from a number of defects. It is not really possible to assume that a particle occupies a differential distance dx or that all of the particles are conveniently lined up as shown in Figure 2.2. Furthermore sophisticated methods are required to take a finite ensemble of objects and produce a viable differential equation. A small typical volume must be chosen and analyzed by means of the Navier-Stokes equation taking account of colloidal effects and the interaction of p_s and the local porosity of the bed. It can be assumed that there is a three-dimensional confluence of fluids at the entrance and exit of the chosen element of volume which yields one-dimensional pressures equal to p_n and p_{n+1}. These pressures then represent the boundary conditions for the element. The resulting difference equation can be solved to yield p_n as a function of n. By letting n approach infinity, a macro differential equation can be obtained.

Although it has been popular to adopt a Kozeny solid as the model for filter cakes, doubt exists as to the validity of the process for compressible materials. At the present time, empirical differential equations must be employed as a basis for practical solutions.

2.4 Compression Permeability Cell

Compression-permeability (C-P) cells Figure 2.3
are used to determine the relationship between porosity, permeability and applied pressure. The cell is a device in which
a mechanical load can be applied to a fixed bed while the permeability and porosity are determined independently. The load
is applied to a movable piston while liquid percolates through
the bed under low hydrostatic head. Both permeability and
porosity (fraction and void space) are obtained as functions of
the applied mechanical pressure. It is generally assumed that
the mechanical pressure produces the same effects as the cumulative frictional drag of liquid passing through the cake. When
the frictional drag divided by the cross-sectional area in a
filter cake equals the mechanically applied pressure in the C-P
cell, the porosity and permeability in the filter cake and C-P
cell are assumed to be identical. Although that assumption is
fundamental to present filtration theory, it is only an approximation which requires careful analysis and study in practice.

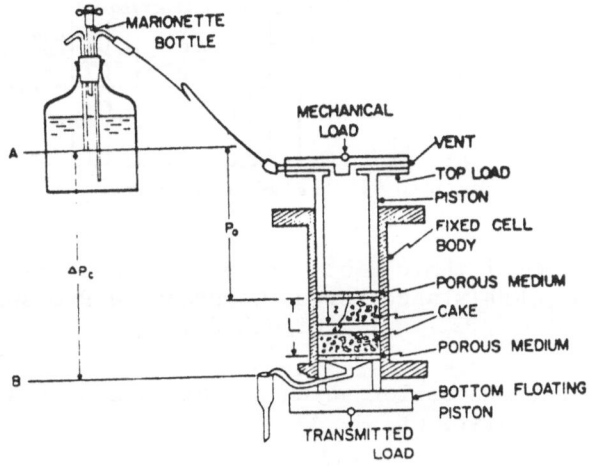

Compression-Permeability Cell
Figure 2.3

The porosity decreases and filtration resistance increases
with increasing p_s. In general, it is necessary to relate α
and ε to p_s by means of empirical formulas. In Figure 2.4, a
typical logarithmic plot of α and ε vs. p_s is illustrated. It
is assumed that the filtration resistance and porosity assume

constant values α_i and ε_i at some low pressure p_i. Above p_i, a power function relation is utilized in the following form:

$$\varepsilon = Ep_s^{-\lambda} \qquad p_s \geq p_i \qquad (2.7)$$

$$\varepsilon = \varepsilon_i = Ep_i^{-\lambda} \qquad p_s \leq p_i \qquad (2.8)$$

$$\alpha = ap_s^n \qquad p_s \geq p_i \qquad (2.9)$$

$$\alpha = \alpha_i = ap^n \qquad p_s \leq p_i \qquad (2.10)$$

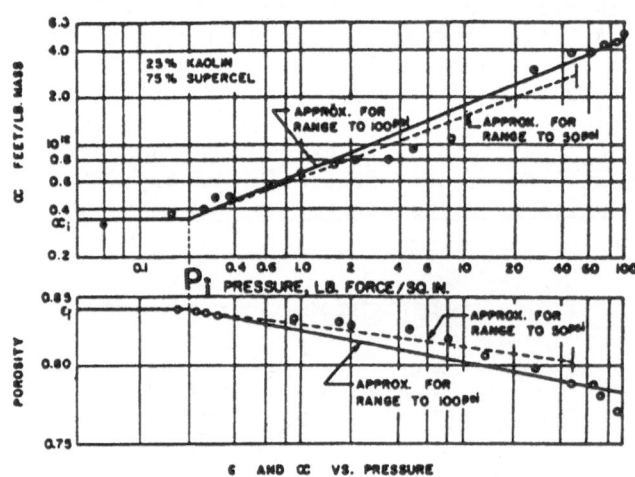

Logarithmic Plots of Porosity and Filtration
Resistance Versus Compressive Pressure
Figure 2.4

An equivalent expression which can be used over a wider pressure range consists of

$$\alpha = \alpha_i + a_1 p_s^n \qquad (2.11)$$

This formula is equivalent to a combination of (2.9) and (2.10). However, it suffers drastically in that it cannot easily be integrated in the equation $\int dp_s/\alpha$ which occurs frequently. Another integral which appears regularly is $\int K \, dp_s$ or its equivalent $\int dp_s/\alpha(1 - \varepsilon)$ (See Equation (1.5) relating K to α). It is useful to have a logarithmic plot of K vs. p_s or equivalently $1 - \varepsilon$. In Figure 2.5, such a graph is shown. It is possible to

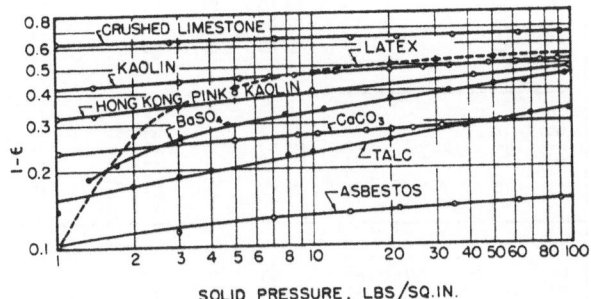

SOLID PRESSURE, LBS/SQ.IN.

Logarithmic Plot of 1 - ε vs. Solid Pressure
Figure 2.5

represent 1 - ε in a manner similar to that used in Equations
(2.7) and (2.8), thus

$$1 - \varepsilon = Bp_s^{\beta} \qquad\qquad p_s \geq p_i \qquad\qquad (2.11)$$

$$1 - \varepsilon_i = Bp_i^{\beta} \qquad\qquad p_s \leq p_i \qquad\qquad (2.12)$$

The permeability can be obtained by utilizing the relationship
between α and K as follows:

$$K = 1/\rho_s(1 - \varepsilon)\alpha = \frac{p_s^{-n-\beta}}{\rho_s Ba} \qquad\qquad (2.13)$$

Small changes in porosity may produce large increments in
filtration resistance. Eliminating p_s between Equations (2.7)
and (2.9) produces

$$\alpha = aE^{n/\lambda}\varepsilon^{-n\lambda} \qquad\qquad (2.14)$$

Differentiation and rearrangement yields

$$\frac{d\alpha}{\alpha} = -\frac{n}{\lambda}\frac{d\varepsilon}{\varepsilon} \qquad\qquad (2.15)$$

The percent change in ε is multiplied by n/λ to get the corresponding percent variation in α. Reference to Table 2.1 indicates that n/λ varies from about 5-10. However, there are
cases (fine silica) where it bas approximately equalled 60.

TABLE 2.1

Typical Values for the Contents in Equations 2.7 - 2.12

Substance	ε_i	p_i	a	n	E	λ	B	β	n/λ
Asbestos	0.902	0.9	-	-	0.90	0.017	0.115	0.057	-
Calcium Carbonate	0.771	0.9	-	0.19	0.77	0.034	0.235	0.063	5.6
Celite	0.872	0.9	-	0.14	0.90	0.017	-	-	8.2
Crushed Limestone	0.375	1.0	-	-	0.375	0.015	-	-	-
Gairome Clay	0.800	1.3	282	0.60	0.815	0.091	0.26	0.13	6.6
Ignition Plug Clay	0.78	0.6	-	0.56	0.75	0.07	0.27	0.128	8.0
Kaolin	0.698	0.03	-	-	0.59	0.045	0.42	0.054	-
Kaolin, Hong Kong pink	0.72	0.5	101	0.33	0.70	0.059	-	0.005	5.6
Solkofloc	-	-	0.0024	1.01	-	-	-	-	-
Talc	-	-	8.66	0.51	0.86	0.054	0.155	0.203	9.4
Titanium Dioxide	-	-	32	-	-	-	-	-	-
Zinc Sulfide	-	-	14	0.69	-	-	-	-	-
General Range	0.4-0.95	0.01-1.5	10^4-10^{14}	0-1.2	0.4-0.95	0-0.1	0.1-0.5	0-0.25	-

2.5 Modified Compressive Pressure

Discussion among researchers in flow through compressible porous media sometimes reflect on the adequacy of Equation (2.3) which yields $dp_L = -dp_s$. Some arguments lead to the incororation of a porosity term in Equation (2.3) which could be expressed generally as

$$dp_L + f(\varepsilon)dp_s = 0 \qquad (2.16)$$

As it is assumed that α, K, and ε are functions of p_s, dp_L is normally eliminated from Darcy's law which can then be placed in the form

$$-\frac{dp_s}{dx} = \frac{\mu q}{f(\varepsilon)K} \qquad (2.17)$$

Experimentation yields the quantity $f(\varepsilon)K$ which is then related to p_s by empirical equations. From a practical standpoint, addition of a function $f(\varepsilon)$ would not alter equations presently in use.

3. INTERNAL FLOW RATE

3.1 Conventional Formulas

Most work in the field of filtration has been based upon the simplified (and at times incorrect) formula

$$q = \frac{\Delta p}{\mu(\alpha w + R_m)} \qquad (3.1)$$

where αw and R_m are respectively the cake and medium resistances. The quantity w is the mass of dry solid deposited per unit area. In constant pressure filtration, it is assumed that boty q and α are constant. It has been demonstrated 1,2.3 that q is not constant and that the velocity of the solids must also be considered. In practical cases, variation in flow rate within a cake is of importance when the quantity of liquid retained in the cake is comparable to the filtrate. Short filtration with concentrated slurries favor variation of internal flow rate. Thus, modification of (3.1) is primarily of importance in filtration carried out with drums and discs.

3.2 Internal Flow Rate Variation

In Figure 3.1, porosity is plotted against x at different times. It can be seen that the porosity decreases at each point in the solid; and, consequently, liquid is squeezed out of the cake as it is compressed. This squeezing action results in an increase of flow rate as the supporting medium is approached.

The fraction of solids $(1 - \varepsilon)$ is also plotted against x at constant time intervals. While the porosity decreases with time at any fixed distance from the medium, the fraction of solids increases. The solid flows toward the medium displacing liquid.

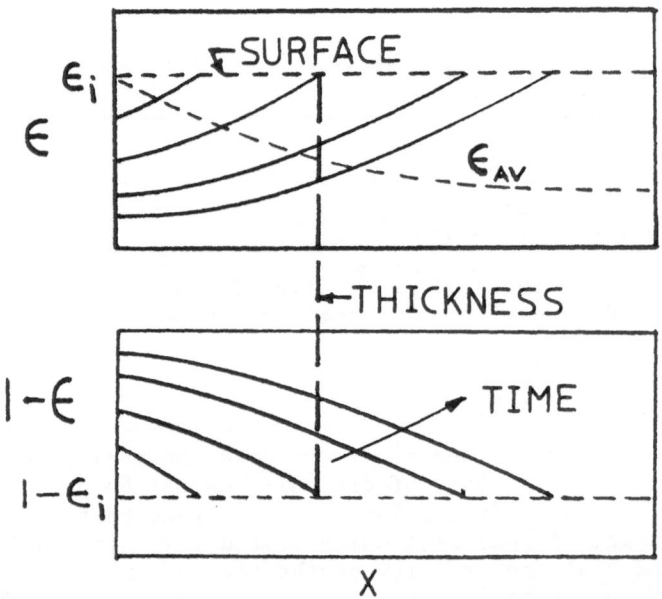

Fraction Voids (ε) and Fraction
Solids $(1 - \varepsilon)$ vs. Distance at Different
Values of Time.

Figure 3.1

When the cake is deposited at the surface, it has a relatively high porosity and large liquid content. As fresh layers
build up, the surface passes into the cake interior; and the
porosity begins to decrease. Then liquid that is first deposited
is later partially removed.

The total volume of liquid per unit area contained in the
distance zero to x is given by

$$\text{liquid volume/area} = \int_0^x \varepsilon dx \qquad (3.2)$$

The liquid volume/area in the cake is represented by the area
under the ε vs. x curve. A material balance over the portion
of the cake lying between 0 and x of the cake will give the
difference between the flow rate in q, and the flow rate out,
q_1, (See Figure 2. 1).

The flow rate into cross section x minus the flow rate out
at the medium equals the rate of change of (3.2) with respect
to time, thus

$$q - q_1 = \frac{\partial}{\partial t} \int_0^x \varepsilon dx \qquad (3.3)$$

The variable of integration in Equation (3.3) can be changed
by noting that

$$\varepsilon dx = d(x\varepsilon) - xd\varepsilon \qquad (3.4)$$

Substituting in (3.3) yields

$$q - q_1 = \frac{d}{dt} [x\varepsilon - \int_{\varepsilon_1}^{\varepsilon} xd\varepsilon] \qquad (3.5)$$

The values of x, ε, and q at different positions in the cake
are given by

Varible	Value at medium	Value at surface
x	0	L
ε	ε_1	ε_i
q	q_1	q_i

Substituting the limits for x = 0, L gives

$$q_i - q_1 = \frac{d}{dt} [L\varepsilon_i - \int_{\varepsilon_1}^{\varepsilon_i} xd\varepsilon] \qquad (3.6)$$

In Figure 3.2 a plot of ε vs. x is shown in relation to
the integral $\int xd\varepsilon$ as it occurs in previous equations. If there
were no squeezing and compressive action within the cake, there

338

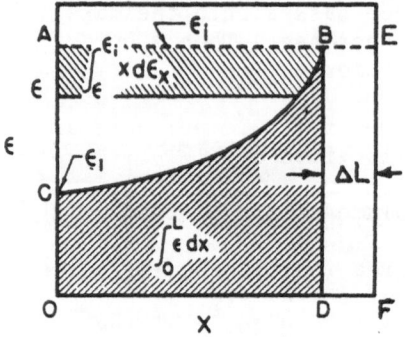

Porosity Relationships
Figure 3.2

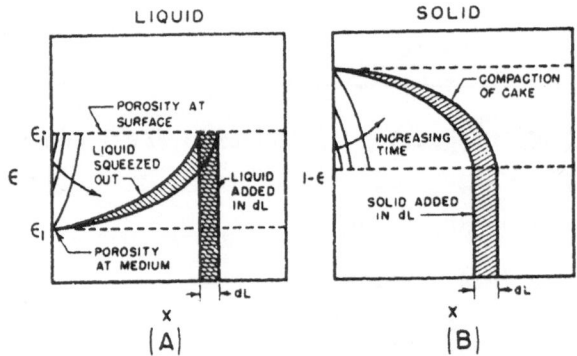

Porosity vs. Distance, Increasing Cake Thickness
Figure 3.3

would be a uniform porosity equal to ε_i, the value in an infinitesimal surface layer. The area in ABEFO in Figure 3.2 is proportional to the volume of liquid per square foot of cross-sectional area which would remain in the cake if there were no compression. The area CBDO, which equals $\int_0^L \varepsilon \, dx$, represents the the volume of liquid retained in the interstices of the cake, and consequently the area ABC + BEFD equaling $\int_{\varepsilon_1}^{\varepsilon_i} x \, d\varepsilon + \varepsilon_i \Delta_L$ is proportional to the liquid which was first deposited and then squeezed out as the cake was compressed during filtration.

The changes accompanying an incremental increase in cake thickness are illustrated in Figure 3.3A and 3.3B. In Figure 3.3A the liquid squeezed from the cake during a differential time dt is represented by the singly cross-hatched area. The doubly cross-hatched area represents the liquid added to the cake which is larger than the liquid removed. In Figure 3.3B $(1 - \varepsilon)$ is plotted against x. The integral $\int_o^L (1 - \varepsilon) \, dx$ corresponds to the volume of solid material.

When the cake increases dL in thickness, a layer dL' is layed down as shown in Figure 3.4. The differential dL' is larger than dL because the cake is being compacted while fresh solid is laid down. At the same time the thickness increases by dL, the original cake is compacted to thickness L'. The amount of cake laid down in time dt equals dL'. The mass of cake deposited is

$$dw = \rho_s (1 - \varepsilon_i) dL' \qquad (3.7)$$

dL' must be used as dw is laid down over the distance dL' rather than dL.

Since $w = \rho_s (1 - \varepsilon_{av}) L$, dw is also given by

$$dw = \rho_s d \, [(1 - \varepsilon_{av})L] = \rho_s (1 - \varepsilon_{av}) dL - \rho_s L d\varepsilon_{av} \qquad (3.8)$$

Equating (3.7) and (3.8) one obtains

$$dL' = \frac{1 - \varepsilon_{av}}{1 - \varepsilon_i} \, dL - \frac{L}{1 - \varepsilon_i} \, d\varepsilon_{av} \qquad (3.9)$$

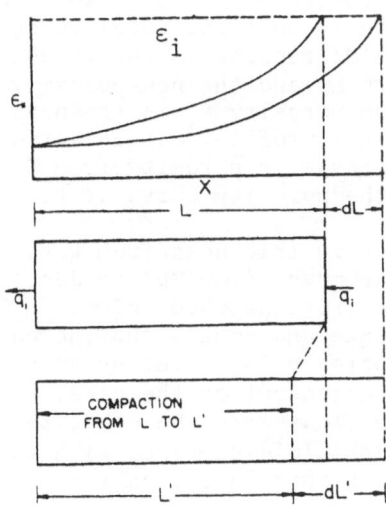

Composition of
Cake
Figure 3.4

The fraction of freshly deposited cake which flows out of dL into the cake is given by

$$\frac{dL' - dL}{dL'} = \frac{(\varepsilon_i - \varepsilon_{av})dL - Ld\varepsilon_{av}}{(1 - \varepsilon_{av})dL - Ld\varepsilon_{av}} = \frac{d[(\varepsilon_i - \varepsilon_{av})L]}{d[(1 - \varepsilon_{av})L]} \qquad (3.10)$$

If ε_{av} is constant as occurs in constant pressure filtration with negligible medium resistance Equation (3.10) reduces to

$$\frac{dL' - dL}{dL} = \frac{\varepsilon_i - \varepsilon_{av}}{1 - \varepsilon_{av}} \qquad (3.11)$$

It is also possible to write

$$dL' = \frac{d[(1 - \varepsilon_{av})L]}{1 - \varepsilon_i} \qquad (3.12)$$

3.3 Flow Rate at Surface of Cake

When calculating the flow variation in a cake, it is necessary to know the value of q_i, the flow rate in that portion of the cake near the surface with a porosity of ε_i. If is assumed that $\varepsilon = \varepsilon_i$ when $p_s = p - p_L < p_i$. It is possible for the porosity ε_i to extend a considerable distance into the cake beyond dL' if there is a very small accumulative drag near the surface. The porosity will have a value of ε_i in thickness dL or dL'.

In Figure 3.5, a physical picture of the flow rate relations is illustrated. At time t the porosity is shown as curve OABCD. The area under that curve represents the total voids in the cake. The companion curve O'A'B'C'D' represents the volume of solids. At time t + dt, L changes to L' and the new porosity curve is given by OEFGH. The maximum porosity ε_i is shown as extending from C to B along the original profile.[1] In an actual case, there would be a slight decrease in porosity from C to B and no sudden change in slope of the porosity curve at B.

When the cake increases in thickness from L to L' the original surface of the cake moves from DD' to JJ'. The solid which was originally in the cross-hatched volume JK'C'D moves into the main body of the cake and equals the increased volume of solids A'B'F'E'. That increase is offset by an equal volume of liquid ABFE which is squeezed out of the cake. When the solid represented by JK'C'D flows into the cake, the accompanying liquid JKCD (equal to J'K'C'D') combines with the liquid from the slurry and passes on through the cake.

341

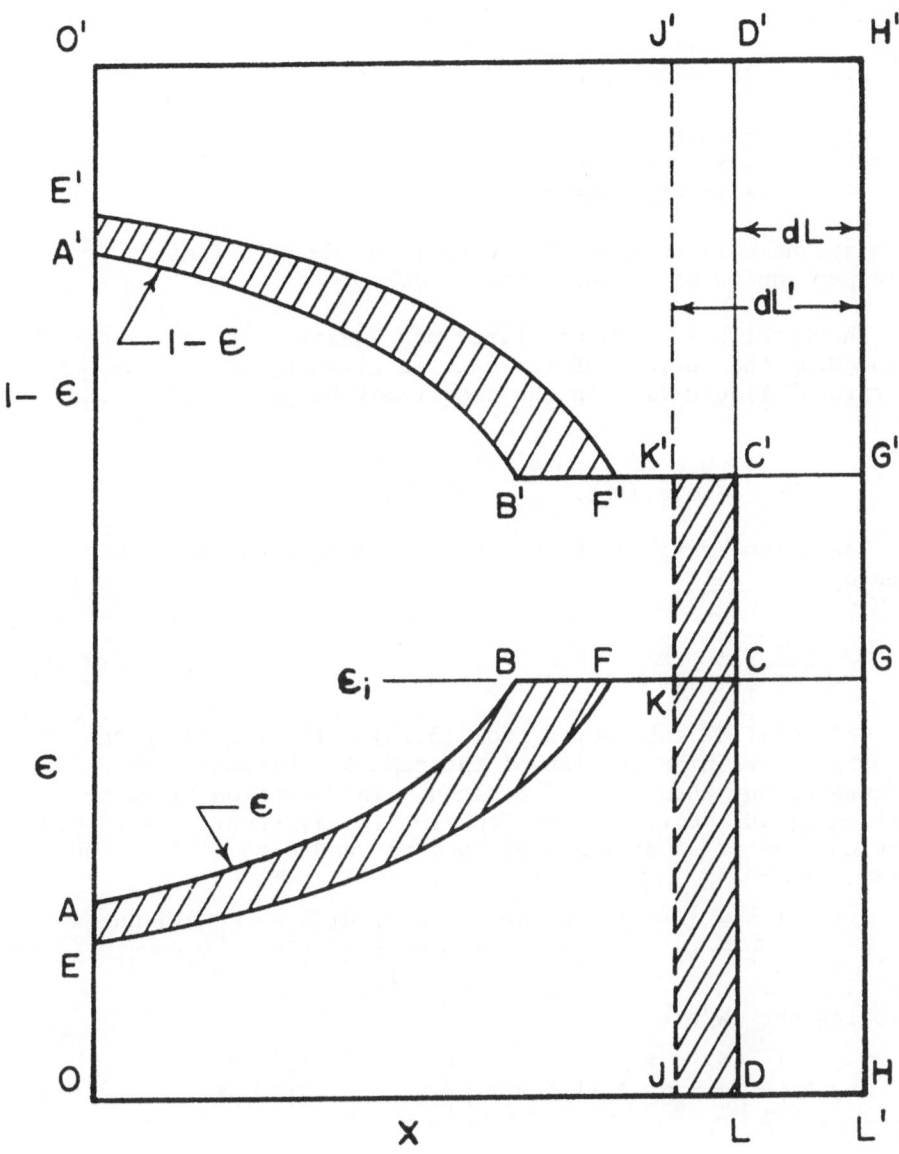

Conditions at Cake Surface
Figure 3.5

The flow rate will be constant and equal to q_i throughout that part of the cake where the porosity equals ε_i. It is convenient to calculate q_i at a distance dL' from the surface of the cake. Then that rate will be the same wherever the porosity equals ε_i. The rate of liquid flowing past the point $(L - dL')$ may be calculated as follows:

q_i = rate of liquid arriving at cake surface
        - rate of liquid deposited in dL'
        + liquid squeezed out of dL' (JKCD)

Calculations will be made on the basis of dw kilograms of dry solid per square meter deposited in dL'.

The total kilograms of slurry per square meter per second approaching the surface of the cake is given by $(1/s)dw/dt$, and the rate of liquid flow in the slurry may be presented in the form

$$q_o = \frac{\text{cu.m of liquid}}{(\text{sq.m})(\text{sec.})} = \frac{1 - s}{\rho s} \frac{dw}{dt} \tag{3.13}$$

The volume rate at which liquid is deposited in dL' is given by

$$\frac{1 - s_i}{\rho s_i} \frac{dw}{dt} \tag{3.14}$$

This will be subtracted from (3.13). The liquid which was originally in the portion of the cake of thickness $(dL' - dL)$ is squeezed out when dw is deposited. The fraction given by Equation (3.10) multiplied by Equation (3.14) equals the liquid squeezed from $dL' - dL$ (JJ'D'D), and is added to (3.14) minus (3.13). Thus

$$\rho q_i = \frac{1 - s}{s} \frac{dw}{dt} - \frac{1 - s_i}{s_i} \frac{dw}{dt} + \frac{1 - s_i}{s_i} \frac{d[(\varepsilon_i - \varepsilon_{av})L]}{d[(1 - \varepsilon_{av})L]} \frac{dw}{dt} \tag{3.15}$$

Factoring out dw/dt

$$\rho q_i = \frac{1}{s_i} \left[\frac{s_i - s}{s} + (1 - s_i) \frac{d[(\varepsilon_i - \varepsilon_{av})L]}{d[(1 - \varepsilon_{av})L]} \right] \frac{dw}{dt} \tag{3.16}$$

Eliminating dw by use of 3.7 gives

$$\rho q_i = \rho_s \frac{s_i - s}{s_i s} \frac{d[(1 - \varepsilon_{av})L]}{dt} + \rho_s \frac{1 - s_i}{s_i} \frac{d[(\varepsilon_i - \varepsilon_{av})L]}{dt} \tag{3.17}$$

$$= \rho_s \frac{1 - s}{s} \frac{d[(1 - \varepsilon_{av})L]}{dt} - \rho_s \frac{(1 - s_i)(1 - \varepsilon_i)}{s_i} \frac{dL}{dt} \tag{3.18}$$

The value of s_i is related to ε_i by

$$s_i = \frac{\rho_s (1 - \varepsilon_i)}{\rho_s(1 - \varepsilon_i) + \rho \varepsilon_i} \tag{3.19}$$

Substituting in (3.18) yields

$$\rho q_i = \rho_s \frac{1 - s}{s} \frac{d[(1 - \varepsilon_{av})L]}{dt} - \rho \varepsilon_i \frac{dL}{dt} \tag{3.20}$$

In terms of w, Equation (3.20) can be changed to

$$\rho q_i = \left[\frac{1 - s}{s} - \frac{\rho}{\rho_s} \frac{\varepsilon_i}{(1 - \varepsilon_{av})} \right] \frac{dw}{dt} - \frac{\rho}{\rho_s} \frac{w}{(1 - \varepsilon_{av})^2} \frac{d\varepsilon_{av}}{dt} \tag{3.21}$$

These equations for q_i can be combined with similar expressions for q_1 and substituted in (3.6).

3.4 Flow Rate at Medium

To relate q_1 to dw/dt an overall material balance can be written over the slurry, cake, and filtrate as follows:

Mass of slurry = mass cake + mass of filtrate

$$w/s = w/s_c + \rho v \tag{3.22}$$

where s and s_c are respectively the mass fractions of solids in the slurry and cake, and v is the volume of filtrate. Solving for v yields

$$v = \frac{1 - s/s_c}{\rho s} w \tag{3.23}$$

Differentiating with respect to time leads to q_1

$$q_1 = \frac{dv}{dt} = \frac{1 - s/s_c}{\rho s} \frac{dw}{dt} + \frac{w}{\rho} \frac{1}{s_c^2} \frac{ds_c}{dt} \tag{3.24}$$

The average mass fraction s_c of solids in the cake is related to the average porosity ε_{av} by

$$s_c = \frac{\rho_s (1 - \varepsilon_{av})}{\rho_s (1 - \varepsilon_{av}) + \rho \varepsilon_{av}} \qquad (3.25)$$

The reciprocal of s_c equals the ratio m of mass of wet to dry cake

$$m = 1/s_c = 1 + \frac{\rho \varepsilon_{av}}{\rho_s (1 - \varepsilon_{av})} \qquad (3.26)$$

Differentiation with respect to t yields

$$\frac{dm}{dt} = -\frac{1}{s_c^2} = \frac{\rho}{\rho_s} \frac{1}{(1 - \varepsilon_{av})^2} \frac{d\varepsilon_{av}}{dt} \qquad (3.27)$$

Substituting (3.25), (3.26), and $w = \rho_s (1 - \varepsilon_{av})L$ in (3.24) leads to

$$\rho q_1 = \left[\frac{1 - s}{s} - \frac{\rho}{\rho_s} \frac{\varepsilon_{av}}{1 - \varepsilon_{av}} \right] \frac{dw}{dt} - \frac{\rho L}{(1 - \varepsilon_{av})} \frac{d\varepsilon_{av}}{dt} \qquad (3.28)$$

Substracting (3.21) from (3.28) yields

$$(q_1 - q_i) = \frac{1}{\rho_s} \frac{\varepsilon_i - \varepsilon_{av}}{1 - \varepsilon_{av}} \frac{dw}{dt} \qquad (3.29)$$

The ratio q_i/q_1 is a useful quantity and can be written as

$$\frac{q_i}{q_1} = \frac{\left[\dfrac{1 - s}{s} - \dfrac{\rho}{\rho_s} \dfrac{\varepsilon_i}{(1 - \varepsilon_{av})} \right] \dfrac{dw}{dt} - \dfrac{\rho}{\rho_s} \dfrac{w}{(1 - \varepsilon_{av})^2} \dfrac{d\varepsilon_{av}}{dt}}{\left[\dfrac{1 - s}{s} - \dfrac{\rho}{\rho_s} \dfrac{\varepsilon_{av}}{(1 - \varepsilon_{av})} \right] \dfrac{dw}{dt} - \dfrac{\rho}{\rho_s} \dfrac{w}{(1 - \varepsilon_{av})^2} \dfrac{d\varepsilon_{av}}{dt}}$$

$$(3.30)$$

In a filtration with constant pressure drop across the cake, the average porosity is constnat; and reduces to

$$\frac{q_i}{q_1} = 1 - s \frac{\rho (\varepsilon_i - \varepsilon_{av})}{(1 - s)\rho_s (1 - \varepsilon_{av}) - S\rho \varepsilon_{av}} \qquad (3.31)$$

The maximum possible concentration in the slurry occurs when $s = s_i$. That also corresponds to the minimum value of q_i/q_1. Substituting $s = s_i$ in (3.31) yields

$$(q_i/q_1)_{min} = \varepsilon_i \tag{3.37}$$

For the general Equation (3.30), the effect of the additional term involving $d\varepsilon_{av}/dt$ increases the value above ε_i. Thus even where ε_{av} is decreasing with time q_i/q_1 will always be greater than ε_i^{av}.

3.5 Flow of Solids[1]

As the cake is compressed, the solids actually flow toward the medium. The displacement of liquid by the solids causes the flow variation of the liquid as calculated in Paragraphs 3.3 - 3.4. Thus whenever flow variation of the liquid is important, it will be necessary to consider the flow of the solid. As a practical matter, flow variations need only be considered when s approaches s_c or s_i, i.e., the slurry is highly concentrated. Normally such slurries are filtered in a rotary, belt, or disc filter where the filtration time is short (frequently in seconds).

The volume of liquid displaced by the solid in time dt is illustrated in Figure 3.2. Similarly the solid which flows into the void space is shown in Figures 3.3 and 3.5. The apparent solid velocity is expressed as r cu.ft. solids/(sq.ft.)(sec.). The total volume of solids per unit area in the cake from 0 to x is given by

$$w_x/\rho_s = \text{vol. solids/area} = \int_0^x (1 - \varepsilon)dx \tag{3.33}$$

where w_x is the mass of dry solids contained in the distance 0 to x. The rate of increase of the volume of solids in distance x is given by the rate of flow in minus the rate out. The rate out at the medium is assumed to be zero, assuming no particles flow into the filtrate. Differentiating (3.33) with respect to time gives

$$\frac{1}{\rho_s}\frac{dw_x}{dt} = r = \frac{\partial}{\partial t}\int_0^x (1 - \varepsilon)dx \tag{3.34}$$

Differentiating under the integral sign

$$r = -\int_0^x \frac{\partial\varepsilon}{\partial t} dx \tag{3.35}$$

Differentiating with respect to x yields

$$\frac{\partial r}{\partial x} = - \frac{\partial \varepsilon}{\partial t} \qquad (3.36)$$

Differentiating Equation (3.3) with respect to x and combining with (3.36), one obtains

$$\frac{\partial \varepsilon}{\partial t} = \frac{\partial q}{\partial x} = - \frac{\partial r}{\partial x} \qquad (3.37)$$

Focusing on q and r

$$\frac{\partial q}{\partial x} + \frac{\partial r}{\partial x} = \frac{\partial}{\partial x} (q + r) = 0 \qquad (3.38)$$

At any instant of time, integration of (3.38) yields $q + r = C$. Application of the condition that $q = q_1$ and $r = 0$ at the medium where $x = 0$ produces

$$q + r = q_1 \qquad (3.39)$$

In the surface layer where q has its minimum value q_i, r has its maximum value

$$r_i = q_1 - q_i \qquad (3.40)$$

Relative Velocity

In the Darcy flow equation, the velocity term should be the velocity of liquid relative to the solid or

$$\text{relative velocity} = u - u_s \qquad (3.41)$$

where u and u_s are respectively the true average velocities of liquid and solid. In relation to q and r

$$u = q/\varepsilon \qquad (3.42)$$

$$u_s = r/(1 - \varepsilon) \qquad (3.43)$$

and

$$u - u_s = \frac{q}{\varepsilon} - \frac{r}{1-\varepsilon} = - \frac{1}{\varepsilon} (q - \frac{\varepsilon}{1-\varepsilon} r)$$

$$\qquad (3.44)$$

$$= \frac{1}{\varepsilon} (q - er)$$

where e is the void ratio.

In Figure 3.6 the physical explanation of the quantities in (3.44) are shown. The curve AB and OD show respectively the variation of q and r. The value of er is given by OC. The difference between AB and OC yields the quantity in brackets in Equation (3.44). When points B and C coincide, B is at a point corresponding to $q_i/q_1 = \varepsilon_i$.

In Figure 3.7, calculated values for the flow variation in ignition (spark) plug material and latex are shown as a function of the distance through the cake. The limiting value of the slurry concentration is given by $s_i = 0.4767$ and limiting values of q/q_1 is 0.78.

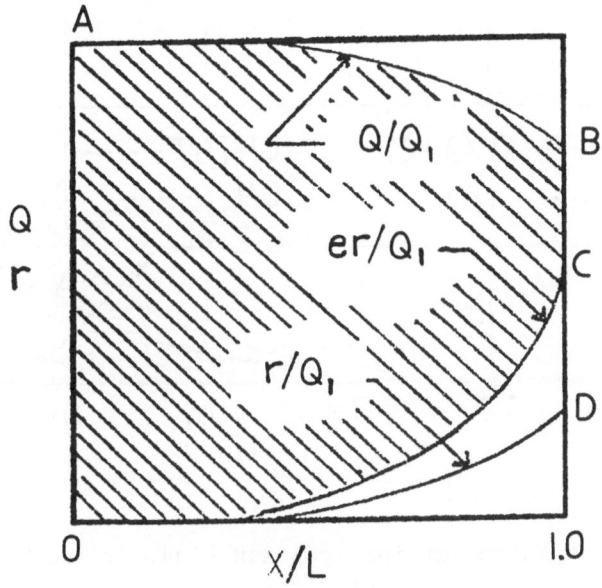

Flow Rate Variations
Figure 3.6

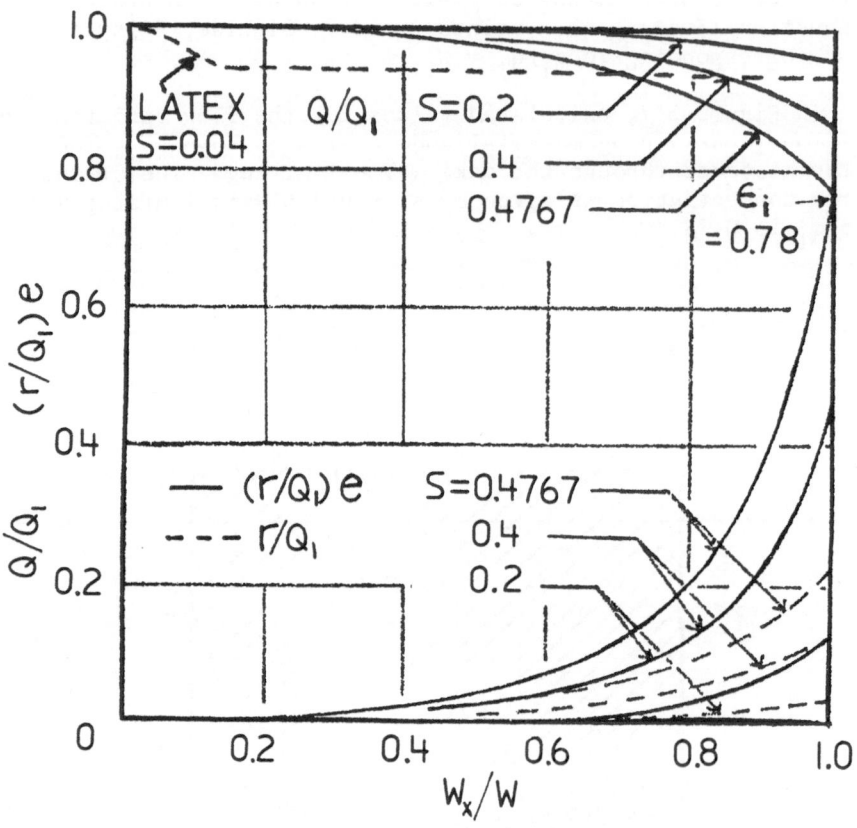

Flow Variation for Ignition Plug, Pressure
Figure 3.7

References

1. Shirato, M., H. Kato, M. Sambuichi, and T. Aragaki, "Internal Flow Mechanism in Filtration", AIChE J., 15, 405, 1969.

2. Tiller, F.M. and H. R. Cooper, The Role of Porosity in Filtration, IV, Constant Pressure Filtration, AIChE J., 6, 595, 1960.

3. Tiller, F. M. and M. Shirato, "The Role of Porosity in Filtration, VI, New Definition of Filtration Resistance, Ibid, 10, 61, 1964.

4. POROSITY AND HYDRAULIC DISTRIBUTION IN COMPRESSIBLE CAKES

4.1 Introduction

In filter cakes the variation of porosity and hydraulic pressure distance from the cake surface is important from both theoretical and industrial viewpoints. Porosity and hydraulic pressure play fundamentals roles in their relation to flow rates, overall pressure drop, and other parameters involved in the differential equations of flow through compressible, porous media. Porosity variation determines the average porosity and liquid content of the filter cake in commercial operation. Since a dry cake is frequently desired, it is important for design purposes to know how the average porosity varies with total pressure. It will be shown that increasing pressure has widely different effects on liquid content. With some materials the average porosity is hardly affected by increasing pressure; with others substantial decreases are involved. After the internal variations of porosity and pressure are found, it is possible to fit the external restraints due to filters and pumps to the boundary conditions and obtain solutions for operating processes.

At this time (1973) it is not customary for designers or users of filtration equipment to employ the methods suggested in this chapter. Unreliability of basic data and theoretical weaknesses contribute to difficulties in accurately predicting experimental relations among basic variables. However, the greatest obstacle to utilization of the methods to be presented is simply ignorance of theoretical principals on the part of engineers involved in filtration. There is much which could be gained if a more analytical approach to filtration were taken in industry.

Few investigations of porosity variation have been reported in the literature. Hutto[2] developed a unique banding technique

in which a colored material was introduced at equal mass intervals and incremental volumes were measured visually. An extensive experimental analysis of porosity and hydraulic pressure
variation was made by Shirato and Okamura[6]. Those authors
graphed porosity ε vs. fractional distance x/L through solids
as well as the average porosity ε_{av} and the ratio of mass of
wet to mass of dry cake as functions of total pressure. They
showed that ε_{av} varied approximately as a power function of the
applied pressure p.

4.2 Local Hydraulic Pressure and Porosity

Formulas 1,3,4,8 for porosity and hydraulic pressure depend
upon the basic equations for flow through compressible porous
media and the relation between local porosity and applied compressive pressure. Approximate equations utilizing empirical
relations will be developed.

The Shirato[7] equation for flow through compressible porous
media is given by

$$\frac{dp_L}{dw_x} = -\frac{dp_S}{dw_x} = \alpha \, \mu (q - e \, r) \qquad (4.1)$$

where dw_x represents the mass of dry solids in distance dx. The
differential can be eliminated by use of the expression $dw_x =
\rho_S (1 - \varepsilon)dx$. Substituting in (4.1), one obtains

$$\frac{dp_S}{dx} = -\mu\rho_S (1 - \varepsilon)\alpha (q - e \, r) \qquad (4.2)$$

In terms of permeability K

$$\frac{dp_S}{dx} = -\frac{\mu}{K} (q - e \, r) \qquad (4.3)$$

Under certain conditions[7] q and r vary markedly throughout
the solid and may not be treated as constants*. For relatively
dilute slurries and long filtration cycles, however, q may be
considered as approximately constant and r = 0. Solving for
dx in (4.3) and placing limits on the integrals one obtains
(with $q = q_1$, the exit rate)

---

*
For a numerical treatment of variable q and r see Chapter 5,
reference 10.

$$\int_0^x dx = x = \frac{1}{\mu q_1} \int_{p_s}^{p-p_1} K dp_s \tag{4.4}$$

where the lower limit on p_s is related to the hydraulic pressure at distance x from the medium by

$$p_s = p - p_L \tag{4.5}$$

The quantity p_1 represents the pressure required to overcome the resistance of the medium[1]. It will be assumed to be zero in the derivations which follow. Equation (4.4) becomes on integration from 0 to L

$$L = \frac{1}{\mu q_1} \int_0^p K dp_s \tag{4.6}$$

Dividing Equation (4.4) by (4.6) one gets

$$\frac{x}{L} = \frac{\displaystyle\int_{p_s}^p K dp_s}{\displaystyle\int_0^p K dp_s} = \frac{\displaystyle\int_{p_s}^p \frac{dp_s}{\alpha(1-\epsilon)}}{\displaystyle\int_0^p \frac{dp_s}{\alpha(1-\epsilon)}} \tag{4.7}$$

In Equation (4.7) the integral in the denominator is constant at any given total pressure p. The integral in the numerator is a function of the limits of integration, and consequently Equation (4.7) defines a relationship between x/L and p_s. The parameters ϵ, α, and K are functions of p_s. They may be obtained from compression-permeability cell measurements or from p-v-t data for actual filtrations.

As both ϵ and p_L are functions of p_s, Equation (4.7) can be considered as defining the relationship among porosity, hydraulic pressure drop, and the fractional distance through the cake. If no solids are being deposited, the thickness L is constant. On the other hand, if a cake is being formed, both L and p will be functions of the time. The use of Equation (4.7) for filtration in which pressure is varying carries the implicit and somewhat weak assumption that transient effects in passing from one pressure to another may be neglected.

Shirato[5] developed a method for finding the hydraulic pressure variation using specially designed manometers connected to probes within a cake. In Figure 4.1 the types of data which were obtained are illustrated for an ignition plug slurry having a mass fraction of solid equal to 0.225 and filtered at a cons-

tant pressure of 71.8 lb./sq. in. (496 kn/m^2). As long as a
probe remained outside of the cake, the pressure was constant
and equal to the applied filtration pressure. As soon as the
cake increased in thickness sufficiently to envelope a probe,
the pressure began to fall. The thickness of the cake could
be determined accurately as a function of time or volume filtered by observing the time at which the pressure on a given
probe began to fall.

Based upon experiments similar to the one illustrated in
Figure 4.1 plots of p_L/p vs. x/L were prepared as shown in
Figure 4.2. It is apparent that time does not appreciably affect the relationship of the hydraulic pressure drop as a
function of the fractional distance through the cake in accord
with the prediction of Equation (4.7).

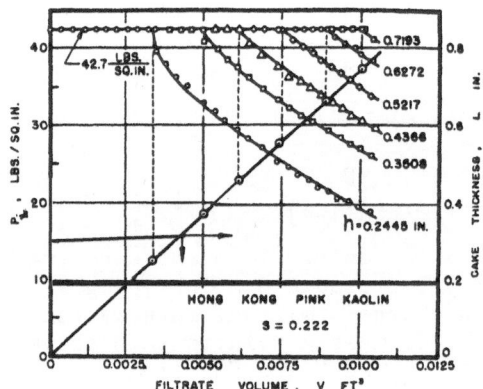

**Hydraulic pressure and cake thickness vs.
filtration volume.**

Figure 4.1

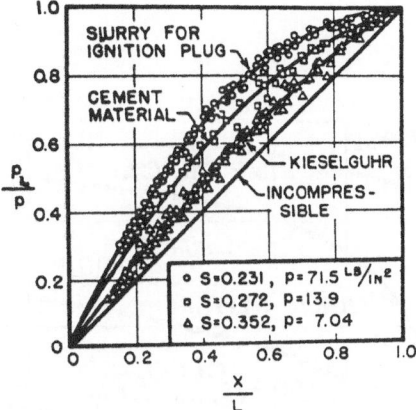

Hydraulic pressure vs. x/L.

Figure 4.2

In Figure 4.3 porosity variations with fractional distance through the cake are illustrated. With the exception of experimental values for ignition plug (clay slurry for spark plug manufacture) all curves were calculated for relatively dilute slurries.

For highly porous and compressible material like polystyrene latex and fine silica, the porosity remains constant throughout most of the cake. However, there is a sharp decrease close to the medium, indicating that a large part of the cake resistance is concentrated in a compact layer next to the supporting medium. The pressures (10 psi) marked on the fine silica and polystyrene latex curves are hydraulic pressures calculated with Equation (4.7). The value of 10 psi on the silica curve indicates that the hydraulic pressure has dropped from 100 to 90 psi. Thus, a loss of 10% of the original pressure does not occur until 90% of the cake has been traversed by the liquid. The effect is even more pronounced with the latex where a 10% loss is not reached until the liquid has passed through 98% of the cake.

In Figure 4.4 the effect of total pressure on the average porosity is illustrated for a number of substances. Filtration pressure has its greatest effect on porosity in the low pressure range. Surprisingly, for the highly compressible latex, porosity tends to remain constant with increasing pressure above 10 psi. The explanation can be seen in Figure 4.3 where increasing total filtration pressure has little effect on the flat portion of the latex porosity curve. Because frictional drag affects only a small part of the porosity curve, increasing

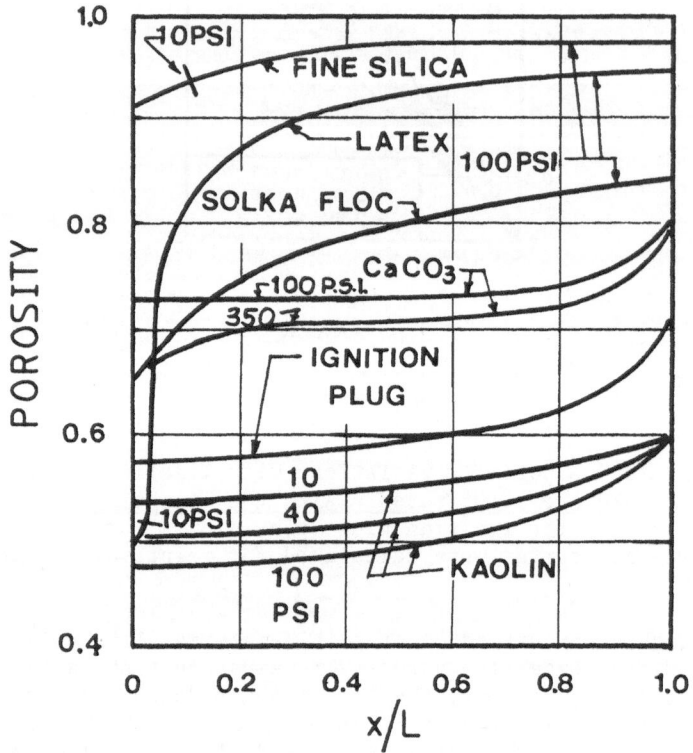

Porosity vs. Fractional Distance
Figure 4.3

pressure does not change average porosity much. However, the
initial decrease in average porosity represents a large percentage reduction in the liquid present.

4.3 Average Porosity

     In process calculations, the average liquid.content is of
more interest than the curves of porosity variation with distance. The average porosity may be obtained by numerical integration of the curves in accordance with the equation

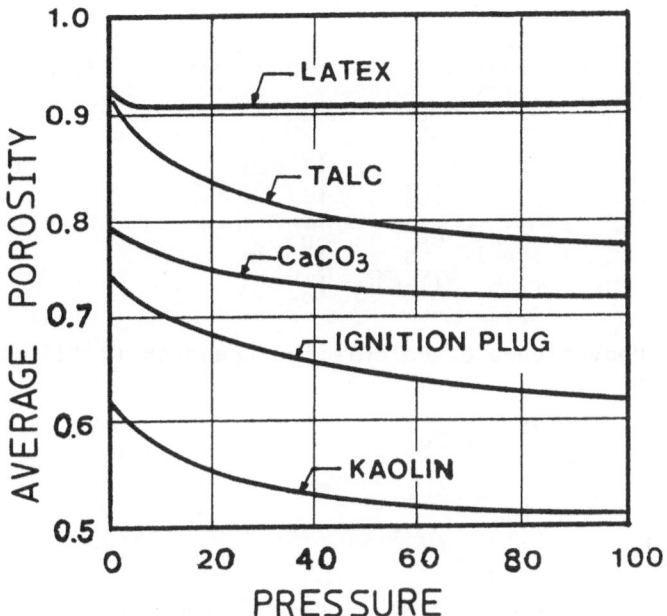

Average Porosity vs. Applied Pressure
Figure 4.4

$$\epsilon_{av} = \frac{1}{L} \int_0^L \epsilon dx \qquad (4.8)$$

On changing the variable of integration one obtains

$$\epsilon_{av} = \frac{1}{L} \int_0^{p-p_1} \epsilon \frac{dx}{dp_s} \, dp_s \qquad (4.9)$$

The term dx/dp_s can be obtained from Equation (4.2) and substituted in (4.9) to yield

$$\epsilon_{av} = \frac{1}{\mu\rho_s L} \int_0^p \frac{\epsilon dp_s}{\alpha(1-\epsilon)(q-er)} \qquad (4.10)$$

or on assumption that $q = q_1$, $r = 0$, and $p_1 = 0$, (4.10) becomes

$$\varepsilon_{av} = \frac{1}{\mu\rho_s q_1} \frac{1}{L} \int_0^{p_1} \frac{\varepsilon dp_s}{\alpha(1-\varepsilon)} \qquad (4.11)$$

The thickness of the cake can be obtained from (4.6) and substituted in (4.11)

$$\varepsilon_{av} = \frac{\int_0^p \frac{\varepsilon dp_s}{\alpha(1-\varepsilon)}}{\int_0^{p-p_1} \frac{dp_s}{\alpha(1-\varepsilon)}} = \frac{\int_0^p \varepsilon K dp_s}{\int_0^p K dp_s} \qquad (4.12)$$

For convenience of calculation Equation (4.12) may re-written as

$$\varepsilon_{av} = 1 - \frac{\int_0^p dp_s/\alpha}{\int_0^p dp_s/\alpha(1-\varepsilon)} \qquad (4.13)$$

Use of Equation (4.13) instead of Equation (4.12) permits a simplification in that the integral of $\varepsilon\, dp_s/\alpha(1-\varepsilon)$ does not have to be obtained.

The fraction of solids s_c in the cake can be related to ε_{av} by

$$s_c = \frac{\rho_s(1-\varepsilon_{av})}{\rho_s(1-\varepsilon_{av}) + \rho\varepsilon_{av}} \qquad (4.14)$$

The ratio m of wet to dry cake frequently encountered in the literature is simply the reciprocal of s_c, i.e., $m = 1/s_c$.

4.4 Analytical Formulas for Average Porosity and Hydraulic Pressure Variation

The curves of Figure 4.4 indicate that a priori prediction of the effect of pressure on liquid content is difficult and not entirely dependent on the compressibility of the solid. An analytical expression relating ε_{av} to p is desirable as it will assist in determining the relative effects of p in the integrals shown in Equation 4.20. The power function approximations previously developed can be conveniently utilized for relating both ε and $(1-\varepsilon)$ to p_s.

In order to find the functional relationship between ε and x/L, the two integrals of Equation (4.7) must be evaluated. Beginning with the integral in the numerator, which will be called I, one can substitute for α and $1-\varepsilon$ in the following manner:

$$I = \int_0^p \frac{dp_s}{\alpha(1-\varepsilon)} = \int_0^{p_i} \frac{dp_s}{\alpha_i(1-\varepsilon)} + \int_{p_i}^{p_s} \frac{dp_s}{aBp_s^{n+\beta}} \qquad (4.14)$$

where α_i and ε_i are considered constants. Integrating and substituting limits one gets

$$I = \frac{p_s}{\alpha_i(1-\varepsilon_i)} + \frac{1}{aB} \frac{p_s^{1-n-\beta} - p_i^{1-n-\beta}}{n + \beta - 1} \qquad (4.15)$$

Substituting $\alpha_i = ap_i^n$ and $(1-\varepsilon_i) = Bp_i^\beta$ and rearranging Equation (4.15) one obtains

$$I = \frac{1}{aB} \frac{p_s^{1-n-\beta} - (n+\beta)p_i^{1-n-\beta}}{1 - n - \beta} \qquad p > p_i \qquad (4.16)$$

The integral of the denominator of Equation (4.7) has the same form as (4.16) except that the total pressure p replaces p_s.

It is also necessary to obtain a value for the integral of dp_s/α_x.

$$\int_0^{p_s} \frac{dp_s}{\alpha} = \int_0^{p_i} \frac{dp_s}{\alpha_i} + \int_{p_i}^{p_s} \frac{dp_s}{ap_s^n} \qquad (4.17)$$

Integrating and combining terms yields

$$\int_0^{p_s} \frac{dp_s}{\alpha} = \frac{p_i^{1-n}}{a} + \frac{p_s^{1-n} - p_i^{1-n}}{a(1-n)} \qquad (4.18)$$

$$= \frac{p_s^{1-n} - np_i^{1-n}}{a(1-n)}$$

If n is 0.5 or less, the term in p_i can be neglected for total pressure above 10 p.s.i. (70 kN/m²). However, if n is as large as 0.7, the p_i term must be included. When n becomes large, the power function approximation is less accurate, and numerical methods should be employed for the integration.

Substitution in Equation (4.7) yields

$$1 - \frac{x}{L} = \frac{p_s^{1-n-\beta} - (n+\beta)p_i^{1-n-\beta}}{p^{1-n-\beta} - (n+\beta)p_i^{1-n-\beta}} \tag{4.19}$$

$$= \frac{p_s^{1-n-\beta} - \text{const}}{p^{1-n-\beta} - \text{const}} \tag{4.20}$$

When p_i is sufficiently small, the constant term in the numerator and denominator can be eliminated to give

$$1 - x/L = (p_s/p)^{1-n-\beta} = (1 - p_L/p)^{1-n-\beta} \tag{4.21}$$

Elimination of p_s in favor of ε in (4.19) yields

$$1 - \frac{x}{L} = \frac{(E/\varepsilon)^{(1-n-\beta)/\lambda} - (n+\beta)p_i^{1-n-\beta}}{p^{1-n-\beta} - (n+\beta)p_i^{1-n-\beta}} \tag{4.22}$$

or for the more simple form in which p_i is neglected

$$1 - x/L = (\varepsilon/\varepsilon_1)^{-(1-n-\beta)/\lambda} \tag{4.23}$$

where ε_1 is the porosity at the medium where p_s reaches its maximum value p (if $p_1 = 0$).

Figure 4.4 compares (4.22) and (4.23) as applied to the data of Shirato and Okamura[6] for Hong Kong pink kaolin. Values used in making the comparisons are as follows: p = 42.2 lb./sq. in. (292 kn/m²), ε_i = 0.55 (experimental), E = 0.695 (calculated), n = 0.332 (calculated by Shirato and Okamura[6]), λ = 0.586 (calculated by Tiller and Cooper[8]), β = 0.095 (calculated by Tiller and Cooper), and p_i = 0.5 lb./sq.in Equation (4.22) and (4.23) are about equally good for representing the data through most of the filter solid. As the cake surface is approached however, Equation (4.22) gives better results, as would be expected since the p_i term was not neglected.

To obtain an analytical expression for the average porosity as a function of total pressure, substitution in (4.13) leads to

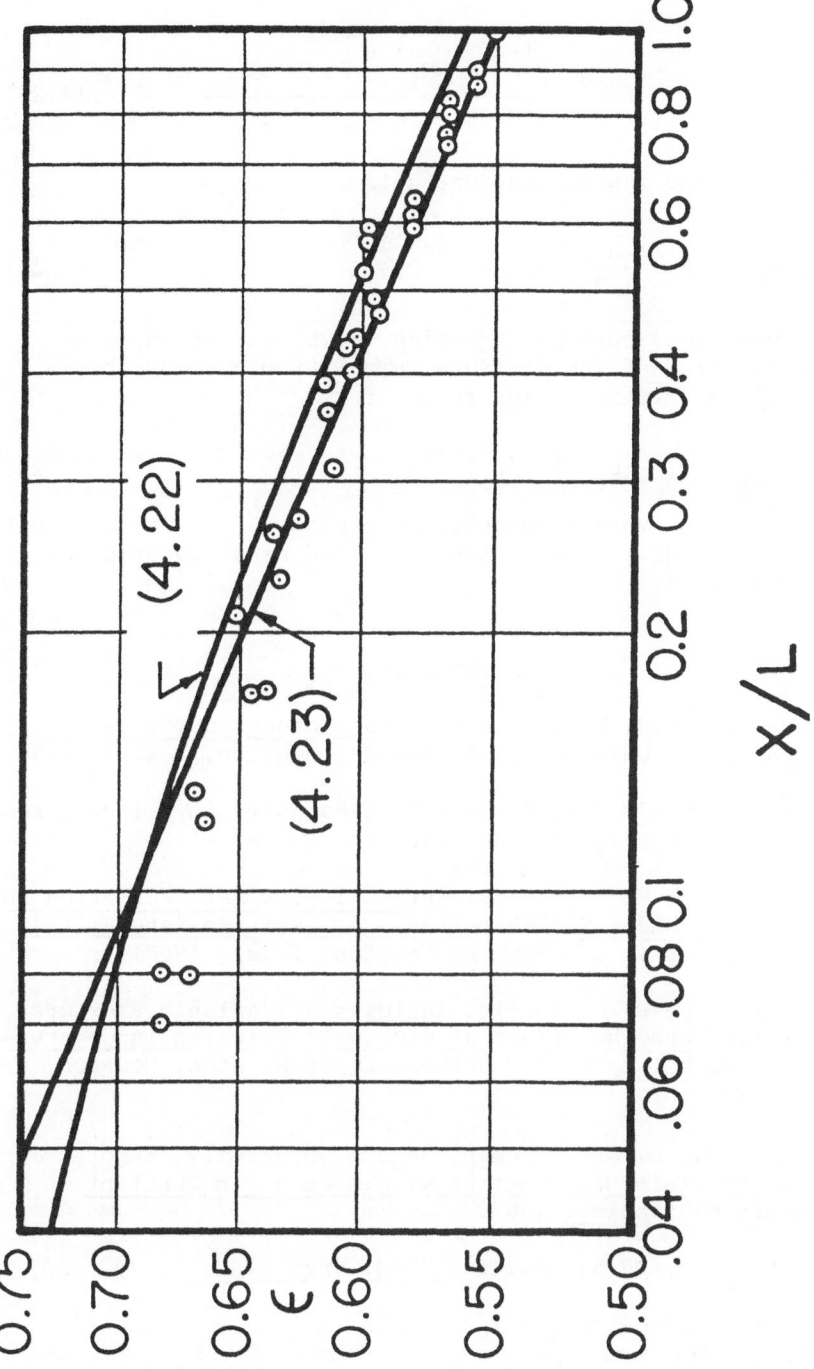

Figure 4.4 Comparison of Experimental
and Theoretical Values

$$\varepsilon_{av} = 1 - B \; \frac{1-n-\beta}{1-n} \; \frac{p^{1-n} - np_i^{1-n}}{p^{1-n-\beta} - (n+\beta)p_i^{1-n-\beta}} \qquad (4.24)$$

Neglecting p_i and rearranging one gets

$$1 - \varepsilon_{av} = B \; \frac{1-n-\beta}{1-n} \; p^{\beta} \qquad (4.25)$$

For practical purposes Equation (4.25) can be improved if the multiplier of the pressure term is replaced by an experimentally determined value in the form

$$1 - \varepsilon_{av} = (1 - \varepsilon_{avo})(p/p_o)^{-\beta} \qquad (4.26)$$

where ε_{avo} is an average porosity at pressure p_o. The exponent β may be derived from porosities obtained as a function of loading pressures.

References

1. Cooper, Harison, M. S. Thesis, Flow Through Compressible Porous Media, University of Houston, Houston, Texas, 1958.

2. Hutto, F. B., Jr., Distribution of Porosity in Filter Cakes, Chem. Eng. Progr., 53, 328 1957.

3. Lu, Wei-Ming, M. S. Thesis, Internal Flow Rate Variation in Filter Cake Under Variable-Pressure, Variable-Rate Filtration, University of Houston, Houston, Texas, 1964.

4. Lu, Wei-Ming, Ph.D. Thesis, Analysis of Variable Pressure Filtration and the Effect of Side-wall Friction in Compression-Permeability Cells, University of Houston, Houston, Texas, 1968.

5. Shirato, M., D. Eng. Thesis, Nagoya University, Nagoya, Japan, Liquid Pressure Distribution within Cakes in Constant Pressure Filtration, 1960.

6. Shirato, M., and S. Okamura, Chem. Eng. (Japan), 19, 105, 1955.

7. Shirato, M., M. Sambuichi, H. Kato, and T. Aragaki, Internal Flow Mechanism in Filter Cakes, AIChE J. 15, 405, 1969.

8. Tiller, F. M. and Cooper, Harrison, The Role of Porosity in Filtration, V, Porosity Variation in Filter Cakes, ibid, 8, 445, 1962.

9. Tiller, F. M. and M. Shirato, The Role of Porosity in Filtration, VI, New Definition of Filtration Resistance, ibid, 10, 61, 1964.

10. Tiller, F. M., Chapter 5, Theory and Practice of Solid-Liquid Separation, F. M. Tiller and P. J. Lloyd, editors, Chemical Engineering Department, University of Houston, Houston, Texas 77004.

5. CAKE FILTRATION

5.1 Introduction

This chapter will treat filtration processes involving cake formation. A variety of conditions will be considered including constant pressure, constant rate, and variable rate, variable-pressure operations. In general, the theory which will be developed will be applicable to plate and frame, between leaf Nutsch, belt, rotary vacuum, and other filters in which one-dimensional cakes are formed.

5.2 Types of Filtration Processes According to Pumping

The previous sections have dealt with the fundamentals of flow through filter cakes, concentrating on what takes place inside the filter. One must also consider the external restaints when calculating filtrate volume per unit area v and applied pressure p as functions of time t. External conditions imposed by the geometry of the filter and the pumping mechanism must be harmonized with an analysis of the internal portion of the cake.

This chapter will concentrate on the most common case of uni-dimensional, linear flow in compressible cakes. Approach to design and analysis of filtration depends upon whether or not the internal flow rate can be considered constant. When conditions are such that q undergoes negligible changes (favored by low concentration in slurry, long filtration periods, high values of ε_i), calculations are considerably simplified. However, in short filtrations with rapid buildup of cake, errors of 5 - 15% may result because of the assumption of constant internal flow rate.

When flow is radial or non-unidimensional, transmission of drag forces is considerably modified, thereby complicating relationships between permeability, flow resistance, and compressive pressure. There are no published experimental data showing the effect of non-linear accumulated drag on α and ε. Much theoretical and experimental investigation remains to be done.

The rate of flow from the cake, $q_1 = dv/dt$, is determined by combining the overall cake and medium resistance with the pressure-flow rate characteristics of the pumping mechanism. Once the rate is determined as function of the amount of cake deposited and the volume of filtrate, the time may be obtained by numerical integration. It is possible to develop analytical expressions for constant-rate and constant-pressure operations.

For purposes of mathematical treatment, filtration processes are classified according to the variation of the pressure and flow rate with time. Generally, the pumping mechanism determines the flow characteristics and serves as a basis for division into the following categories:

1. Constant pressure filtration. Actuating mechanism is compressed air maintained at a constant pressure.

2. Constant rate filtration. Positive displacement pumps of various types are employed.

3. Variable-pressure; variable-rate filtration. The use of a centrifugal pump results in the rate varying with the back pressure on the pump.

4. Stepped pressure. For experimental purposes, it is possible to manually increase pressure during a filtration and simulate various pumping conditions.

Flow rate vs. pressure characteristics for the four types of filtration are illustrated in Figure 5.1. Arrows drawn on the curves point in the direction of increasing time. The constant pressure curve is represented by a vertical line, the downward arrows indicating that the rate decreases with time. Drawn horizontally, the constant-rate filtration curve has arrows pointing to increasing pressure with time. The rate for a filter actuated by a centrifugal pump will follow the downward trend of the variable-pressure; variable-rate curve. Depending upon the characteristics of the centrifugal pump, widely differing curves may be encountered. If the first portion of the curve is nearly flat, the pump will produce a filtration which is almost at constant rate. The dotted curve is approximately

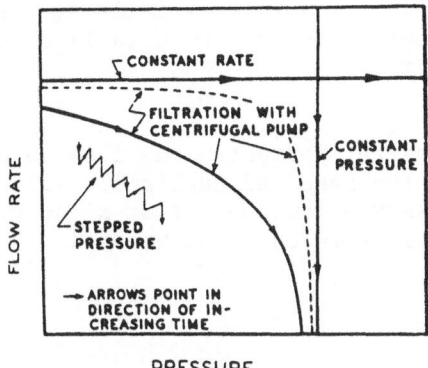

Figure 5.1 Relation of Flow Rate to Pressure
for Different Methods of Operation

equivalent to a filtration carried out first at constant rate
and then at constant pressure. In the literature, by far the
greatest attention has been focused on constant-pressure filtration. While a small amount of effort has been directed
toward the more significant constant-rate filtration, the in-
dustrially important area of variable-pressure, variable-rate
filtration has been virtually untouched.

5.3 Filtration Resistance

In general the filtration resistance α_{av} of Ruth has been
defined as

$$\alpha_{av} = \frac{p - p_1}{\int_0^{p - p_1} dp_s/\alpha} \tag{5.1}$$

where the pressure p_1 required to overcome the medium resistance
is given by

$$p_1 = \mu q_1 R_m \tag{5.2}$$

The value of q_1 is the flow rate at the exit of the cake
where $x = 0$. If p_1 is small, the value of α_{av} depends only on
the total pressure p, and the volume vs. time relation will be
parabolic for constant pressure filtration. Frequent lack of
good ($\pm$ 20%) agreement and occasional wide variation in calculated

and experimental value of α are partially accounted for by weaknesses in the definition given in (5.1) as well as by experimental difficulties.

The definition of average filtration resistance in (5.1) is based upon a constant instantaneous flow rate (but varying with time) through the cake and negligible velocity of the solids. If q is assumed independent of x and equal to q_1 and r = 0, then the basic flow equation becomes

$$\frac{dp_s}{dw_x} = - \mu \alpha q_1 \tag{5.3}$$

Rearranging and integrating

$$\mu q_1 \int_0^w dw_x = \mu q_1 w = \int_0^{p-p_1} dp_s/\alpha \tag{5.4}$$

Dividing each side of (5.4) into $p - p_1$ leads to

$$\alpha_{av} = \frac{p - p_1}{\int_0^{p-p_1} dp_s/\alpha} = \frac{p - p_1}{\mu q_1 w} \tag{5.5}$$

Substituting (5.2) for p_1 in (5.5) and solving for q_1 produces

$$q_1 = \frac{dv}{dt} = \frac{p}{\mu(\alpha_{av} w + R_m)} \tag{5.6}$$

To develop a better definition of an average α based on variable q and r, Shirato, et. al.[6,7], started with

$$\frac{dp_s}{dw_x} = - \mu \alpha (q - er) \tag{5.7}$$

This equation is placed in the form:

$$\int_0^{p-p_1} dp_s/\alpha = \mu \int_0^w (q - er) dw_x \tag{5.8}$$

The right-hand side of (5.8) can be placed in the form:

$$\mu q_1 w \int_0^w \left[\frac{q}{q_1} - e \frac{r}{q_1} \right] d(w_x/w) \tag{5.9}$$

The value of the integral is the area under the curve of the integrand in Equation (5.10) plotted against w_x/w from zero to one and is called J_s which will be less than one because the integrand in Equation (5.11) is less than unity. Thus, Equation (5.8) can be placed in the form:

$$\int_0^{p-p_1} dp_s/\alpha = \mu J_s q_1 w \tag{5.10}$$

Dividing both sides into $(p - p_1)$, replacing p_1 by (5.2) and solving for q_1, yields:

$$q_1 = \frac{dv}{dt} = \frac{p}{\mu(J_s \alpha_{av} w + R_m)} \tag{5.11}$$

which is the customary form for presenting the filtration equation, except that α_{av} is multiplied by J_s. A further modification of α will be made later with respect to surface area growth in leaf filters.

An experimental determination of average filtration resistance finds the factor $J_s \alpha_{av}$ rather than α_{av} as commonly reported in the literature. The factor J_s is always less than unity but only significantly so for highly concentrated slurries and filtrations lasting a short period of time.

The idea of a concentrated slurry is of importance. Frequently, certain ranges of slurry concentration are said to favor particular equipment and process operations. However, the percentage of solids which forms a "thick" or "concentrated" slurry necessarily depends upon the percent solids in the cake. The average percent solids is s_c, and the value at the surface $(x = L)$ is s_i. A slurry becomes "solid" when the slurry concentration equals s_i and mechanical dewatering is the operation rather than filtration. Thus, the values of s/s_c or s/s_i are the factors which determine whether or not the slurry is concentrated.

While there is not a great deal of information available at the present time, it appears that when s/s_i is less than 0.6 the slurry can be treated as "dilute." In Figure 5.2, a plot of J_s[6,7] as a function of s is shown. Up to concentration of 0.4, the J_s curve is relatively flat.

5.4 Dimensions of the Filtration Resistance

The dimensions of the specific resistance α and the filter

366

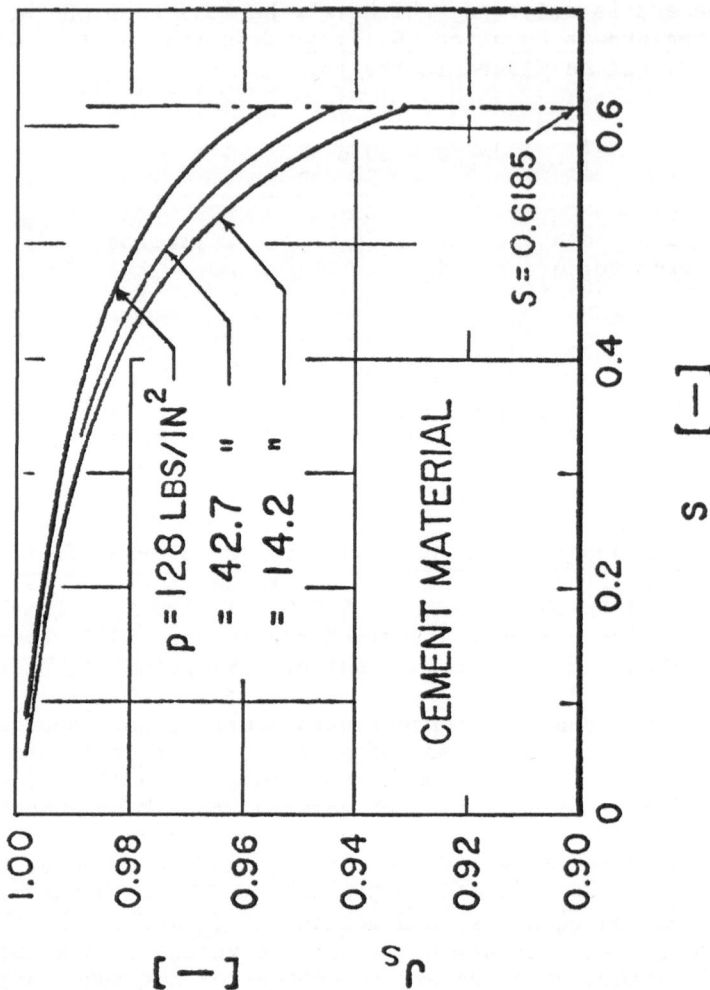

Figure 5.2 J_s vs. Concentration

medium resistance R_m are of interest. The medium resistance
has the dimension of reciprocal length and α has the dimensions
L/M. The order of magnitude of α in either m/kg or ft/lbm is
approximately 10^{10}

Considerable confusion has arisen in the literature concerning the units and numerical values of α. Many authors
have employed inconsistent units and have reported values which
are not in accord with the best engineering practice. Because
of this confusion it is sometimes rather difficult to compare
the numerical results of different investigators.

It is not infrequent that the specific resistance has been
expressed in terms of T^2/M. With such dimensions, w must be
expressed as weight instead of mass. Generally authors using
the T^2/M basis have not made a clear distinction between force
and mass. Gale[2] discussed the units of sec.^{2}/gram normally
used in the field of sewage sludge. He pointed out that
$\alpha(sec.^2/g)$ must be multiplied by g(981 cm./sec.^{2}) to obtain
$\alpha(cm./g.)$.

A summary of units and conversion factors follows

$$\alpha(ft/lbm.) = 14.9 \ \alpha(cm/g) = 1.49 \ \alpha(m/kg)$$

$$= 14,612 \ \alpha \ (sec^2/g)$$

In Figure 5.3B a filter cake is illustrated in which flow
of fluid takes place from left to right, and distance is measured from the medium. As the liquid flows frictionally through
the compressible, porous media, p_L drops until it reaches the
value p_1 at the interface of the cake and supporting medium.
The relationship of time to pressure p_1 at the medium and the
pressure drop across the cake $(p-p_1)$ is illustrated in Figure
5.3B for talc filtered at a constant pressure of 5 lb./sq.in
(34.5 kN/m^2). Initially when there is no cake, the entire
pressure drop is across the medium and $p=p_1$. As dv/dt decreases
with time, p_1 falls in accord with Equation (5.2) and the pressure drop $(p-p_1)$ across the cake builds up.

In general a medium should be chosen to give a minimum resistance consistent with the production of satisfactory clarity.
Grace[3] has indicated that a filter medium exhibiting a resistance
equivalent to no more than 0.01 in. of cake can usually be
selected with the result that the pressure drop across the medium
becomes a negligible portion of the total pressure drop for
a major portion of the filter cycle.

368

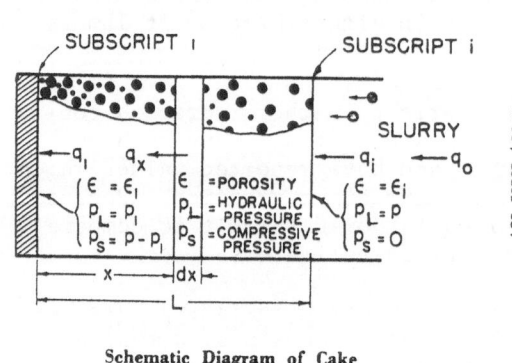

Schematic Diagram of Cake

Figure 5.3A

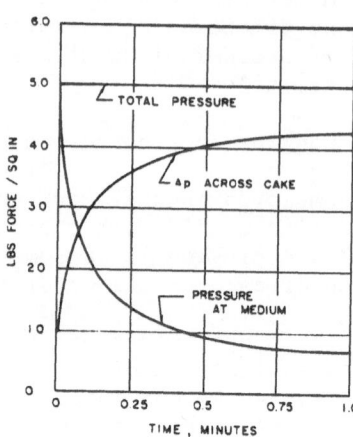

Variation of pressure drops across cake
and medium.

Figure 5.3B

The thickness of the cake is given by the mass of dry
cake per square foot divided by the dry cake density $\rho_s(1-\varepsilon_{av})$.
The ratio R_m/α represents the mass of dry solids per square
foot equivalent to the resistance of the medium. Consequently
the equivalent thickness of the medium is given by

$$L_{eq} = \frac{R_m}{\alpha \rho_s (1-\varepsilon_{av})} = \frac{R_m}{\alpha w} L \qquad (5.12)$$

As ε_{av} decreases with increasing time, the equivalent cake
thickness is a maximum at the start of filtration; however it
rapidly decreases to an approximately constant value. For
the example illustrated in Figure 5.3B the initial equivalent
medium thickness is 0.41 in., while the ultimate value approached
is 0.05 in.

In Figure 5.4 calculated values of the specific resistance
are plotted against the time for the constant pressure filtration of talc. The curve for average α starts at a low value,
corresponding to zero compressive pressure in the first layer
of cake, and rises very rapidly. The limiting value of the average α corresponds to the condition inwhich p_1 has reached
a negligible value.

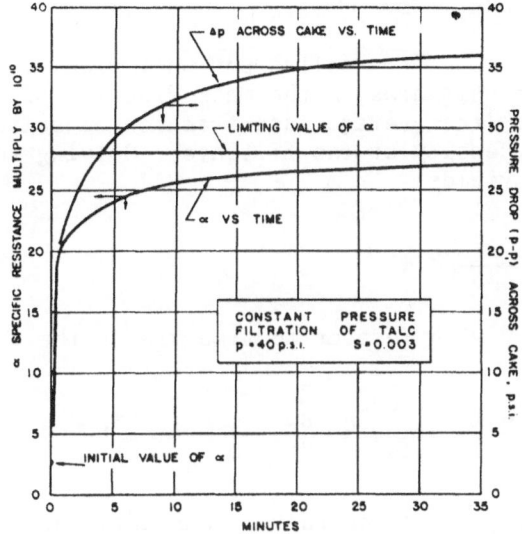

Figure 5.4   Filtration Resistance vs. Time
in Constant-pressure Filtration

5.6   Simplified Constant Pressure Filtration

Before treating the general case in which the specific
filtration resistance is considered variable, the more simple
situation will be treated in which the porosity is considered
constant. If the porosity is constant, α in Equation (5.11)
will be constant; and J_s = 1. Simplified equations frequently
encountered in the literature result from these assumptions.
However, it should be recognized that there are limitations
which have seldom been recognized in the analysis of experimental data. The limitations will be discussed after the conventional equations have been derived.

Both internal and external material balances must be
written relating the rate of flow to other variables. In most
derivations, it is assumed that the rate of liquid flow through
the cake is constant and that the solids velocity is zero or
negligible. Both assumptions are valid for dilute but not
concentrated slurries.

From an over-all viewpoint, a material balance can be
written on a unit area basis in the form:

Mass of slurry = mass of cake + mass of filtrate:

$$w/s = w/s_c + \rho v \tag{5.13}$$

where w is the mass of dry solids per unit area, v the filtrate volume per unit area, s the mass fraction of solids in the slurry, s_c the average mass fraction of solids in the cake, and ρ the density of the filtrate. Solving for v in Equation (5.13) yields:

$$v = \frac{1 - s/s_c}{\rho s} w \tag{5.14}$$

Differentiating v with respect to time yields the flow rate q_1 of filtrate[10]

$$q_1 = \frac{dv}{dt} = \frac{1 - s/s_c}{\rho s} \frac{dw}{dt} + \frac{w}{\rho} \frac{1}{s_c^2} \frac{ds_c}{dt} \tag{5.15}$$

For the most part, s_c is considered constant, and ds_c/dt is placed equal to zero. In general, s_c varies continuously in any filtration in which the pressure continuously rises. In constant-pressure filtration of talc, latex, and calcium carbonate, s_c or m reached a constant value in less than 0.5 min.[8]. For long batch filtration, a variation over a short period of time would be of little significance and could be neglected. However, in a rotary drum filtration where the filtration time for a 120-deg. submergence at 0.5 rpm. is 40 sec., it is not possible to assume s_c or m constant. In general, sophisticated numerical methods are required for continuous rotary filtration.

It is possible to eliminate w by combining (5.11) with (5.14). Thus

$$\frac{dv}{dt} = \frac{p}{\mu \left[\frac{\alpha s \rho}{1 - s/s_c} v + R_m \right]} \tag{5.16}$$

Separating the variables and integrating produces

$$\frac{\mu R_m}{p} v + \frac{\alpha s \rho \mu}{p(1-s/s_c)} \frac{v^2}{2} = t \tag{5.17}$$

This equation may be conveniently written as

$$bv + v^2/K = t \tag{5.18}$$

The constants in Equation (5.17) may be determined from the equation in the form

$$b + v/K = t/v \qquad (5.19)$$

It is also possible to work directly with (5.16) in the form

$$b + 2v/k = dt/dv \qquad (5.20)$$

Equation (5.20) is superior to (5.19) in that dt/dv measured at the midpoint $(t_1 + t_2)/2$ of the interval in question can be replaced by $\Delta t/\Delta v$. For parabolic relations the derivative taken at the midpoint is precisely equal in slope to the secant regardless of the size of interval. Equation (5.19) depends upon knowing absolute values of time and volume while (5.20) involves only increments. As it is somewhat difficult to decide upon the point at which t = 0 in some experiments, it is convenient to have an expression which is independent of the starting point.

5.7 Neglecting Medium Resistance in Constant Pressure Filtration

Taking Equation (5.17) as representing the volume vs. time relationship for a constant pressure filtration, it can be noted the v^2/k term will be small in the initial stages compared to the linear bv term. This is to be expected since the latter quantity contains the resistance of the septum which controls the filtration at the beginning of the process. However, as time becomes greater the quadratic term in v increases rapidly and becomes controlling in determination of the time. As the cake continues being deposited, the resistance of the medium becomes a decreasing part of the total resistance. Thus if a filtration is carried out over a sufficiently long period of time, the medium resistance R_m, may be neglected leading to the following expression as an approximation:

$$v^2 = Kt = \frac{2p(1-s/s_c)}{\mu\rho s\alpha} \, t \qquad (5.21)$$

In terms of the mass deposited per unit area w, the equation becomes

$$w^2 = \frac{2p\rho_s}{\alpha\mu(1-s/s_c)} \, t \qquad (5.22)$$

Another important equation can be written in terms of the thickness as a function of time

$$L^2 = \frac{2p}{\alpha\mu\rho_s(1-\varepsilon_{av})^2(1-s/s_c)} \, t \qquad (5.23)$$

w is related to L by

$$w = \rho_s(1 - \varepsilon_{av}) \, L \qquad (5.24)$$

and s_c is related to ε_{av} by

$$s_c = \frac{\rho_s(1-\varepsilon_{av})}{\rho_s(1-\varepsilon_{av}) + \rho\,\varepsilon_{av}} \qquad (5.25)$$

If R_m is neglected, it is apparent that the calculated time of filtration for a definite volume of filtrate will be smaller than the calculated time including the medium resistance. Subtracting Equation (5.22) from an equation in w equivalent to (5.17) gives the error which results from neglecting R_m. Thus

$$\text{error in } t = (\frac{\mu R_m}{p}) \, \frac{1-s/s_c}{\rho_s} \, w \qquad (5.26)$$

The fractional error is given by the error in Equation (5.26) divided by t. Thus

$$\text{fractional error} = \frac{1}{1 + \frac{w}{2}\frac{\alpha}{R_m}} \qquad (5.27)$$

If the percentage error is to be less than 5%, then $\alpha w/2R_m$ must be approximately 20 or greater; therefore

$$w > 40 \, R_m/\alpha \qquad (5.28)$$

The term R_m/α has dimensions of M/L^2. Grace[3] has pointed out that except for very thin cakes an approximation can be made in which $R_m/\alpha = 0.5$ kg/m^2 if the proper type of medium is employed. An analysis of the relatively meager data presented by Grace indictaes that R_m/α might be closer to 0.1 - 0.2. Since the septum resistance varies greatly with age and treatment, care must be observed in making this approximation. However assuming that R_m is one half of α, Equation (5.28) states that w should be greater than 20 kg/m^2 if R_m is to be neglected. For a solid with a specific gravity of 2.5 and a cake porosity of 0.5, $\rho_s(1 - \varepsilon_{av}) = (2500)(0.5) = 1250$ kg/m^3. Dividing 20 by 1250 gives 1.6 cm. of cake as the thickness necessary to reduce the error in using Equations (5.21) - (5.23). Caution should be used in employing the approximations.

5.8 Constant Pressure Filtration With Variable Filtration Resistance

Throughout filtration literature, experimentors have assumed that α was constant during a constant pressure filtration; and consequently, both α and R_m could be obtained from a straight-line plot of t/v vs. $\bar{V}$. Neither α or s_c is constant except when $R_m = 0$; and commonly accepted theory must be carefully evaluated. Returning to Figure 5.3, it is apparent that the pressure drop across the cake changes with time. As α depends upon the pressure drop across the cake and not the total pressure drop, it is obvious that α must vary. A typical case is illustrated in Figure 5.4 where α is shown to vary markedly with time. Clearly it is necessary to review conventional theory.

If ε_{av} (and s_c) vary, it is necessary to modify (5.14) utilizing the value of ε_{av} as given by Equation (4.12) or (4.13). The quantity $(1 - s/s_c)$ can be changed in Equation (5.14). First finding s_c in terms of ε_{av}

$$1/s_c = m = 1 + \frac{\rho \varepsilon_{av}}{\rho_s (1 - \varepsilon_{av})} \tag{5.28}$$

where m is the ratio of the mass of wet to mass of dry cake. Substituting (4.13) for ε_{av} in (5.28) yields

$$m = 1 - (\rho/\rho_s) + (\rho/\rho_s) \frac{\int_o^{p-p_1} dp_s/\alpha \,(1 - \varepsilon)}{\int_o^{p-p_1} dp_s/\alpha} \tag{5.29}$$

This equation assumes that J_s will have little effect on m. For constant pressure filtration p is constant, but p_1 varies. With the upper limit, $p - p_1$, variable in both integrals, s_c must also vary. Eliminating w between Equations (5.10) and (5.14) and solving for v

$$v = \frac{(1 - s/s_c)}{J_s \mu \rho s q_1} \int_o^{p-p_1} dp_s/\alpha \tag{5.30}$$

374

Substituting (5.29) in (5.30) yields

$$J_s v q_1 = \frac{\sigma - s(\sigma - 1)}{\mu \rho_s s} \int_0^{p-p_1} \frac{dp_s}{\alpha} - \frac{1}{\mu \rho_s} \int_0^{p-p_1} \frac{dp_s}{\alpha(1-\varepsilon)} \qquad (5.31)$$

where σ is the specific gravity of the dry solids. This equation[10] can be viewed as an improvement over the usual differential equation presented for constant pressure filtration. The last term in Equation (5.31) represents a correction factor for the variation of the average liquid content of the cake. When the cake porosity approaches a limiting average value, the last term combines with the first to yield the conventional filtration equation. Calculations[10] with the exit rate and based on (5.31) are illustrated in Figure 5.5 for talc at 5.0 and 15 lb./sq.in. and polystyrene latex at 10 lb./sq.in. The curves are plotted in the form of dt/dv vs. v; and according to conventional theory, such curves should yield straight lines if α, s_c, and R_m were constant. Variations in filtration resistance and average porosity during the initial period cause a marked deviation from straight-line plots. If the linear portions of the curves are extrapolated to the vertical axis, incorrect medium resistances will result. In the case of latex it would appear that the resistance was zero from an extrapolation. In evaluating medium resistances from experimental data great care should be employed in extrapolating dt/dv vs. v data to zero volume.

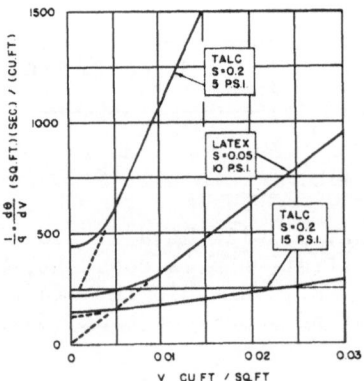

Figure 5.5   Reciprocal Rate vs. Volume

Porosity varies throughout the cake as a function of time and thickness as illustrated in Figure (5.6), where x for constant values of t. Data used for the graph represent calculated values and are purely illustrative. The first infinitesimal layer at the cake surface has a porosity and specific resistance corresponding to zero compressive pressure. At each instant of time the porosity drops throughout the cake until the medium is reached where it has its least value. The cake thickness increases, and at a given distance from the medium the porosity decreases as time goes on. As the pressure drop across the cake increases, the porosity at the medium decreases and eventually reaches a minimum value at B equal to a porosity determined by the maximum applied pressure. Obviously as ε_{av} decreases, s_c will increase.

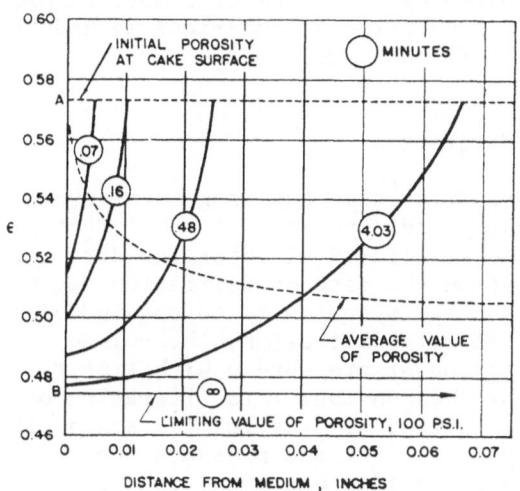

Figure 5.6 Porosity vs. Distance from Medium

5.9 Analytical Equations for Constant Pressure Filtration

Equation (5.30) and (5.31) can be integrated if α and $(1-\varepsilon)$ are assumed to be power functions of p_s in accord with Equations (2.7) - (2.12). Substitution of Equation (4.18) in (5.30) yields

$$\frac{dv}{dt} = \frac{(1 - s/s_c)}{\mu s \rho a (1 - n)} \frac{(p - p_1)^{1-n} - np_i^{1-n}}{J_s v} \qquad (5.32)$$

As p_l is a function of dv/dt, Equation (5.32) cannot be directly integrated. For the case in which p_i and p_l are negligible, it can be integrated to give

$$\frac{v^2}{2} = \frac{(1 - s/s_c)}{\mu s \rho a J_s} \frac{p^{1-n}}{1-n} t \qquad (5.33)$$

Equation (5.31) can be integrated using (4.16) and (4.18). If p_l and p_i are neglected, there results

$$J_s v q_1 = \frac{\sigma - s(\sigma-1)}{\mu \rho_s s} \frac{p^{1-n}}{a(1-n)} - \frac{1}{\mu \rho_s} \frac{p^{1-n-\beta}}{aB(1-n-\beta)} \qquad (5.34)$$

The left hand side can be integrated to give $J_s v^2/2$ which then equals the RHS multiplied by t. In general more complex analytical formulas are not justified as numerical methods are more appropriate for more difficult problems.

5.10 Constant Rate Filtration

Constant pressure filtration has long been favored in both university teaching and experimental investigations, presumably because of the simplicity of obtaining data. In constant pressure filtration the rate of filtration decreases as the cake builds up, and it is customary to obtain volume vs. time data. In constant rate filtration, the volume is linear in time; and the variation of pressure p with time t is studied. In Figure 5.7 data for the constant rate filtration of two paper slurries are exhibited. In the initial stages the pressure increased slowly, but after the pressure reached 15 lb. force/sq.in. the rate of buildup was much greater. In Figure 5.8 constant rate filtrations for a number of different laboratory investigations are illustrated. The curves are similar to those shown in Figure 5.7 except that the time is much shorter. In general, in a constant rate filtration, the initial linear pressure rise is followed by a curve of increasing upward curvature. With slightly compressible materials, this approximate linearity may extend over a considerable pressure range; while with highly compressible materials, there may be a large curvature in the pressure vs. time curve from the beginning.

The volume is related to the time by

$$v = q_1 t \qquad (5.34)$$

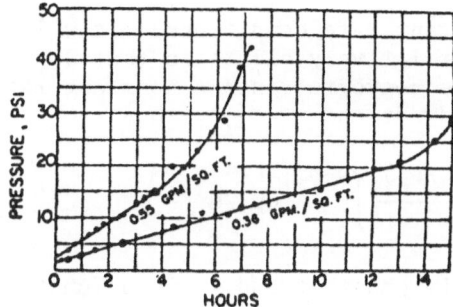

Figure 5.7 Pilot Plant Runs for Paper Slurry

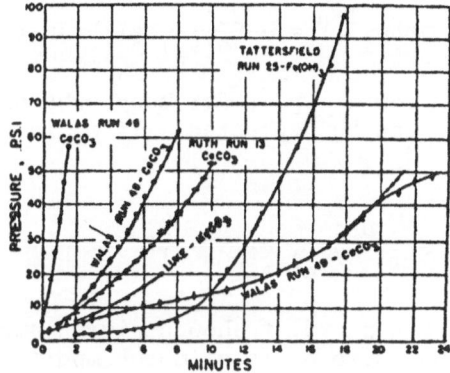

Figure 5.8 Pressure vs. Time for Laboratory
Constant Rate Filtrations

If v is eliminated from (5.16) and dv/dt is replaced by q_1, there results

$$p - p_1 = \frac{\alpha\mu\rho s}{1 - ms} q_1^2 t \qquad (5.35)$$

As J_s = 1 when α is constant, it has been eliminated. Equation (5.35) indicates a linear variation of p with t. Although the assumption of constant α is legitimate for constant pressure filtration when R_m is small, the same assumtion for constant rate filtration is not justified as indicated by the data in Figures 5.7 and 5.8. Therefore it is necessary to utilize either (5.30) or (5.31) in developing analytical equations for filtration under constant rate conditions. It will be assumed that the solution is relatively dilute so that $J_s = 1$ and Equation (5.30) can be converted into a constant rate equation by substituting $q_1 = dv/dt$ and $v = q_1 t$, thus

$$q_1^2 t = \frac{(1 - s/s_c)}{\rho s \mu} \int_0^{p-p_1} dp_s / \alpha \qquad (5.36)$$

In this equation, p_1 is constant and s_c is a function of p. The time t is a function of p both because it appears in the upper limit of the integral and because s_c depends on p.

Combining Equation (5.36) with (4.18) yields

$$(p - p_1)^{1-n} - n p_i^{1-n} = \frac{\mu s \rho}{(1-s/s_c)} a(1-n) q_1^2 t \qquad (5.37)$$

This equation is valid for $p \geq p_i$. For many substances where n is less than 0.5, p_i can be neglected. If p_i is omitted (5.37) becomes

$$(p - p_1)^{1-n} = \frac{\mu s \rho}{(1-s/s_c)} a(1-n) q_1^2 t \qquad (5.38)$$

A logarithmic plot of $p - p_1$ vs. t should give a straight line. At the low pressure end where p_i is of more importance, the plot should lie above the straight line predicted by (5.38)

The average specific resistance with $J_s = 1$ is defined by

$$\frac{p - p_1}{\alpha_{av}} = \int_0^{p-p_1} dp_s / \alpha \qquad (5.39)$$

If the integral is replaced by Equation (4.18), α_{av} can be placed in the form

$$\alpha_{av} = \frac{a(1-n)(p-p_1)}{(p-p_1)^{1-n} - n p_i^{1-n}} \qquad (5.40)$$

If p_i can be neglected, then

$$\alpha_{av} = a(1-n)(p-p_1)^n \qquad (5.41)$$

Equations (5.37) and (5.38) can be rewritten in exactly the
form of Equation (5.35) <u>provided</u> α_{av} is used and is taken as
a function of $p - p_1$. The value of α_{av} is precisely the same
for constant pressure and constant rate filtration when the
pressures are equal.

In Figure 5.9 data are shown for a number of constant
rate runs made with aqueous slurries of kaolin. The percentage of solids was varied from 0.002% to 1.0%, and the rate
of filtration ranged from 2 - 11 cu.ft./(hr.)(sq.ft.). Final
pressures ranging from 30 to 100 lb. force/sq. in. were recorded with the total filtration time ranging from 10 min.
to 2 hr. In general it was possible to obtain reasonable
checks on the data, except when the time required to reach
100 lb. force/sq.in. was less than 10-15 min. Because of the
difficulty of obtaining p_1 (ranging from 0.1 to 0.5 lb. force/
sq.in.) and the initial time exactly, the logarithmic plots
are not too reliable in the region below 10 psi and 10 min.
When the log $(p - p_1)$ vs. log t plot is used, the p_1 term in
Equation (5.38) is neglected when a straight line is drawn.
Inclusion of the p_1 term would yield a negative correction
to the pressure which might be significant at low pressure.
In the low pressure range, all the data lie above the straight
line, thereby indicating that a negative correction would
being the points closer to the line predicted by theory. The
data of Figure 5.9 can be represented by the equation

$$(p - p_1)^{0.56} = 6.98(10)^9 s\mu q_1^2 t \tag{5.42}$$

when the slurry is dilute and the pressure difference is above
approximately 8.0 lbf./sq.in. Equation (5.42) utilizes pressure in p.s.i. To change to lbf./sq.ft., the right-hand side
must be multiplied by 7.73.

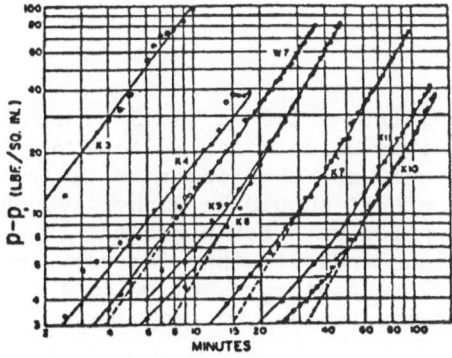

Figure 5.9 $p-p_1$ vs. Time

Above an n of 0.5 - 0.7 the approximations used in deriving
Equation (5.42) lose accuracy. Nevertheless, useful qualitative information can be obtained by manipulation of the equations for highly compressible materials for which n may be
greater than unity. The time t_i required to reach a pressure
drop across the cake of $\Delta p_c = p - p_1 = p_i$ can be calculated
using (5.37) and assuming $\bar{\alpha} = \alpha_i$. The ratio of the time t
to t_i is given by

$$\frac{t}{t_i} = \frac{\Delta p_c^{1-n} - n p_i^{1-n}}{(1 - n) p_i} \tag{5.43}$$

Plots of Δp_c vs. t/t_i at constant values of n are illustrated
in Figure 5.10 and 5.11.

The curves of Figure 5.10 are similar to those previously
shown in Figure 5.7 and 5.8. For n < 0.1, a straight line
may be used to approximate the data over a considerable pressure range. When n increases to 0.3, it is difficult to represent the pressure vs. time as a linear function beyond
5-10 psi. When n reaches values near 0.7, the curves are
characterized by an initial flat portion followed by a rapidly
increasing slope. For n greater than unity (5.43) is best
placed in the form

$$\frac{t}{t_i} = \frac{1}{(n - 1) p_i} \left[\frac{n}{p_i^{(n-1)}} - \frac{1}{\Delta p_c^{(n-1)}} \right] \tag{5.44}$$

A plot of Δp_c vs. t/t_i for n > 1.0 is shown in Figure 5.11.
The curves, although qualitative in nature, indicate an asymptotic approach to infinite pressure at a finite time. In
Equation (5.44), as Δp_c approaches infinity, the last term becomes zero, and the ratio of t/t_i is given by

$$\frac{t}{t_i} = \frac{n}{n - 1} p_i^{-n} \tag{5.45}$$

For example, if n = 2.0, t/t_i = 200 in (5.45), indicating
that the pressure will rise asymptotically to a very large
value when the time reaches two hundred times the value required to reach p_i. Again it should be emphasized that the
calculations are only approximate for n greater than 0.7.
Practically, the equations indicate that for very compressible
materials filtered at a constant rate, there is a finite time
at which the pressure rises rapidly to large values.

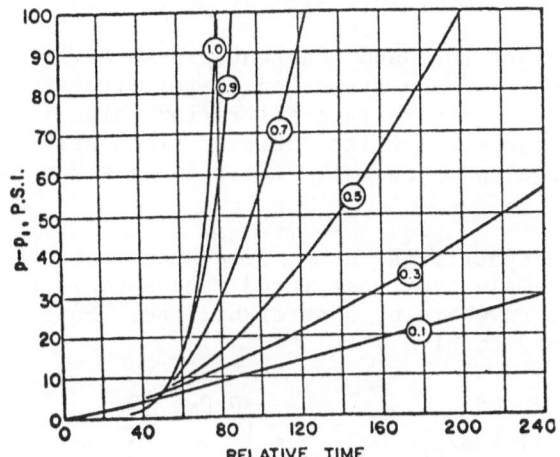

Figure 5.10
Effect of Compressibility on Pressure vs. Time
Curves for Constant Rate Filtration

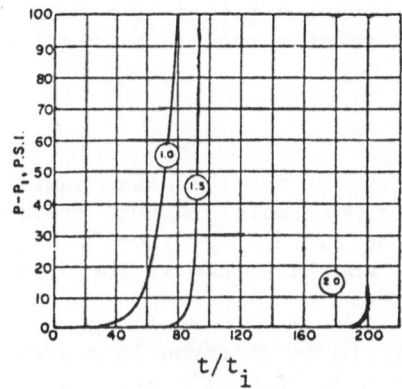

Figure 5.11
Pressure Rise vs. Ratio of Filtration
Time to Time Required to Reach p_i

5.11 Variable-Pressure, Variable-Rate Filtration

While constant rate and constant pressure filtrations have received most of the attention in the literature, it is variable pressure-variable rate filtration which is of most practical importance. A filter unit is frequently operated with a centrifugal pump in which the rate is a function of the back pressure.

Consider a centrifugal pump having the characteristics of Figure 5.1. Only over restricted ranges could the curve be approximated anywhere by a straight line. Equation 5.10 can be rearranged to give

$$^w q_1 = \frac{\rho s}{1 - s/s_c} ^v q_1 = \mu J_s \int_0^{p-p_1} \frac{dp_s}{\alpha} \qquad (5.46)$$

Equation (5.46) states that the product of the mass of dry solids per unit area multiplied by the rate must equal the integral on the right-hand side, which is itself a function of both applied pressure p and pressure at the septum p_1. When graphical or numerical methods are used to solve Equation (5.46), the volume is first obtained as a function of q_1 and p. Then knowing v in relation to q_1, t is gotten from:

$$t = \int dv/q_1 \qquad (5.47)$$

Equations (5.46) and (5.47) are the basic equations needed to solve variable-rate, variable-pressure filtration problems. They also reduce to the formulas previously developed for constant pressure and constant rate if the appropriate restrictions are utilized.

Example. Talc is to be filtered in a press with a centrifugal pump. The pump characteristics, filtration resistance, and parameters necessary to the solution are given in Table 1. Other values necessary to the solution are

$$
\begin{aligned}
\mu &= 0.001 && \text{lb. mass/(ft.)(sec.)} \\
\rho &= 62.4 && \text{lb. mass/cu.ft.} \\
\rho_s &= 167 && \text{lb. mass/cu.ft.} \\
s &= 0.003 && \text{fraction solids in slurry} \\
R_m &= 2.0(10^{10}) && \text{ft.}^{-1} \\
m &= 2.5 && \text{(approximate average value)}
\end{aligned}
$$

TABLE 1

Pump Characteristics			Filtration Resistance	
q_g gal./(min.)(sq.ft.)		lb./sq.in.	α ft./lb.mass	lb./sq.in.
0		43.5	$1.057(10^{11})$	1.10
0.1		42.6	1.39	2.23
0.2		38.4	1.97	4.50
0.3		32.3	2.51	7.9
0.4		24.1	3.03	11.3
0.5		10.3	3.67	17.1
0.57		0	4.19	22.7
			4.78	28.3
			6.46	47.9

The values chosen for the medium resistance is of a reasonable order of magnitude and corresponds to values which might be encountered in practice. It is convenient to change from seconds to minutes, cubic feet to gallons, and pounds force/square foot to pounds force/square inch as follows:

$$t = 60\, t_m \tag{5.48}$$

$$q = q_g/(60)(7.48) = 0.00226 q_g \tag{5.49}$$

$$v = v_g/7.48 \tag{5.50}$$

$$p = 144\psi \tag{5.5.}$$

Substituting in (5.46)

$$\frac{v_g}{7.48} = \frac{(32.17)(1 - 0.0075)(144)}{(0.001)(0.003)(62.4)(0.0026)q_g} \int \frac{d\psi_s}{\alpha} \tag{5.52}$$

$$v_g = \frac{8.13(10^{10})}{q_g} \int_o^\psi \frac{d\psi_s}{\alpha} \tag{5.53}$$

A plot of ψ_s vs. $1/\alpha$ is shown in Figure 5.12. The pressure p_1 required to overcome the resistance of the medium is obtained from

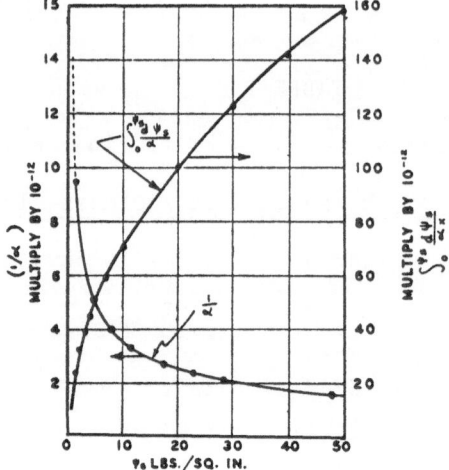

Figure 5.12 ψ_s vs. $1/\alpha$

$$p_1 = \mu q_1 R_m \tag{5.54}$$

or in terms of psi

$$= \frac{(0.001)(2)(10^{10})(0.00226)}{(32.17)(144)} q_g$$

$$\tag{5.55}$$

$$= 9.75 q_g$$

The lowest pressure for which a value of α is known is 1.10 lb./sq.in. As it is necessary to determine the area under the $1/\alpha$ curve starting at $p_s = 0$, an extrapolation as indicated by the dotted lines in Figure 5.12 is necessary. In Equation 2.9 with $n = 0.506$ and $a = 8.66(10^{10})$ is extrapolated to 0.1 lbf./sq.in, α becomes 2.7 (10^{10}) ft./lb. mass, a value that issued in constructing the curves in Figure 5.12. The choice of $p_i = 0.1$ lbf./sq.in. amounts to little more than a conservative, educated guess. The area under the $1/\alpha$ curve between 0 and 1.0 lb./sq.in. is a large fraction of the total area, amounting to about 15% of the area from 0 to 50 lb./sq.in. For materials more compressible than talc, the area between 0 and 1.0 lb./sq.in. could become the major portion of the integral. For filtrations carried out to pressure of 100 to 150 lb./sq.in., the area under the 0 to 1.0 lb./sq.in. is not so significant for compressibility coefficients, n, less than 0.5 to 0.7. However, for low-pressure filtrations as effected by rotary vacuum operation, the area under the 0 to 1.0 lb./sq.in. portion of the curve might become a significant portion of the total area.

While it is difficult to obtain compression-permeability
data at very low pressures, it is important to carry experiments to as low a pressure as possible to miniffize the need
for excessive extrapolation.

In order to obtain v_g in (6.99) as a function of the
pressure ψ, it is necessary to find ψ_1 and q_g for various times.
In Figure 5.12 the pump rate is plotted against both the pressure ψ_1 at the entrance to the medium and the filtration pressure ψ in pounds/square inch. At the instant that flow is
started, the entire pressure drop will be across the medium,
and the initial pressure and rate can be found at A, where
the two curves intersect. As time progresses, the rate will
drop; the pressure at the surface of the cake will increase;
and the pressure at the medium ψ_1 will decrease. In general
the pressure drop across the cake will be given by lines like
BC. From Figure 5.13 q_g, ψ, and ψ_1, and the value of the
integral in (5.53) can be taken from (5.12). The volume then
may be calculated as illustrated in Table 1.

To complete the problem, the time must be calculated by
use of Equation (5.47) with a graphical integration of a plot
of the reciprocal of the rate q_g vs. the volume of filtrate
per square foot, v_g. (Figure 5.14). With values of q_g, the
pressure can be found from Figure 5.13 as demonstrated in
Table 2.

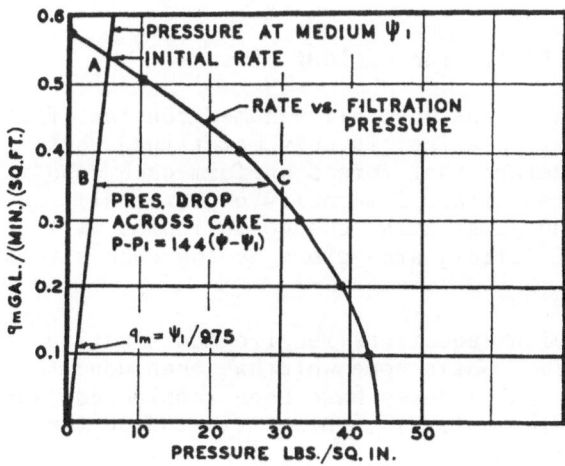

Figure 5.13 Pump Rate vs. p and p_1

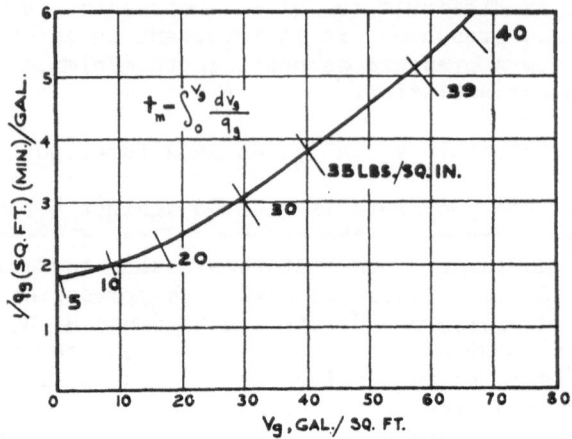

Figure 5.14
Graphical Integration for Determination of Time

Plots of the volume per unit area, filtration pressure, and pressure at the medium vs. the time are shown in Figure 5.15. The medium pressure decreases as the rate drops off. The filtration pressure increases with time and begins to approach a constant pressure toward the end of the curve. The inflection at the first of the curve is caused by the relatively large area under the $1/\alpha_x$ vs. p_s curve in the low-pressure region under 1.0 lb./sq.in.

5.12 Variable Area Filtration of Leaf Filters[1,7]

The previous equations have all assumed constant filter area, which, strictly speaking, is only maintained where there is a containing structure that forces uniform cake deposition. If a cake is deposited either internally or externally on a circular element, any substantial change in radius will vary the area. In a leaf filter, area grows as the cake extends beyond the medium.

All of the previous equations require modification if area growth is appreciable. While some work has been done to account for this, only ideal cases have been considered, and no rigorous solutions are available. Brenner[1] studied the growth of a cake in a three-dimensional filtration on a circular leaf. Amplifying that work, Shirato and his co-workers[5] developed a term, "effective filtration area factor." For two and three-

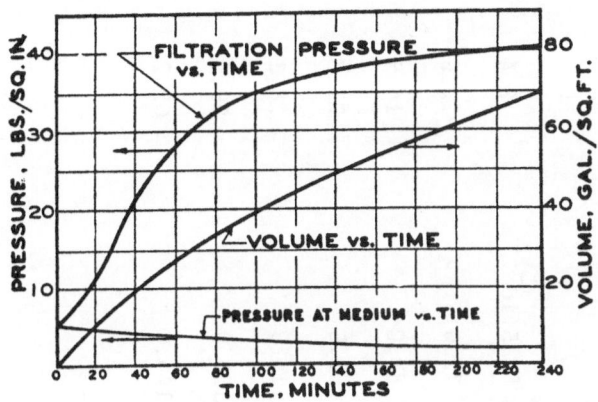

Figure 5.15   Pressure and Volume vs. Time

TABLE 2

Filtration Pressure ψ, lb./sq.in.	Pressure at Medium ψ_1 lb./sq.in.	Pressure Difference, lb./sq.in	Rate of Filtration q_g, gal./(min.)(sq.ft.)	Value of Integral Figure 5.12	v from (5.53) gal./sq.ft.
5	5	0	0.540	0	0
10	4.9	5.1	0.506	$52.5(10^{-12})$	8.4
15	4.6	10.4	0.470	72.0	12.5
20	4.2	15.8	0.434	88.0	16.5
25	3.8	21.2	0.391	102.5	21.3
30	3.4	26.6	0.336	115.5	28.1
35	2.5	32.5	0.263	129.0	39.9
40	1.7	38.3	0.172	140.0	65.2

TABLE 2 Continued

Filtrate Volume v_m, gal./sq.ft.	Value of Integral t, min. Equation (5.47)	Rate of Filtration q_g, gal./ (min.)(sq.ft.)	Filtration Pressure, lb./sq.in	Medium Pressure, lb./sq.in.
0	10.0	0.540	5.0	5.0
10	19.2	0.500	10.3	4.9
20	41.6	0.420	21.5	4.1
30	69.0	0.322	31.0	3.1
40	103	0.263	35.0	2.6
50	145	0.218	37.5	2.1
60	195	0.186	39.4	1.8
70	253	0.160	40.3	1.6

dimensional filtration involving a circular or square plate
of radius or half length r, the area factors j_{II} and j_{III}
were found experimentally for four different materials to be:

$$j_{II} = 1 + 0.801 \, (L/r)^{1.22} \qquad (5.56)$$

$$j_{III} = 1 + 1.47 \, (L/r)^{1.20} \qquad (5.57)$$

The experimental values were slightly lower than those predicted theoretically. The area factor is incorporated in
the basic differential equation as:

$$q_1 = \frac{(p - p_1)}{\mu\alpha w/j} \qquad (5.58)$$

where j can be replaced by j_{II} or j_{III}. Both q_1 and w are
based on the original area. For an L/r ratio of 0.1
(a 2-in. cake on a 40-in. leaf), the actual experimental
correction factors, both two and three-dimensional, varied
from 1.03 to 1.10.

Incorporating the old definition of filtration resistance
α_{av} along with the correction factors J_s for slurry concentration and j for area growth into equations for variable area
filters leads to

$$\alpha = (J_s/j)\alpha_{av} \qquad (5.59)$$

As $J_s < 1$ and $j > 1$, the actual resistance would be less
than that calculated by use of α_{av} alone. The correction may
yield values of α varying up to 15 to 20% from α_{av}

5.14 Local Resistance and Porosities Derived from Overall
      Values

Although it is possible to utilize either constant-pressure or constant-rate filtration for laboratory investigations
if the proper precautions are observed, variable-pressure
variable-rate operation offers some advantages. It is more
simple than constant rate as no controls are required to
maintain the rate at a fixed value. It is better than constant pressure in that resistances are obtained for a variety
of pressures in a single run.

Basically all that is needed is a tank with a stirred connected to a filter press by a pump. While any pumping device can be used, it is essential to avoid changing the nature of the precipitate during the pumping and stirring operation. A high-speed centrifugal pump may develop shearing force on the particles with resulting increase in filtration resistance. Practically it is difficult to find centrifugal pumps with sufficiently small capacity for the size of filter convenient to laboratory testing. A small constant rate pump (piston, peristaltic, etc.) can be used with a bypass to simulate a centrifugal pump[11]. Based upon experimental data, an empirical relationship between the α_{av}, it is possible to calculate the local α as a function of p_s without having to resort to its experimental determination in a permeability-compression cell. Rearranging Equation (5.1) gives

$$\int_0^{p_o} \frac{dp_s}{\alpha} = \frac{p_s}{\alpha_{av}} \tag{5.58}$$

Differentiating with respect to p_s yields

$$\frac{dp_s}{\alpha} = \frac{\alpha_{av} dp_s - p_s d\alpha_{av}}{\alpha_{av}^2} \tag{5.59}$$

Solving for α results in

$$\alpha = \frac{\alpha_{av}}{1 - \dfrac{p_s d\alpha_{av}}{\alpha_{av} dp_s}} = \frac{\alpha_{av}}{1 - \dfrac{d\ln\alpha_{av}}{d\ln p_s}} \tag{5.60}$$

If a plot of $\ln \alpha_{av}$ vs. $\ln p_s$ yields a straight line of slope n, Equation (5.60) reduces to

$$\alpha = \alpha_{av}/(1 - n) \tag{5.61}$$

Equation (5.61) is a satisfactory approximation when n is less than approximately 0.5 and 0.7 and p_i can be neglected.

A similar technique can be used for obtaining the local porosity[5]. Rewriting (4.13) and replacing the numerator by p_s p_s/α_{av}, one obtains

$$\varepsilon_{av} = 1 - \frac{p_s/\alpha_{av}}{\displaystyle\int_0^{p_s} dp_s/\alpha(1-\varepsilon)} \qquad (5.62)$$

Solving for the integral

$$\int_0^{p_s} \frac{dp_s}{\alpha(1-\varepsilon)} = \frac{p_s}{\alpha_{av}(1-\varepsilon_{av})} \qquad (5.63)$$

Differentiating with respect to p_s

$$\frac{1}{\alpha(1-\varepsilon)} = \frac{1}{\alpha_{av}(1-\varepsilon_{av})} - \frac{p_s}{\alpha_{av}^2(1-\varepsilon_{av})^2} \frac{d[\alpha_{av}(1-\varepsilon_{av})]}{dp_s} \qquad (5.64)$$

Multiplying by $\alpha_{av}(1-\varepsilon_{av})$ and replacing quantities of the form dx/x by $d \ln x$ leads to

$$\frac{\alpha_{av}(1-\varepsilon_{av})}{\alpha(1-\varepsilon)} = 1 - \frac{d \ln[\alpha_{av}(1-\varepsilon_{av})]}{d \ln p_s} \qquad (5.65)$$

$$= 1 - \frac{d \ln \alpha_{av}}{d \ln p_s} - \frac{d \ln(1-\varepsilon_{av})}{d \ln p_s} \qquad (5.66)$$

In Chapter 4, it was shown that the last term in Equation (5.66) has a value of β when (2.11) is valid. Under those curcumstances

$$\frac{\alpha_{av}(1-\varepsilon_{av})}{\alpha(1-\varepsilon)} = 1 - n - \beta \qquad (5.67)$$

Thus with a knowledge of the variation of α_{av} and ε_{av} with p, it is possible to calculate local values of both α and ε. As yet, the practical utility of Equations (5.61) and (5.67) has not been established.

References

1. Brenner, Howard, Three Dimensional Filtration on a Circular Leaf, AIChE Jour., 7, 666, 1961

2. Gale, R. S., J. Inst. Water Poll Control, No. 6, 1967.

3. Grace, H. W., AIChE J., Structure and Performance of Filtermedia, 2, 307, 1956.

4. Jahreis, Carl, Role of Pumping Equipment in Liquid Filtration, Atlanta Meeting, AIChE, 1960.

5. Lu, W. M., F. M. Tiller, F. B. Cheng, and C. T. Chien, J. Chinese Inst. Chem. Eng., 1, 45, 1970.

6. Shirato, M., M. Sambiuchi, H. Kato, and T. Aragaki, Internal Flow Mechanism in Filter Cakes, AIChE J., 15, 405, 1969.

7. Shirato, M., T. Murase, H. Hirate, and M. Miura, Studies in non-unidimensional Filtration, Kagaku Kogaku, 29, 1007, 1965.

8. Tiller, F. M., Chem. Eng. Progr., The R ole of Porosity in Filtration, Part 2, Analytical Equations for Constant Rate Filtration, 51, 282, 1955.

9. Tiller, F. M., The Role of Porosity in Filtration, Part 3, Variable-pressure, Variable-rate Filtration, AIChE J., 4, 170, 1958.

10. Tiller, F. M. and Harrison Cooper, The Role of Porosity in Filtration, Part 4, Constant pressure filtration, ibid, 6, 595, 1960.

NOMENCLATURE

a	constant in Equation (2.9)
A	cross sectional area, L^2
A_c	area of contact among particles, L^2
b	intercept on the graph of dt/dv vs. v, reciprocal of the initial rate of filtration T/L
B	constant in Equation (2-1) dimensions meaningless
C_{sub}	constants, Equations (1.10) and (1.16)
D	diameter of channel, L
D_p	particle diameter, L
e	void ratio, $\varepsilon/(1 - \varepsilon)$, dimensionless
E	constant in Equation (2.7)
f	Fanning friction factor, see Equation (1.11) dimensionless
f*	modified friction factor, $f\varepsilon^3/(1 - \varepsilon)$, dimensionless
F_s	frictional head loss, L
j	symbol for j_{II} or j_{III}
j_{II}	correction-factor for area growth in two dimensions, dimensionless
j_{III}	same in three dimensions
j_s	correction factor for filtration resistance due to variable flow rate, dimensionless
k	Kozeny "constant", dimensionless
K	permeability, L^2
K	constant in Equation (5.28), L^6/T

L	thickness, L
L	length of bed, L
L_{eq}	thickness of cake equivalent to resistance of medium, L
m	ratio of mass of wet to mass of dry cake, dimensionless, $1/s_c$
m_i	value of m in infinitesimal surface layer of cake
n	constant defined in Equation (2.9), dimensions meaningless
N_R	Reynolds number, see Equation (1.12), dimensionless
N_R^*	modified Reynolds number, $N_R/(1 - \varepsilon)$, dimensionless
p	applied filtration pressure, F/L^2
$p_f, \Delta p$	frictional pressure drop, F/L^2
p_i	pressure below which porosity and filtration resistance assumed constant, F/L^2
p_L	hydraulic pressure at distance x from medium, F/L^2
p_s	solid compressive pressure at distance x from medium, also total compressive pressure, F/L^2
p_1	pressure at interface of medium and cake, F/L^2
p_o	pressure at which $\varepsilon_{av} = \varepsilon_{avo}$, F/L^2
Δp_c	$p-p_1$, F/L^2
q	superficial rate of flow of liquid in cake at distance x from medium, $L^3/L^2 T$
q_g	rate of flow at interface of medium and cake, gal./sq.ft.)(min.)

q_i value of q in infinitesimal surface layer of cake, L^3/L^2T

q_1 value of q at interface of medium and cake, L^3/L^2T

r half-dimension, length or diameter of filter element, L

r rate of flow of solid in cake at distance x from medium, L^3/L^2T

R_c cake resistance, αw, 1/L

R_H hydraulic radius, ratio of flow area to wetted perimeter, L

R_m medium resistance, 1/L

s mass fraction solids in slurry, dimensionless

s_c average mass fraction solids in cake, dimensionless

s_i fraction solids in infinitesimal layer of cake where x = L

s_o Specific surface, surface area/unit volume of pure solid which occupies $1/(1-\varepsilon)m^3$ of space

t time, seconds

t_i time required to reach p_i, sec.

t_m time, minutes

u average liquid velocity in pores, L/T

u_s average superficial velocity of solids, L/T

v volume of filtrate, L^3/L^2

v_g volume of filtrate, gal./sq.ft.

x distance from medium

w total mass of dry solids per unit area, sq.ft.

w_x mass of solids per unit area in distance x from medium

Greek Letters

α	value of specific resistance at distance x from medium where solid compressive pressure is p_s, L/M
α_{av}	average value of α uncorrected for variation of flow rate or area growth, L/M
α_i	value of α at pressure below p_i, L/M
β	exponent in Equation (2.11), dimensions meaningless
ϵ	porosity at distance x from medium, dimensionless
ϵ_i	porosity at pressure below p_i, dimensionless
ϵ_{av}	average porosity, dimensionless
ϵ_{avo}	average porosity at pressure p_o, dimensionless
λ	exponent in Equation (2.7)
μ	viscosity of liquid, FT/L^2
ρ	density of liquid, M/L^3
ρ_s	true density of solids, M/L^3
σ	ρ_s/ρ, dimensionless
ψ	pressure, lb. force/sq.in.
ψ_s	solid compressive pressure, lb. force/sq. in.
ψ_1	pressure at interface between cake and medium lb. force/sq.in.

BRIEF NOTES ON THE WASHING OF FILTER CAKES

A. S. WARD

LOUGHBOROUGH UNIVERSITY OF TECHNOLOGY

WASHING OF FILTER CAKES

This process involves the removal of soluble material contained in the liquid which is retained in the filtercake at the end of the filtration operation. The removal is usually effected by the application of a wash liquid, miscible with the filtrate and a solvent for the material that is to be removed, to the filtercake in situ.

The physical mechanisms which occur within the cake during this operation include displacement, mixing and diffusion. In practical terms it is clear that a highly efficient washing operation will require the cake to be free from cracks with an even thickness and uniform porosity and that the wash liquid should be applied evenly and in such a way that the filtercake structure is not disturbed. It will be seen that these ends are not easily met in the design of equipment.

The process of filtercake washing is important in many industrial processes, e.g., those involving crystallisation or precipitation where a solid phase has to be produced in a highly pure state from the contaminant filtrate.

THEORY

Many workers have been concerned to produce mathematical equations which are to be used to predict the change in solute concentration that might be obtained after washing for a certain time with a given amount of wash liquor.

The general approach has been to use a mathematical model based on the general physical picture described above. The various approaches have included displacement plus mixing/diffusion, longitudinal mixing/diffusion and latterly blind side channel models, which attempt to take into account the solute material entrapped in pores within the solids and in dead spaces in the cake structure. In this case a historical approach can be used to review the field and to discuss broadly these general attempts to develop a theory.

Rhodes (1) postulated an exponential decay form of equation in order to describe the way in which the concentration of solute material in the wash liquor changes with time. The equation,

$$c = c_o \exp \left[- \frac{k F t}{L} \right]$$

where c is the solute concentration after a washing time of t, c_o the initial concentration of soluble material in the filtrate L is the cake thickness, F the volumetric rate of flow of wash liquid per unit area and k is an equilibrium constant, was derived from experimental work and is semi empirical in nature. The filtrate and wash liquor are assumed to be perfectly mixed in the washing operation and therefore no account is taken of the diffusional processes that occur. Not surprisingly, the Rhodes model is only reasonably accurate for short washing times.

Choudhury and Dahlstrom (2) developed Rhodes' ideas and applied his equation to the continuous equipment used in vacuum filtration. The mixing of wash liquor and filtrate was assumed to be a controlling factor. Their equation, developed from material balances, is

$$R = (1 - \frac{E}{100})^n \qquad \text{where R is the}$$

fraction of soluble material remaining in the cake after washing, E is the washing efficiency expressed as a percentage and n is the wash ratio, i.e., the ratio of the volume of wash liquor to the volume of filtrate in the unwashed cake. The experimental data they presented agreed with theory for wash ratios less than 2.1. The experimental value of efficiency was lowered by 10% for full scale applications in order to take account of uneven cake thicknesses and maldistribution of the wash liquor across the filtration area.

Another important relationship studied was that between the time of washing and the wash ratio. Initially the filtercake is characterised by use of the form filtration equation

$$\frac{V_f}{A} = \left[\frac{2 \Delta_p t_f}{u r v} \right]^{\frac{1}{2}}$$

where V_f is the volume of filtrate obtained in time t_f, Δp is the constant pressure difference, μ the liquid viscosity, r the resistance of the filtercake and v the volume of cake deposited by unit volume of filtrate. This equation is obtained by integration of the usual incompressible cake filtration equation with the medium resistance assumed to be negligible. If wash liquor and filtrate are taken to have the same viscosity then the wash rate per unit area is given by

$$\left[\frac{1}{A}\frac{dv}{dt}\right]_W = \frac{A\,\Delta p}{u\,r\,v\,V_f}$$

which should be constant.

If the volume of wash liquid is V_W and the washing time t_W then

$$\frac{V_W}{A} = t_W \left[\frac{1}{A}\frac{dV}{dt}\right]_W = \frac{A.\,\Delta p}{u\,r\,v\,V_f}.\ t_W = \left[\frac{\Delta p}{2urv\,t_f}\right]^{\frac{1}{2}} t_W$$

The volume of residual filtrate, V_m, in the cake is proportional to the quantity of cake and therefore to the volume of filtrate so

$$\frac{V_m}{A} \propto \frac{V_f}{A}$$

and

$$\frac{V_m}{A} = k'\left[\frac{2\,\Delta p\,t_f}{u\,r\,v}\right]^{\frac{1}{2}}$$

Thus the wash ratio $n = V_W : V_m$ can be obtained as

$$n = \frac{V_W}{V_m} = \frac{t_W}{2\,k'\,t_f}$$

where k' is a proportionality constant determined by the cake porosity. Plots of t_W versus n for different values of t_f, the cake formation time, gave straight lines (Figure 1). Many experimental results obtained from vacuum leaf filter tests were presented to support the theories.

A more fundamental approach to the problem due to Taylor (3), (4) stems from consideration of the displacement of one miscible

fluid by another in a single straight tube. The initial shape
of the interface is taken to be flat and a parabolic profile
develops with time such that the velocity n is given by the
equation

$$u = 2 \bar{u} (1 - \frac{r^2}{R^2})$$

where $\bar{u}$ is the average velocity, u is the velocity at radius r
and R is the tube radius.

For a tube of length L the concentration of the displacing
liquid C(t) is zero for times less than the value of $\frac{L}{2\bar{u}}$.

At time $t = L/2 \bar{u}$, the displacing liquid has just reached the
end of the tube.

For times greater than $L/2 \bar{u}$ the concentration C(t) is
given by

$$C(t) = \frac{1}{\pi R^2 \bar{u}} \int_o^r 2 \bar{u} (1 - \frac{r^2}{R^2}) 2 \pi r \, d r.$$

The value of r is found from the equation

$$r = R (1 - \frac{L}{2 \bar{u} t})^{\frac{1}{2}}$$

which is found from the area of the tube occupied by the displacing liquid.

The effect of molecular diffusion can be examined by considering that the concentration of the displacing fluid is a
function of radius r, time t, and x the distance along the tube.
The diffusion equation can be written as

$$D_m \frac{\partial^2 c}{\partial r^2} + \frac{1}{r} \frac{\partial c}{\partial r} + \frac{\partial^2 c}{\partial x^2} = \frac{\partial c}{\partial t} + V_o (1 - \frac{r^2}{a^2}) \frac{\partial c}{\partial x}$$

where D_m is the molecular diffusion coefficient taken to be a
constant, V_o is the tube volume and a the tube diameter. At
the impermeable wall

$$\frac{\partial c}{\partial r} = 0 \quad \text{at} \quad r = R$$

For a long thin tube in which longitudinal transport due to diffusion is neglected in comparison with the convective transport and the time for radial variations in concentration to disappear due to diffusion is small compared to the time for longitudinal convective transport, the net transfer relative to a plane moving with velocity $\bar{u}$ is given by the equation

$$\frac{\partial c_m}{\partial t} = \frac{4\,a^2\,u_m^2}{192\,D_m}\,\frac{\partial^2 c_m}{\partial x^2}$$

where c_m is the mean concentration in a radial plane. This process is thus described by the diffusion equation with a longitudinal diffusion coefficient D given by the expression

$$D = \frac{4\,a^2\,u_m^2}{192\,D_m}$$

Aris (5) extended this analysis to include axial molecular diffusion and showed that the dispersion coefficient for this case was

$$D_m + \frac{4\,a^2\,\bar{u}^2}{192\,D_m}$$

A complete description of mechanisms controlling dispersion in a capillary tube as a function of dimensionless time and Peclet number is available from the finite difference solution of the full convective diffusion equation due to Ananthakrishnan et al. (6). The same arguments have been applied by many workers to the problem of longitudinal mixing in packed beds but perhaps the most concise approach is that due to Brenner (7). This paper is concerned with the behaviour of an initially sharp interface between two miscible liquids having identical dynamical and kinematical properties. The process is described by the equation

$$\frac{\partial C}{\partial T} + \frac{u\,\partial C}{\partial X} = \frac{D\,\partial^2 C}{\partial X^2}$$

where $C = C(X,T)$ is the concentration of soluble material, T is the time from the commencement of the wash, X is the distance from the point of introduction of the wash liquid, and u the average interstitial velocity ($u = Q/A\varepsilon$ where Q is the volumetric flowrate of wash liquid, A is the cake area and ε the cake porosity). The bed thickness is L and D is an axial dispersion coefficient. The initial solute concentration is c_o, which is a constant throughout the bed at time T = 0, and c_f is the solute concentration of the incoming fluid. In order to be able to integrate the above equation, Brenner made the following substitutions

$$c = \frac{C - C_f}{C_o - C_f}, \quad t = \frac{UT}{L}, \quad x = \frac{X}{L} \quad \text{and} \quad Pe = \frac{UL}{4D} \left(\begin{array}{c} \text{Péclet} \\ \text{No.} \end{array} \right)$$

to give

$$\frac{\partial c}{\partial t} + \frac{\partial c}{\partial x} = \frac{1}{4\,Pe} \frac{\partial^2 c}{\partial x^2}$$

and by substituting

$$c(x,t) = v(x,t) \exp (2\,Pe. x - Pe. t)$$

obtained

$$\frac{\partial v}{\partial t} = \frac{1}{4\,Pe} \frac{\partial^2 v}{\partial x^2}$$

which is a one-dimensional equation of the type often discussed in the heat conduction field.

The boundary conditions are of some interest. The first one is concerned with the inlet, i.e., at $x = 0$. Since there should be no loss of soluble material through a plane at which the wash liquor is being introduced as material balance gives

$$UC - D\frac{\partial C}{\partial X} = UC_f \quad \text{for all} \quad T > 0.$$

At the bed exit a similar equation can be written to conserve solute mass, viz

$$Uc - D\frac{\partial C}{\partial X} = UC_e \quad \text{where } C_e = \text{exit concentra-}$$

tion. But the second boundary condition is obtained when it is realised that to avoid the unacceptable conclusion that solute concentration passes through a maximum or minimum value at some point in the interior of the bed, it is therefore necessary to impose

$$\frac{\partial C}{\partial X} = 0 \quad \text{at} \quad X = L \quad \text{for all } T > 0.$$

The initial condition is $C(x,o) = C_o$ = constant for all X.

The average concentration of soluble material in the bed at any time is of considerable interest. This value, $\bar{C}$, is given by the equation

$$\bar{C} = \bar{C}(T) = \frac{1}{L} \int_0^L C(x,T)\,dx.$$

The solutions are developed from general solutions due to Corslaw
and Jaeger (8) and numerical values have been calculated. The
results are presented in graphical form (see Figures 2,3) and
show the relationship between exit solute concentration, C_e, or
average solute concentration $\bar{C}$ versus displacement time, t,
which is termed the number of displacements. The parameter is
Peclet number, which includes the axial dispersion coefficient D.
Taylor has shown that this coefficient D is related to the
molecular diffusion coefficient D_m by the relationship

$$D = \frac{4\,a^2\,U^2}{192\,D_m}$$

where a is the mean pore diameter. By substituting this
expression in that for Peclet number the following expression is
obtained

$$P_e = \frac{12\,L\,D_m}{a^2\,U}$$

Brenner's work, when applied to filtercake washing, is
inadequate in that it does not take into account the removal of
soluble material retained in the blind pores and dead spaces in
a drained filtercake. Practically, drained cakes are common,
many examples being found in centrifugal filtration applications.
To deal with this situation a blind side channel model has been
presented by Han (9,10) and developed by Wakeman (11). The
model assumes that the residual filtrate is contained in a system
of blind channels in the cake and that the flow of wash liquid is
along straight channels without any axial mixing. The soluble
material in the trapped filtrate is transferred by a diffusion
process and the solute then transported through the bed by the
wash liquid flowing in a plug flow fashion. The blind channels
were assumed to be wedge shaped. The straight channels are
assumed to be empty of filtrate so that diffusion of soluble
material from the side channel starts when the wash liquid reaches
the point of emergence of the side channel into the main channel.
Han's solution is given as

$$c(t,y) = \frac{2\,C_o\,A_1\,D_m\,L_e}{Q.\ell} \sum_{n=1}^{\infty} \exp\left\{ - \frac{D_m\,\beta_n^{2}(t - L_e/v)}{\ell^2} \right\}$$

for $t > \dfrac{Le}{v}$ where t is the washing time, y is the distance coordinate in the direction of flow of the wash liquid, c_o is the initial solute concentration of the filtrate, Q is the wash flowrate, v is the velocity of the wash liquid through the cake and β_n are the roots of the first kind of zero order Bessel function.

A_1 is the area of the blind side channel, which is obtained from the residual equilibrium saturation volume ratio

$$R_o = \frac{\text{Total volume of side channels}}{\text{Total volume of straight channels}}$$

by the equation

$$A_1 = \frac{2 A \, \mathcal{E} \, R_o}{\ell}$$

where A is the area of the filtercake and $\mathcal{E}$ the porosity. ℓ is the length of the blind side channel. Wakeman uses an equation relating the blind side channel length to the mean particle size, viz,

$$\ell = 0.148 - 7.46 \, d_p \qquad \text{for } 0 < d_p < 0.013 \text{ cm.}$$

L_e is the effective depth of the bed which is greater than the actual bed depth by a tortuosity factor. Some experimental results are shown in Figure 4.

It has been observed (12) that as washing liquid flow rate increases, the graphs of concentration versus time approach more closely to the plug flow displacement condition and that as washing time increases the effects of dispersion/mixing and, subsequently, molecular diffusion become more relevant.

REFERENCES

(1)   Rhodes, F.H.   Ind. Eng. Chem.   26, p. 1331 (1934).

(2)   Choudhury A.P.R., and Dahlstrom, D.A.   A.I.Ch.E.J., 3, p. 433 (1957).

(3)   Taylor, G.I.   Proc. Phys-Soc.   London 67, p. 857 (1954).

(4)   Taylor, G.I.   Proc. Roy. Soc.   A219, p. 186 (1953).

(5)   Aris, R.   Proc. Roy. Soc.   A235, p. 67 (1956).

(6)   Ananthakrishnan, V., Gill, W.N., Barduhn, A.J., A.I.Ch.E.J., 11, p. 1063 (1965).

(7)   Brenner, H.,   Chem. Eng. Sci., 17, p. 229 (1962).

(8)   Carslaw, H.S., and Jaeger, J.C.   "Conduction of Heat in Solids", O.U.P. (1959).

(9)   Han, C.D.,   Chem. Eng. Sci., 22, p. 837 (1967).

(10)   Han, C.D. and Bixler, H.J.,   A.I.Ch.E.J., 13, p. 1058 (1967)

(11)   Wakeman, R.J.   M.S. Thesis, Manchester University 1971.

(12)   Michael, A.S., Baker, W.E., Bixler, H.J., and Vieth, W.R. I.E.C. Fundamentals 6, p. 33 (1967).

BIBLIOGRAPHY

Chapters 8 and 9, in "Flow through Porous Media"   ed. R.J. Nunge. American Chemical Society, 1970.

Chapter 8 in "Flow of Fluids through Porous Materials"   Collins R. Reinhold, 1961.

Chapter 10, in "Dynamics of Fluids in Porous Media".   Bear, J. American Elsevier 1972.

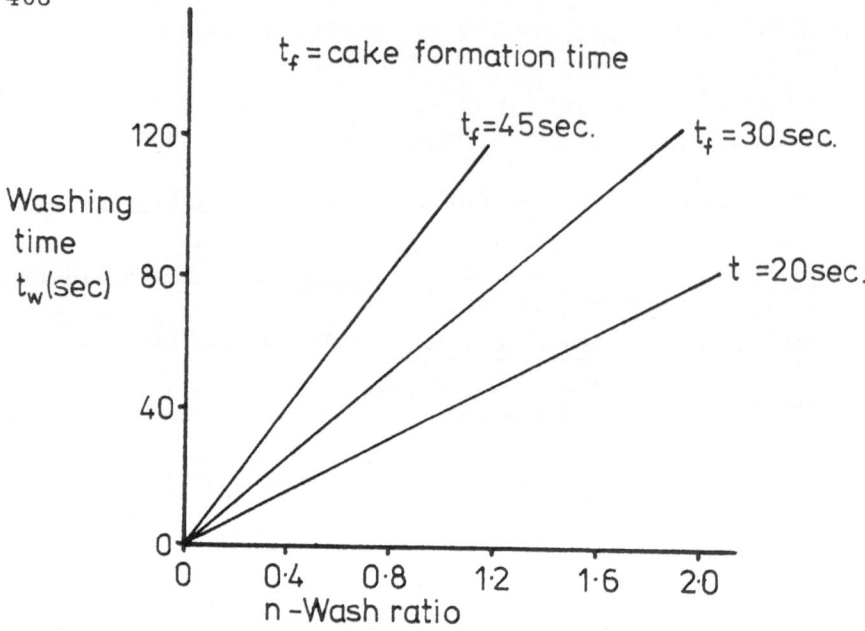

Figure 1 (Ref. 2)
Washing time versus wash ratio with cake formation time as
a parameter.

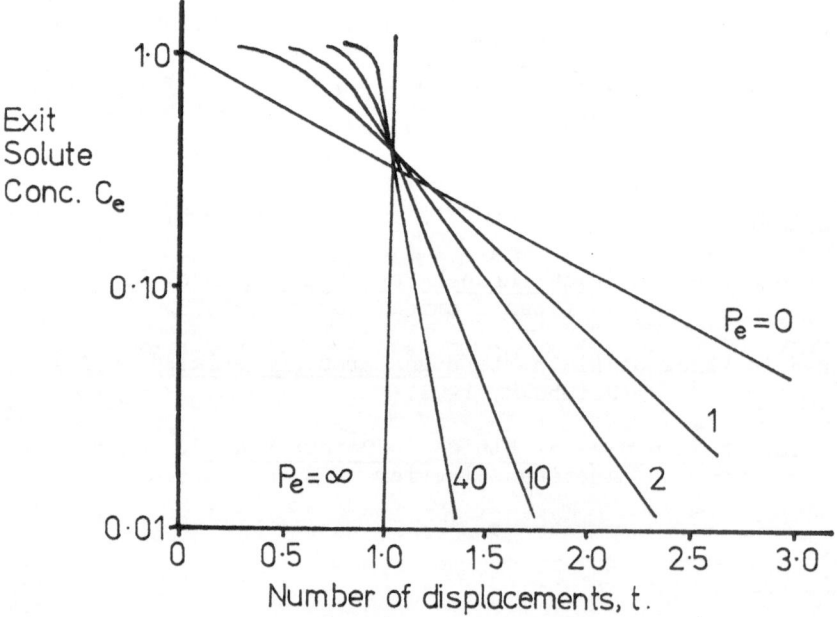

Figure 2 (Ref. 7)
Exit solute concentration versus number of displacements.

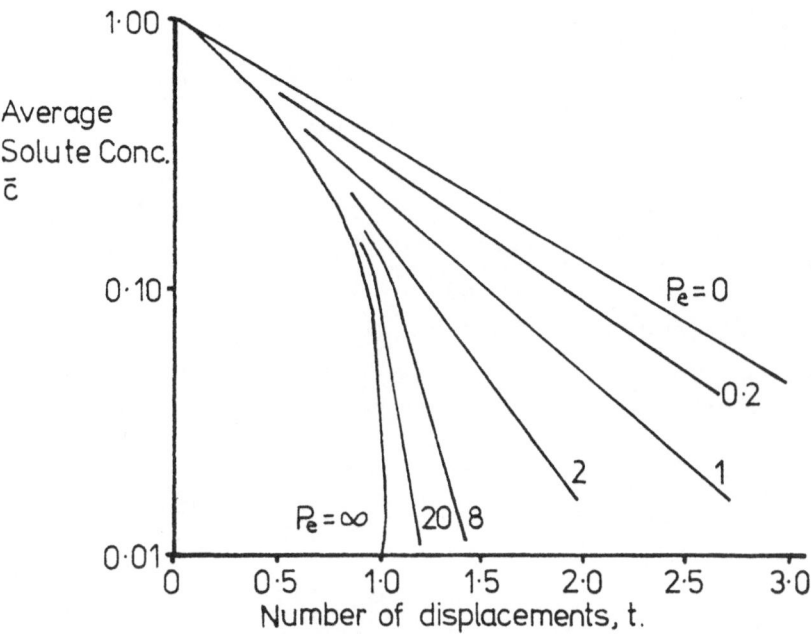

Figure 3 (Ref. 7)
Average solute concentration versus number of displacements.

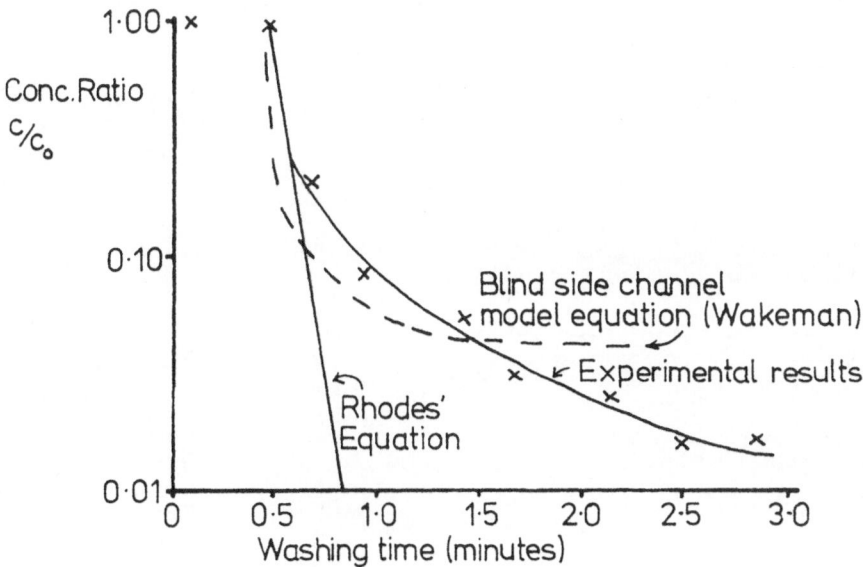

Figure 4 (Ref. 11)
Washing curve (concentration ratio vs time).

PRACTICAL PROBLEMS IN CHOOSING FILTRATION
PROCESS, AND FUTURE DEVELOPMENTS

Professor Dr. -Ing. Christian Alt

Lehrstuhl für Mechanische Verfahrenstechnik
University of Stuttgart

A. The Engineering Progress by using Filtration Theory
   and Practical Selection of Filters

   As shown in the preceding lectures the filtration theory is
directed to the description of the actual process of liquid
separation from solids by means of filtration. However, the
individual proceedings of industrial filtration are comparatively
so complicated that, up to this day, a mathematical approach
could be only successful by simplifying some details which are
more or less important for the assessment of the complete
process.

   The aim of this lecture is to demonstrate as to what in
practice could be done, if both theoretical and special know-
ledges in filtration are to be considered. It is a fact that this
is not practised perhaps too often, particularly in industries.
But the present situation in filtration theory seems to fill up
the gap between what theoreticians can predict and experts
wish to know.

   The peculiarity in the selection of the correct type of filter
and in the assessment of behaviour ensues from the fact that
here, in contrary to other equipments of chemical engineering,
most machines have been introduced some time before filtration theory was developed sufficiently. This results from
empirical efforts for improving existing devices. Thus, many
filters are used with quite different filtration proceedings. The

filtration engineers, particularly those who are at the beginning of their industrial activity, are faced sometimes with very difficult problems of selecting the right type of filter.

Irrespective of special material properties and filter type, some general rules should be observed:

At first, the process requirements are to be defined clearly in considering the performance of the filtration process, which is mostly a part of multistage process within a production of materials or compounds. In this connection, it is to be decided if the filtration can run
a) continously or
b) batchwise and also to know
c) the capacity required.

Then, the suspension should be analysed with regard to some material properties which are important for filtration, for example
a) concentration,
b) particle size, particle distribution, particle shape,
c) porosity and surface or
d) filtration resistance under operating conditions (as minimum).

Unfortunately, the number of occuring suspensions is too extensive to go through such a procedure, preferably in smaller industries which are not in possession of adequate laboratories. On the other hand, the process variables are neither too obvious nor could be disclosed, and this is one of the reasons why in practice, experimental procedures are preferred for filter design. It is evident that this cannot be satisfied all the time. Hence, one divests oneself the possibility to optimize the process by means of calculation and to assess the behaviour of the filter.

In addition to this, a filtration process in a filter actually involves some further operations as it passes through its cycle. For, depending on its nature, the filtration is a batch process irrespective of how it is recognized by the description based on filtration. One exception is given in the case of sieve filtration where the separated solids are discharged immediately from the filter medium.

In the most cases of cake filtration the question is whether
the liquid can be completely separated from solid particles for
economic reasons. This may be done only by several additional
procedures which together form the filtration cycle.

For example these operations may include

1. Cake formation - The liquid is removed by laws of filtration with the exception of that amount which is retained by
   adhesion in the form of liquid film on the surfaces of the
   particles and inside the smaller pores and holes.
2. Cake dewatering - Removal of liquid by contraction of the
   pores and holes due to mechanical compression. This
   assumes certain properties of solids which implies compressibility
   and/or
3. Cake dewatering - Removal of liquid by drag and displacements due to an intensive flow of air or steam across the
   cake.
4. Cake washing - Washing the solids with soluble-free wash
   liquor which occurs as a result of displacement and diffusion in order to reduce the solution in the residual liquid.
5. Cake dewatering - Removal of liquid within the cake to
   minimize the residual moisture.

According to the requirement of the process these stages
can vary in different extensions and successions. Each of them
is based on more or less different principles. The most disclosed is the filtration period. The other functions must be
evaluated mainly by empirical methods.

I. Filtration

We begin with filtration period where the cake is formed.
This is the true filtration period. The basic relationship of
constant pressure filtration is given by the flow through porous
media which is defined in conformity by Darcy's law

$$\frac{d V_1''}{A\,dt} = \frac{\Delta p}{\eta'' \,(\,\bar{m}\,\alpha' + \beta\,)} \qquad (1)$$

with V_1'' = volume of filtrate $[m^3]$
A = area of filter $[m^2]$
t = time $[s]$
p = pressure differential $[N/m^2]$
η'' = dynamic viscosity of filtrate $[Ns/m^2]$
α' = specific filtration resistance of the cake $[m/kg]$

$$= \frac{\alpha}{\varrho' \,(1 - \varepsilon)}$$

α = filtration resistance of the cake $[m^{-2}]$
ϱ' = density of solid $[kg/m^3]$
ε = porosity
$\overline{m}$ = solids per filter area $[kg/m^2]$

$$= \frac{\varrho'' \, V_1'' \, m'}{A \left(1 - \frac{m'}{m'_k} \right)}$$

ϱ'' = density of liquid $[kg/m^3]$
m' = mass rate of solid in the suspension (feed)
m'_k = mass rate of solid in the cake
β = filtration resistance of filter medium $[m^{-1}]$

In many operations the value of β becomes insignificant and can be disregarded. The equation is simplified and shall be taken as a basis in the following considerations. Keeping all terms except $\overline{m}$ as constant for a general assessment, it is apparent that the increase of filtrate per time is a function of $1/V_1''$ with $\overline{m} \sim V_1''$, i.e. the increase is diminishing in the course of filtration. This implies a filtration cycle to be as short as possible for economic reasons. To maximize filtration efficiency short filtering times or periods should be used.

There are about two cases in practice in which this just mentioned conclusion is not correct,

1) when the porosity varies during the course of filtration and along with that the filtration resistance also,
2) when a mechanical cake discharge requires a certain cake thickness.

In the first case, it may be advantageous to apply a higher filtration pressure when the filter aids are used at the same time. As a rule, this occurs in all cases of filtration where the cake is comparatively deformable and the particles are slimy.

The second case depends on the fact that a thin cake cannot be discharged satisfactorily from the filter. This is a matter of the type of particle and discharge mechanism. As a general rule Dahlstrom [1] gives the minimum dischargeable cake thickness for various types of filters as following:

Horizontal belt filter	3	... 6,5	mm
Drum filter	6,5	... 10	mm
Top feed drum filter	12	... 20	mm
Disc filter	12		mm
Pan filter	20		mm

In particular the basic equation of filtration permits to recognize a few variables which affect filtration rates, and thus to obtain the best efficiency.

The material factors ϱ'', η'', α', m'_κ must be accepted and can be changed for the benefit of filtration rates only in a very few cases. Sometimes it may be possible that a smaller particle distribution as a result of a diversity of packing configurations leads to a low filtration resistance. Changes in particle shape will also affect the flow of filtrate through the cake. Particle size is usually found to be unchangeable in any particular process.

Pressure drop along the filter medium and the filtration cycle are the variables which facilitate the manufacturer to design a filter plant. These features are influenced by process variables which include concentration, temperature and particle size of feed slurry. Viscosity of liquids, and also of the filtrate can vary usually with temperature in such a way that it decreases with increasing temperature. Therefore, the feed slurry should be heated sometimes if necessary or should not be cooled, particularly if the liquid is a solution of high concentration.

The problem is now as before to obtain the correct influence of variations in materials. The determination of material factors is widely difficult if a sufficient accuracy should be achieved. This refers to the particle size, the particle size distribution and the specific surface. A screen analysis is insufficient to define the particle size distribution and the

measurment of specific surface is not usually made in industrial laboratories. These facts lead to the conclusion that the
most satisfactory method available, at this time, for the design of filters is from the information obtained by experiments,
if no other data exists.

Once the correlation of filtration cycle, i.e. time of cake
formation is known, the parameters of pressure, temperature
and viscosity can be easily developed. Among these, the
pressure is usually the most effective variable which can be
adapted by the type of filters. In the case of non-compressible
filter cakes, the filtration rate is directly proportional to
pressure. If a low filtration rate is accepted by the nature of
materials, i.e. by very fine particles, the application of a
comparatively high filtration pressure is indicated.

In a following chapter more details on the correlation
between particle size and type of filter can be found.

II.   Dewatering and washing

We continue the considerations of theoretical correlations
with an analysis of dewatering and washing cycles.

IIa.   Dewatering

At the end of filtration cycle, all pores and holes are filled
by liquid and the likelihood of the bubbles getting caught in a
portion of the pores is out of question. If no external forces
have effect on the residual liquid in the cake, the moisture is
a function of porosity alone. It is

$$\varphi_r = \frac{\varrho'' \varepsilon}{\varrho' - \varepsilon(\varrho' - \varrho'')} \qquad (2)$$

with φ_r = residual moisture.

This amount of liquid in the cake is usually too high for
economic reasons. However, there are some possibilities to
reduce this. The most inexpensive method is drainage by
gravity which can take place immediately subsequent to filtration cycle and the result is theoretically caused by a combined action of adsorptive, capillary and wetting forces. If
each of their effect is expressed by an efficiency factor, such
as φ_a, φ_k and φ_s, Batel [2] obtains the residual moisture as

$$\varphi_r = (\varphi_a + \varphi_s)\left(\frac{s - h_s}{s}\right) + \frac{h_s}{s}\,\varphi_K \qquad (3)$$

with s = cake thickness [m] and h_s = capillary rise [m]

Fig. 1 shows a typical plot, from which it can be seen that in a state of equilibrium the upper layers of cake have a lower moisture than those which are below them.

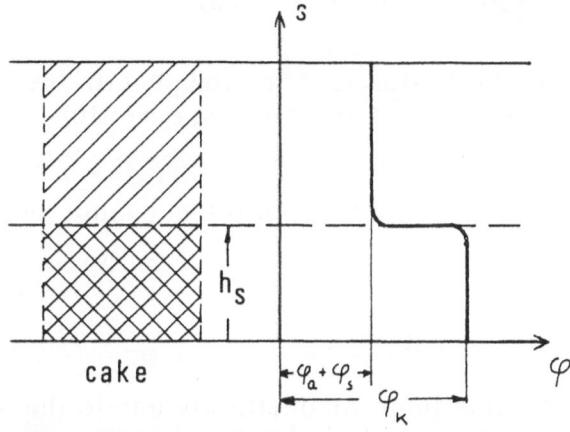

cake

Such a moisture is usually still too high, particularly if the pores of cake are very small, because the higher capillary rise would fill in more pores.

It is now a frequently used practice to reduce the residual liquor extra by blowing or sucking through the cake. To determine the required pressure which can remove the liquid from the pores, the following relationship may be used

$$Eu \cdot We \geqq 1 \qquad (4)$$

wherein the Euler Number $Eu = \dfrac{\Delta p}{\varrho'' w^2}$ and the

Weber Number $We = \dfrac{1 \cdot w^2\, \rho''}{\sigma}$

In the case of packings, it is to be replaced by

$$Eu \cdot We = \frac{\Delta p \;\epsilon\; d_p}{2\sigma \cos\delta\,(1 - \epsilon)\; f_\alpha} \qquad (5)$$

with Δp = differential pressure   [N/m^2]
     ε = porosity of cake
     d_p = particle diameter   [m]

σ = surface tension $[Nm/m^2]$
f_α = form factor (> 1) indicating variation
of particle shape from sphere (=1)

Experiments have shown that Eu . We may have values
from 2 to 3. Assuming for instance Eu . We = 2,5, d_p =
50 μm (= 5 . 10^{-5} m), σ (H_2O, 20^0 C) = 7,3 . 10^2 Nm/m^2,
ϑ (H_2O → Quartz) = 0, ε = 0,4 and f_α = 1, we will
obtain the required pressure differential as

$$\triangle_p = 3,3 . 10^4 \, [N/m^2] = 0,33 \, [bar].$$

The equation (3) indicates the beginning of drainage with the
help of air or gas and the state is called moisture equilibrium
or residual saturation.

The other factors which are important from the economic point
of view are

a) the time for drainage
b) the final cake moisture content.

It is difficult to determine them theoretically due to the
complexity of the intersticial geometry of cakes and the vague
drag effects. Some empirical correlations have been developed,
which when all put together result into a general condition that
straight line relationships can be assumed among all the combinations of the four factors; cake moisture, reciprocal of
viscosity, surface tension and air rate. Observations and a
regressive analysis indicate predominant effects of viscosity
which cause the variation in moisture content to a great extent,
while the contribution of surface tension effects is relatively
unimportant. This agrees with the theory as mentioned above.

Silverblatt and Dahlstrom [3] suggest that it may also be
possible to reduce cake moisture by washing with a comparatively concentrated wetting agent. However, only for agents
with high concentrations the cake moisture content could be
reduced and a minimum air rate could be passed.

The most important factors that affect final moisture
content are the following:

1. Size distribution of solids,
2. Shape and surface characteristics of solids,
3. Concentration of feed,
4. Viscosity of liquid,
5. Dewatering time,
6. Cake thickness,
7. Pressure drop through the cake in dewatering cycles,
8. Air, gas or steam rate,
9. Filter media.

The last five factors listed above are filtration conditions which can be adjusted to increase or decrease the cake moisture and the filtration rate. Unfortunately, optimal conditions for cake dewatering do not correspond to optimal conditions for filtration rate. For example, increase in cake thickness to obtain drier cakes will usually decrease the filtration rate.

The rate at which the filter cake moisture approaches the final residual moisture is a function of an approach factor f_A which can be defined by

$$f_A = f \left(\frac{t_2}{\eta''} \right) \left(\frac{V_L}{m'} \right) \left(\frac{\Delta p_2}{\Delta p_1} \right) \tag{6}$$

wherein t_2 = dewatering time
$\dot{V}_L$ = air (gas) flow rate
m' = mass of the cake per filter area
Δp_2 = pressure differential of dewatering
Δp_1 = pressure differential of filtration

Additional factors which are taken into account are the saturation equilibrium and the nature of particles. The correlation is not yet developed. Plotting according to this equation gives a graph (Fig. 2) as shown by Dahlstrom[1]. It is evident that the dewatering occurs rapidly at low values of the approach factor which is called as "correlating factor". Operating at low values is essentially by evaporative drying and is not practical unless it can eliminate the necessity of an additional moisture removal step.

420

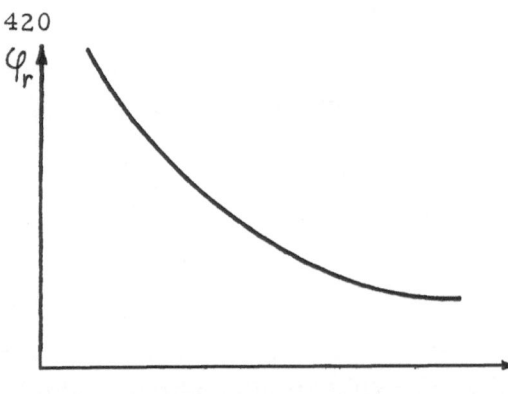

φ_r

f_A

Fig. 2   The approach factor of
dewatering filter cakes

It is also been recommended
by Roskydálek [4] to determine the approach factor by
laboratory measurements,
but the difficulties involved
in these measuring methods
should be pointed out.

An empirical relationship correlating the moisture con-
tents φ_o at the beginning of dewatering and φ_t at the time t
was suggested by Batel [2] as

$$\varphi_t = \varphi_o \exp \left[- K \left(\frac{\Delta p - h_s \, \rho'' \, g}{h_s \, \rho'' \, g} \right) t \right] \qquad (7)$$

wherein the constant K comprises all the material properties
and is to be determined by experiments.

The final moisture content is then formed by adsorptive
liquid $\varphi_a (w)$ and wedge (capillary) liquid $\varphi_z (w)$. Both of
them depend on the air flow rate, w = volume of air/filter
area if drag occurs. For this Batel (l.c.) obtained the
equation

$$\varphi_a + \varphi_z = \frac{\rho''}{\rho'} \left(K_o \, \frac{\sigma \cos \vartheta}{\rho''} \, \frac{O_s}{z \cdot d_p} \right) \qquad (8)$$

which includes O_s = specific surface
$\underline{z}$ = acceleration field (gravitational = 1)
d_p = average particle diameter
K_o = a proportional factor

It was found by experiments that the term within
the parenthesis may vary from 10^2 to $2 \cdot 10^5$ and is a function of material constants only. For example, in the case of
coal and limestone, the term $K_o \sigma \cos \delta / \rho''$ was 3340 cm
and 5180 cm. The factor z also includes centrifugal field and

has influence on the pressure differential when the gas is blowed or sucked.

Furthermore, it can be seen that this moisture content is decreasing with surface tension; strictly speaking, a 50 % reduction in surface tension results into a decrease of 15 % in moisture content. It is obvious that in many practical cases of filtration, it is not economical to use such a considerable amount of agents. A variation in the contact angle is however sometimes more effective. The process of coal filtration is mentioned in support of this where the oil happens to displace the moisture. (Convertol-Process)

The final moisture content varies at different layers of the cake and increases in the direction of the flow of liquid. But this must be considered only in the case of thicker cakes, when the thickness exceeds 10 mm.

In the recent years, a new procedure has been introduced into the practice of filtration which has enabled either to reduce the moisture content or to shorten the dewatering time to some extent. It could be found that some filter cakes discharge a remarkable amount of moisture by mechanical compression. This presupposes a certain deformability of packing. The theories for cake compression and liquid discharge are not developed sufficiently for practice, (but it was noted that the squeezing and filtration take place simultaneously). The peculiarity of this procedure is that the cake is compressed inside of the filter device by elastic filterplates, and their constructions are described in a following chapter.

IIb. Washing

Washing of filter cakes is of distinct importance in practice. Sometimes it can be decisive for selecting the filter process.

Washing is a removal of residual mother liquors (filtrates) from the cakes. Its object is to produce a valuable solution by separating it from the futile solids and at the same time it should be noted that a retained liquor in the cakes might cause impurities in the solids. At the beginning of washing, the residual filtrate undergoes a hydraulic displacement by the wash fluid. This is followed by mixing and diffusion of the two liquids distributed on different concentrations of solute within the voids of the cake.

A mathematical simulation could not yet lead to satisfactory results due to the complexity of a mathematical description of the geometry of cakes as well as the complex flow pattern of wash fluid in the interparticle voids. Trials of different mathematical models, such as that of Wakeman [5], to approach the prediction of washing performance cannot claim general application. An empirical correlation is used by Hackl [6] to determine the rate of wash liquor as a function of cake thickness and wash efficiency and it is as follows:

$$(1 - \eta)^n = \exp \left[\ (V_w/s) - a \ \right] \tag{9}$$

where η_w = wash efficiency
$\dot{V}_w$ = volume of wash liquor $[m^3/m^2]$
s = cake thickness $[m]$
$n; a$ = constants

Besides, the laws of flow through porous media are valid for the determination of rate and pressure drop of wash fluid.

III.   Conclusion

Dewatering and washing time give, in addition to the time of discharge of the cake the so called filtration cycle. The correct assessment of the individual cycle is not possible by theoretical considerations, but needs either some experience or some tests to be carried out. It is to be remembered that the tests or experiments which are mainly used have a limited transferability. The theoretical relations are to be considered if flexibility or adaption is desired in the filter plant.

In practice, it can be found that the true filtration requires short time in comparison with the filtration cycle. Hence the decision of a filter must be done by the assessment of all the cycles.

B. Laboratory tests

In principle, two methods are used for laboratory tests; one for vacuum filtration and the other one for over pressure filtration. Special test devices because of their limitations to general application cannot be considered just like the devices are excluded for deep bed filtration.

The laboratory test has to carry out the operations of filtration as well as dewatering and washing, and all the operations usually occur in the same device. Although the test can give information about the behaviour of the materials to be treated in filters, it would be advantageous to analyse separately the material properties for reasons of systematic evaluations and comparative purposes. They can be of valuable help in identifying the test and should include particle analysis, viscosity and pH-value of liquids, and if necessary the conditions of operation.

It is not intended in this lecture to cover sufficiently all the problems that one would face in these tests. But in all the experiments, some secondary phenomena which are in good support with the experimental results have to be considered. Unfortunately, no general standard test technique exists and in most of the countries different devices and methods are used. At present several Standard Committees are engaged themselves in preparing the standards of filters which can be used in certain applications [7].

If the laboratory test is the basis of design of the filter plant, the following parameters have to be taken into consideration:

a) the sample of the materials to be filtered,
b) the pressure in practice,
c) the filter media (cloth),
d) the actual concentration in all parts of the suspension,
e) temperature,
f) air rate during dewatering,
g) wash liquor,
h) cake discharge, and if necessary
i) precoating.

a) Sample

The sample for tests must be true representative as mentioned above, i.e. all characteristic properties for filtration have to be exactly the same in test as they are in practice. Many discrepancies in filtration can be explained by the results obtained by using the samples. If a proper sample is not obtained, the best test and most reliable method will be of absolutely no help [7]).

The same sample is used frequently for several tests. Such a procedure slightly leads to errors, particularly if the particle sizes are reduced by a repeated agitation in the containers to avoid sedimentation of particles.

b) Pressure

Tests should be performed at expected pressure levels for actual installation. But, from the actual point of view, the selection of the best pressure level depends on the behaviour of the cake during its formation. It is mostly ascertained by experience whether a vacuum or an overpressure level would be optimal. Practical considerations will be demonstrated in a following chapter.

Unlike in the filter plant, it is not in general appropriate to run the filter test at different pressures when the cycle is being partly formed. Mostly the laboratory tests make it possible to apply better vacuum (or pressures) as in practice, and the device is more leakproof. In other words, special pressure regulation facilities indicate operations on the filter plant considerably at equal pressures.

c) Filter media

It is common that the laboratory test at first concentrate in giving information about the filter media. From the considerations of the interactions between the particles and elements of filter medium, it is evident that there are more or less suitable media for special slurries. The most interesting factor here is the blinding of cloths by retained solids. To study blinding, it is necessary to run seven or even more tests although it would be almost impossible to determine the blinding in the laboratory with sufficient accuracy.

The laboratory test can also give information about the cake discharge from the filter media. In cake filtration, it is required either by the operation or the equipment that no cake is retained by the filter media. This is not so in the case of clarifying filtration occurring in deep bed filters where solids are retained completely inside the medium. It is obvious that the mechanisms of discharge of solids in these two cases are quite different, and they are discussed later.

Since the filtration efficiency depends on the effect of separation by the filter medium, reliable measurements of concentration of the feed suspension and the filtrate are absolutely necessary. It is to be noted that a variation in the efficiency is a function of load.

There is also a dependance of filtrate clarity on the filtration time. Normally, a better clarity is achieved when once the particles have bridged across the filter medium. Thus, a poor clarity is caused when the former part of the cake is formed. Longer formation time will therefore usually result in clear filtrates and better filtration efficiencies. Tighter media would be more effective; however, they have a tendency to blinding. In some cases, a recycling of the first part of the filtrate is more effective.

d) Concentration

The above-mentioned measurements of more reliability require normally some expensive techniques. For example, the technique of weighing the components separated by evaporation and drying is time-consuming. It is common to evaluate the concentration by the visual methods but they are not as reliable as quantitative methods.

More reliable techniques of measurements are obtained with the help of scattering lights in combination with photocells which has to be adjusted accordingly to each and every material. Finally, electrical techniques have been developed which supply outstanding values.

A well established laboratory experimental setup should include the measurement of the concentration of the feed suspension and the filtrate. If in addition to these concentrations, the residual moisture is known, a material balance can help to improve the accuracy of the test. It is known that difficulties are involved in the measurement of filtrates with poor concentration.

The measurements of the actual feed concentration is normally a question of sampling. In laboratory tests, special attention must be given to the surety of a homogeneous suspension. This is to observe if the solids tend to settle quickly, preferably in the case of relatively coarse particles. Depending upon the tests made, whether upstream or downstream, different results with different accuracies may be obtained. In the first case, the finer particles remain in the upper layers of the container and cause a change in concentration at the bottom. The acting concentration of the suspension at the filter medium may then be quite different from the mean value. In the second case, a cake may be built by coarser materials at first, and the filtration rate does not agree with the expected one.

To avoid sedimentation and change in concentration, an agitation of the suspension is recommended. Dahlstrom[)prefers agitation by using hand because he found results in agreement with his experimental data. To achieve this settling, no other procedure is as quick and reliable as hand agitation. If a slurry cannot be kept in suspension, this general laboratory test is inpracticable and should be replaced by special tests in connection with full scale investigations.

e) Temperature

As seen above it may be advantageous to filter at higher temperatures. In running equivalent tests it would be the best always to branch off the suspension for test from the process directly. In the most cases where this will be impossible, care must be taken in heating the suspension and maintaining the temperature.

f) Air rate in dewatering

Dewatering in laboratory tests may be simulated by discharge of the slurry from the container or by taking out the filter device. In these cases normally the rate of air flow is not controlled. The consumed energy of the pump will be unknown although this is very important from the economic point of view. It is suggested that a gas meter should be used to measure the volume of air which has passed through during the test.

Precautions like considering the gas laws have to be taken for the determination of the air rate. Attention is drawn to their dependence on temperature and pressure.

The laboratory test of dewatering will indicate the likelihood of cake cracking which is absolutely not suitable for the following steps like washing or drying. The investigator will try even to make the cake by a spatula to find out the possibility of using a roller in practice.

g) Wash liquor

The process of washing in the laboratory tests should be carefully carried out. In any case, a procedure similar to that of filtration like dipping the filter leaf into the wash liquor is not useful according to the principle of different processes in practice, where the wash liquor normally comes from above owing to a free flow.

The efficiency of washing is indicated by the dilution of mother liquid as well as by the amount of wash liquor consumed and washing time. All the factors are affected by the design of filter. Therefore, a general rule for washing cannot be set up by considering extremely different types of washing mechanisms and the operations in different filters. The washing test is mostly an optimal approach to reality in the direction of improvement of efficiency. One just thinks of the conditions in pan, drum or disk filters. It is common to determine the actual washing efficiency expected in practice at least in a small-scale setup which is in proportion with the presumed full-scale filter.

h) Cake discharge

The laboratory test may give only limited chances of simulation of the cake discharge in practice. Although in the case of leaf filtration tests, the discharge by scrapers or strings with or without blowing may be observed in principle, the design in practice, as a rule is quite different. The experts will be able to draw the corresponding conclusions only from laboratory test. Up to date, there are no investigations made on the correlations between structures of filter cakes and discharge devices. Apart from rheological problems, the compression, resp. consolidation and abrasion must be considered.

The special types of discharges like the belt, roll and centrifugal discharges can be clearly tested in practice or in small scale setups.

The type of discharge affects the filtration cycle. As mentioned above, the cake thickness is connected with the type of filter because of the used discharge device. In addition to the above listed data, a minimum cake thickness for each of the discharge devices of drum filters is given as follows.[1]

Scaper blades	6 to 10 mm
Strings	5 mm
Belt discharge	1 to 2 mm
Roller	0,7 mm

i) Precoating

The techniques of precoating and its physical basis are certainly the main objects of a preceding lecture. At this point, it should be remembered that precoating may also be tested in experiments which give information about the permeability and separation in the same way as in the experiments of filtration.

Now, a well practised laboratory setup for vacuum filtration will be described as an example, where the points from the practical point of view can be demonstrated.

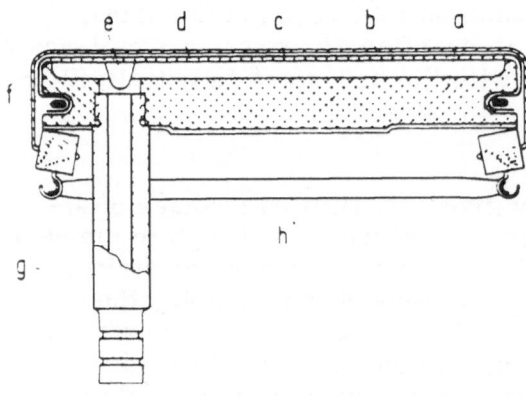

Fig. 3 Handfilter plate

The suitable filter is a rectangular leaf of ca. 10 to 15 cm lateral length. As shown in Fig. 3, it consists of a base plate of hard rubber a, which has grooves at the side of the filter medium for the outlet of filtrate, and a connecting piece g to draw off the filtrate by vacuum. The filter medium c is stretched on the plate by the device f and is supported by a perforated metallic sheet d. A string device b is preferably used for cake discharge.

In Fig. 4 an useful arrangement for handfilter tests is shown. The vacuum is here produced by a jet of water from the device e which is sufficient for most of the tests. It must be measured unconditionally by a manometer d. Two flasks c serve as collecting containers for filtrate and wash liquor. The suspension is in the container b where the filter may be dipped.

The procedure of the test is specified by numerous observations. According to these, it is recommended to move the filter plate slightly in the slurry to incite a required amount of agitation. If it is not possible to avoid settling of the solids, a top feed system can be provided. It is obvious that both the clarity of the filtrate and its volume are functions of the time t and the amount of filtrate is collected in the flask c. Generally the time-study is of an outstanding importance because of its effect on all the filter cycles. The filtration process can be stopped whenever it is required and this is obviously done when the filtration rate is decreased. For dewatering, the plate is taken out of the suspension and turned in such a way that the cake is ahead. As the drainage takes place, it has to be observed if the cake starts cracking. A cracked cake makes it impossible to have a homogeneous dewatering and also a washing which follows subsequently. In practice, the cracks (or punctures) may be closed by rollers.

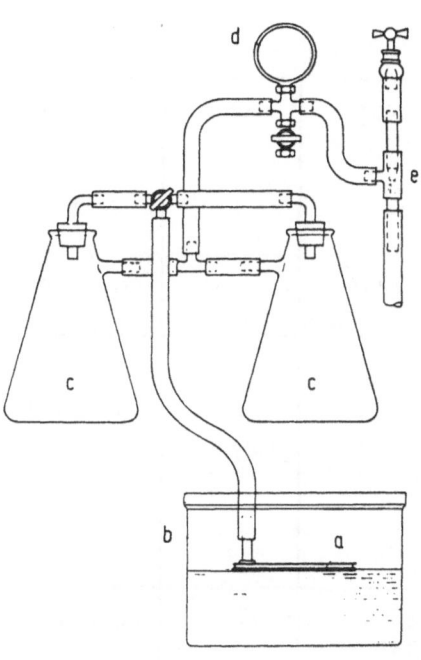

Fig. 4   Laboratory experimental setup

C. A classification of filters

The enormous number of filter types is to be divided into groups to facilitate the filter selection. It can be found that a group of filters are more or less suitable for a certain type of slurries.

In practice, the following classification (table I) may be useful.

Test data form the basis for the following factors:

1) filter cycles,
2) arrangement of filters like submerging of drums, disks, etc.,
3) speed of rotation.

Sufficient records cannot be given in detail.

Table I [8]

Group of suspension Filtration characteristics	Fast filtering	Medium	Slow	Dilute	Very dilute
Cake formation rate	2 to 4 cm/s	1,2 to 5 cm/min	1,2 to 6 mm/min	1 mm/min	no cake
Normal Concentration %	> 20	10 to 20	1 to 10	< 5	< 0,1
Settling rate	rapid diffi- cult to sus- pend	medium	slow	slow	
Leaf test rate kg/m^2h	> 2500	250 to 2500	25 to 250	< 25	no test
Filtrate rate m^3/m^2h	> 10	0,5 to 10	0,0025 to 0,05	0,025 to 0,05	0,025 to 0,05
Typical slurry	Cry- stallic solids	Salts	Pig- ments	Waste water	Water

It appears that filter types can be coordinated with one of the above-mentioned groups, and some to more. Since the sedimentation and therewith the classification of filters depend on the particle size, a graph can be drawn (Fig. 5) from which the range of application for general type of filters can be read.

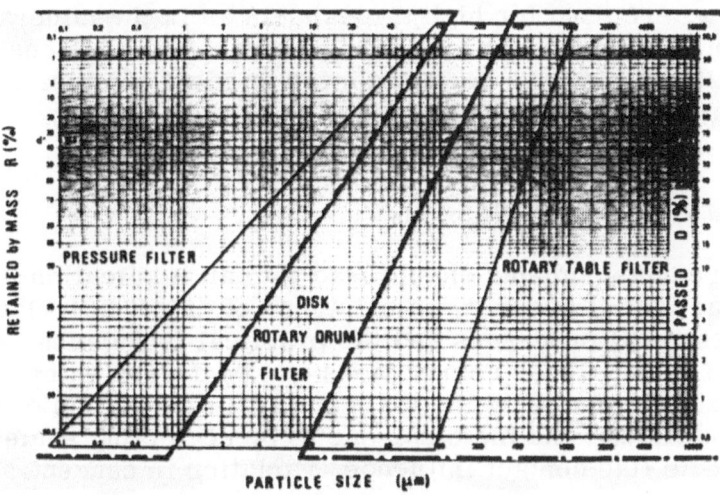

Fig. 5   The range of application

In addition to this, it can be seen that the range of application is a function of the particle spectrum; the suspension with finer particles requires pressure filters while the coarse suspension will be treated preferably in pan or table filters. Further, the amount of different particle sizes in the case of broad size distribution affects the application of a certain filter. For example, an amount of 10 % of retained mass (by sieve) of 10 µm requires in general a pressure filter while an amount of 70 % of the same size may make a rotary filter useful.

This graph contains only general types of filters, since a graph of this type cannot give more general information to a filtration engineer, an entire supplement with all the filter types does not appear to be very useful. Moreover, some filters have a range of application for different suspensions.

In principle, a classification of filter types depending on their operations like vacuum, pressure etc. and also with the combination of the type of filter element is hardly possible. Therefore, more efforts have to be taken for the classification of filters on the basis of filter elements. In practice, the following elements are mainly used.

Bed (Deep bed)   Disk   Nutsch
Belt   Drum   Pan
Candle (Cartridge) Leaf   Plate

Most of them have been introduced into practice since a long
time. Some are used for high throughputs as in the mineral industry, others are preferred for high separating efficiencies.
In general, efforts have been taken for an ideal combination.
New developments are described in the next chapter.

D.   Future Developments

Future developments should lead into an implication for
use in practice. It assumes an existence of direct correlations
between physical processes and the practical techniques. As
mentioned above, these correlations can be developed by two
simplifications; one is by the mathematical description of materials and fluids, and the other one is by neglecting some additional effects like contact influence, variation in concentration
etc.

The implication for control of filtration systems was the
object of the first chapter. Future developments are required
for the correlations between the actual properties of the materials and the mathematical simulations which obviously means
that different materials have to undergo the process of filtration
and many investigations have to be carried out. But it can be
hoped that in future several typical materials will be found
which may be used as a basis for the efficiency of the filters.

Experiences have concluded that the design of the filters
should make it feasible to recognize the optimal conditions
given by the laws of filtration, dewatering and washing. Moreover, optimization assumes the application of mathematical
relations which are applicable only if the design of the filter
permits it.

From the basis of filtration theory, it can be found that
the problematic filtration of the finer suspensions with high
separating efficiencies and low residual moisture content can
be listed for present and future developments of filter designs
as following in Table II.

Table II

Suspension of fine Particles	Operation				Filter Type
	Pressure	Washing	Compression	Continuous (c) automatic (a)	
		x		c	Belt
x		x	x	c + a	Belt with Compression
x	x	x	x	a	Nutsch
		x		c	Tilting Pan
x	x			a	Leaf
x				c	Drum Belt
x	x	x		c	Pressure Drum
x	x	x	x	a	Filter Press with Compression
x	x	x	x	c	Belt Filter Press
x	x		x	a	Tube Press
x	x			c	Rotating Filter Press

The filtration of coarser suspensions which is not so problematic is not considered, excepting such cases where an excellent washing is required for an inexpensive design like the belt filter and tilting pan.

Some typical designs of these types are;

1. The Belt filter [9]

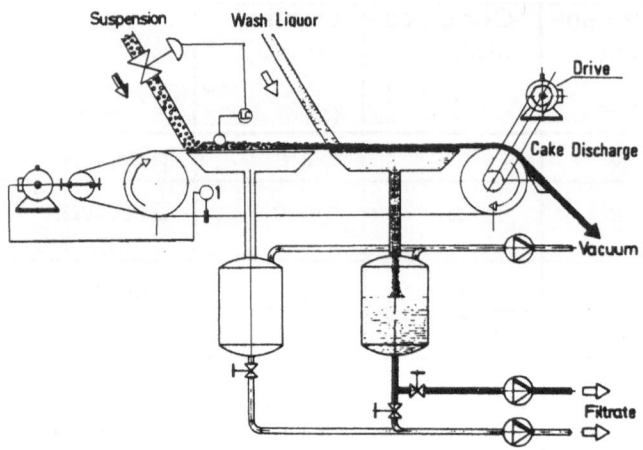

Fig. 6   Washing on Belt Filter

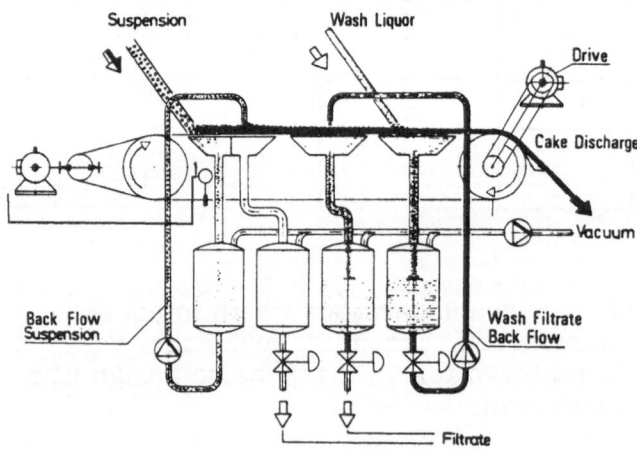

Fig. 7   Reverse flow of wash liquor

In this particular type of filter, the endless belt passes over the vacuum chambers. This type has two advantages, the first is the correct control period of filtration cycles during which an excellent separation can be obtained and the second one is the washing of filter medium when it runs back.

The excellent separation of liquids along the belt permits a reverse flow of wash liquor as shown in Fig. 7. In this way, minimizing the wash liquor is possible, and further from the economic point of view the filtration process is improved.

The frictional drag between the belt and the vacuum chambers which is present in these equipments is eliminated by introducing an unit of reciprocating chambers; movable vacuum trays are attached to the belt during the time of filtration, and then they are released and brought back to their original points.

Belt filter includes two kinds of designs of compression; the semi-continous filter and the belt filter with rollers (Fig. 8 and 9).

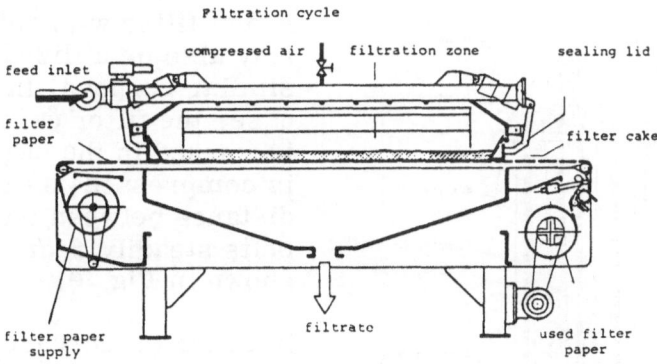

Fig. 8   The semicontinous belt filter

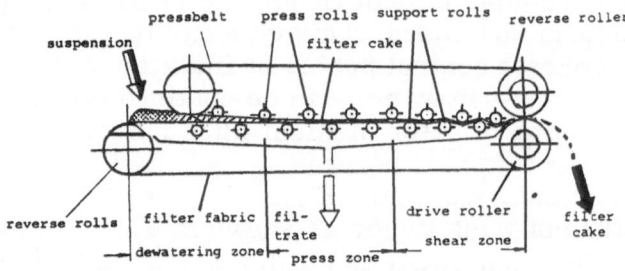

Fig. 9   Pressure belt filter with press rollers

The semi-continous belt filter has a belt with a stepwise movement; the suspension is pumped into the filtration chamber where the filtration occurs when the belt becomes stationary. The filter cake can be dewatered with compressed air. In the other type, the dewatering results from hydraulic compression. In order to discharge the cake, the flaps have to be raised and the filter medium has to be moved. The cake is thrown off when it passes through a deviating roller. This semi-continous belt filter has an application in the filtration of waste water where slimy cakes are obtained.

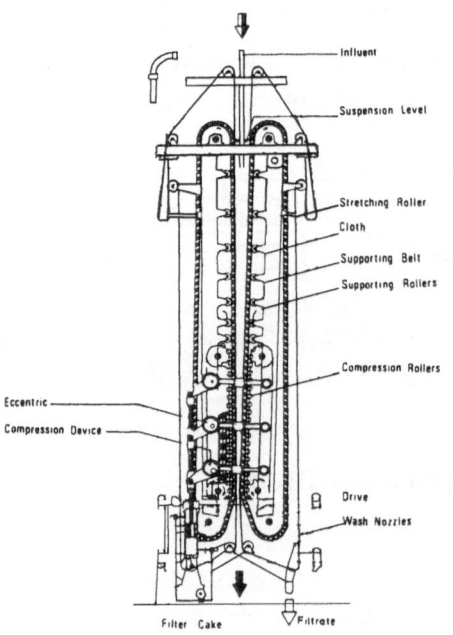

A belt filter with rollers may also be utilized for a similar purpose. In this case, the filter cake which is formed in the beginning is compressed, as the distance between the two belts steadily decreases as shown in Fig. 9.

In order to increase the filtration rate before the cake gets compressed, a vertical belt filter was invented. (Fig. 10). The slurry is fed into the filter from above and then filtration occurs in gravitational field while the coarser particles are settling on the cloth at lower parts which

Fig. 10   The Tower Belt Filter

may reduce the blinding. After this settling, the two belts come together and are pressed by rollers. The pressure is regulated by means of a hydraulic system. The application of this type may be once again found in the case of waste water.

2. Automatic Nutsch Filters

The frequently used Nutsch filters are preferred because of their extreme adaptibility to various filtration cycles, provided if they could be applied to different materials. To use these filters in modern filter plants, many efforts have to be taken to automate the operation. One of the main problems of Nutsch filters is the cake which is being automatically discharged. With the intention of solving this problem, various designs have been shown in Fig. 11a - 11c. All these filters are automatically controlled. The first two have plough bit or screw discharge of the cake, while the last has been provided with a piston which also serves as a device for the mechanical compression of the cake.

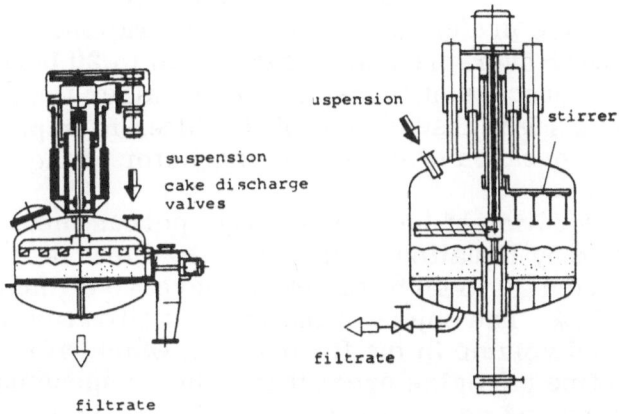

Fig. 11a The CIBA Nutsch   Fig. 11b the Rosemund Nutsch
         filter

During filtration, cracks might occur in the filter-cake which can be smoothened out either by agitation or by discharge device.

Regulation and control of the filtration cycles are performed by control panel methods.

438

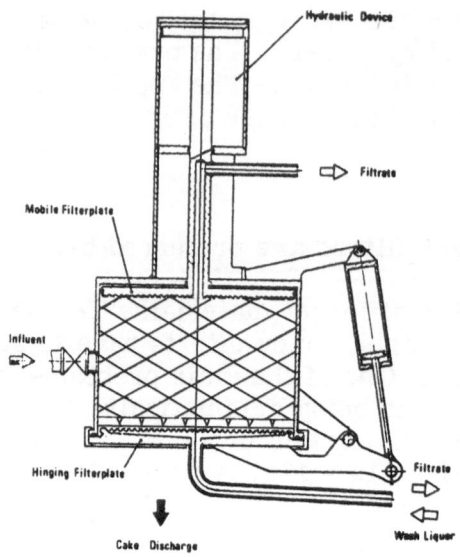

Fig. 11c   The Double Nutsch Filter

The Double Nutsch Pressure Filter, Fig. 11c has a cylindrical filtration chamber with a sleeve at the side which serves as an inlet for the suspension and the two plates built inside serve as filter elements. The upper plate can be moved hydraulically and can empty the chamber by compressing the cake.

3.   Leaf Filters

Leaf filters are mostly used as clarification filters for suspensions with poor concentration and are also preferred for precoat filtration. As a rule they operate under over-pressure which is up to 20 bars or even more. The process of filtration obeys rather the theoretical relations; a right assessment of the filter is in practice always a failure because of the slimy character of the materials treated in here.

The basic equation of the flow through porous media, viz. the Carman-Kozeny Equation indicates that an increase of the cake porosity from 20 to 40 % cause an increase of the flow rate of some 70 %. The object of the precoat filtration is to increase the void volume in the filter cakes which will be low because of the fine particles present in some suspensions like juice, wine, beer and so on. At the same time, the precoat filtration should prevent not only the puncture of the finest particles in a suspension which cannot be retained practically by the most compact filter cloth, but also the blinding of the filter medium.

According to these two objects, two different methods of precoat filtration are used. In the first case, the precoat is dosed into the feed suspension continuously in such a way that the cake is disaggregated by the precoat materials with comparatively high porosities of more than 80 %. In the second case, the precoat layer with a thickness of few millimeters is formed at the beginning of filtration cycle. This layer takes over the function of the filter medium which is now less susceptible to blinding because of the property of deep bed filtration. As mentioned above, only the so called first procedure can be considered by the theoretical relations, if sufficient data of materials are available.

In practice, the main object of leaf filters is the cake discharge which is not quite simple basically for a pressure filter. Several new discharge devices and procedures are being developed for a long time. Centrifugal or vibrating device is one of the most successful for cake discharge. Fig. 12a and

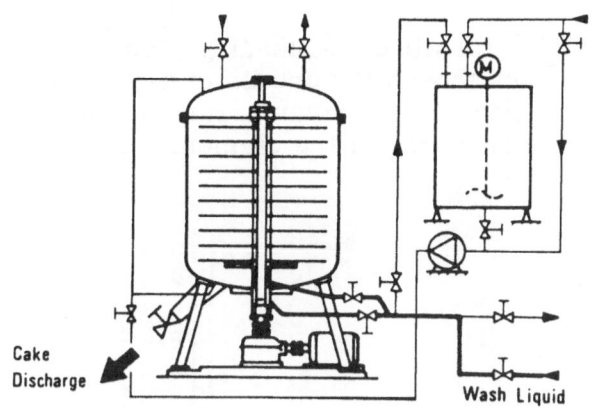

Fig. 12a   Pressure Leaf Filter with Horizontal Cake Discharge

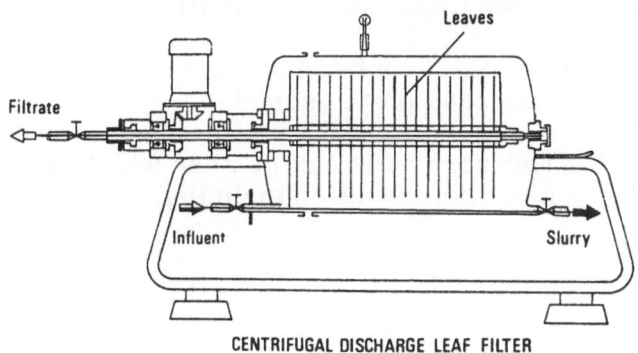

Fig. 12b   Pressure Leaf Filter with Vertical Cake Discharge

Fig. 12b show the horizontal and vertical arrangements of centrifugal cake discharge systems. These filters are mainly used for the filtration of dilute suspensions, as classified in table I.

4.   Filter Types including Cake Compression

At the end of the chapter on dewatering, it was mentioned that under the supposition of some characteristic properties of materials, the squeezing of cakes which takes place subsequently to the building up of the cakes can mainly cause a decrease in its residual moisture content. But the theories do not lead to general conclusions.

The principles for all the squeezing elements in situ are almost similar even though the details of arrangements may be a little different. Fig. 13 shows an arrangement within a common filter press consisting of the elements, plate and frame.

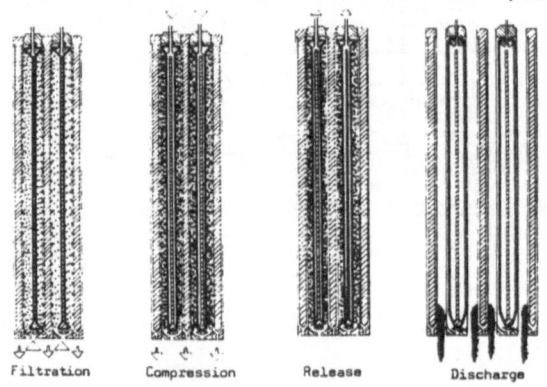

Filtration   Compression   Release   Discharge

Fig. 13   Cake Compressions in situ

In an analogous way, this can be made clear in the case of a filter press with recessed plate assembly.

In Fig. 14 a new apparatus can be seen which may be suitable for an automatic operation including cake discharge.

Industrial Filtration of Liquids

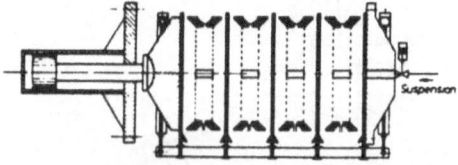

a. initial open position

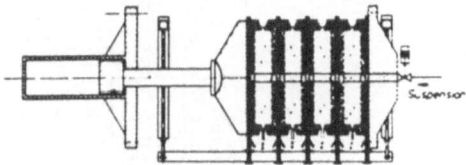

b. during the compression stage

Fig. 14   The Plate Press Filter

Fig. 15 shows the British tube press where the marked
Fig. 9 and Fig. 10 indicate the compression cycle.

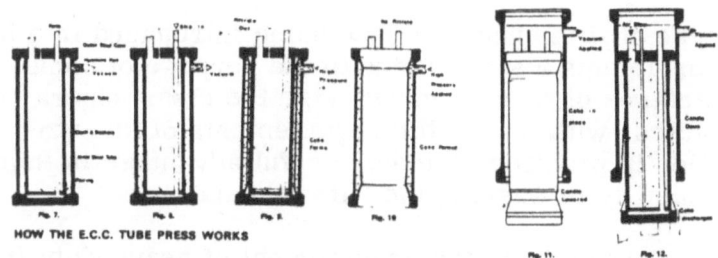

HOW THE E.C.C. TUBE PRESS WORKS

Fig. 15   The Tube Press

A more automated filter press which operates semi-conti-
nously is shown in Fig. 16a and Fig. 16b. The first figure out-
lines the filter elements where the elastic elements for squeezing which are called as diaphragms and their operation can
be seen. The second figure shows how the cake discharge
occurs; i.e. an endless belt passes over the filter elements
always periodically whenever the dewatering resp. squeezing
ends. It is appreciated from the practical point of view that the
filter medium can be washed from both the sides.

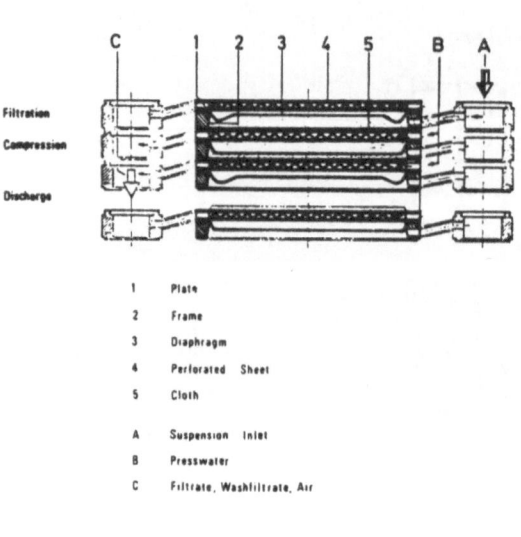

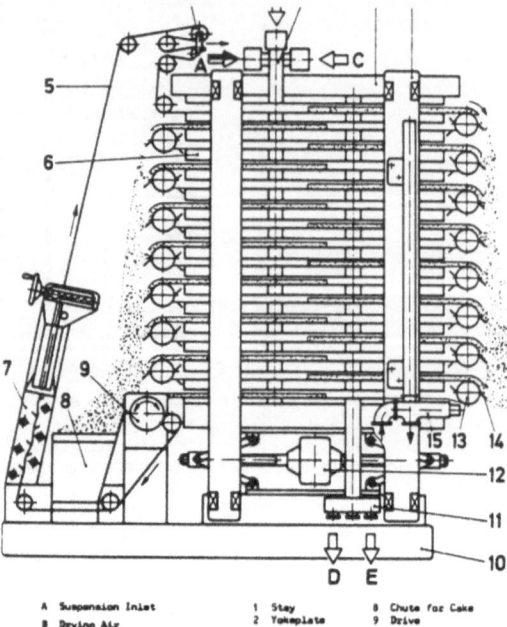

Fig. 16a The Filter Element of
the Automatic Filter
Press

Fig. 16b Scheme of Automatic
Filter Press

In this type of filter which is now being introduced into the
industry, experiments were made with the purpose of combining the advantages of filterpresses, viz. the clear separation of the cycles, with that of the requirements of an automatic operation in practice. Another useful advantage in this
is the practicability in washing the filter media.

One of the last steps in the development of heavy-duty filters is the continously operating filter press as shown in
Fig. 17. This press has stationary and rotating filter elements,
d and e, which are supposed to undertake the process of filtration, and which of course will be done in different quantities.
The improvement in filtration rate which is surprisingly found
in many investigations is probably caused because of the higher
permeability in a moving filter cake. The physical background
is not yet analysed and this may constitute a part of future development.

In Fig. 17, a is the tube for influent of the suspension by
pressure, b is the outlet opening periodically by a valve, c is
the main shaft, d and e are the stationary and rotating filter

plates. This filter press is not yet common in industries and more knowledge about the mechanism is necessary for the estimation of efficiency.

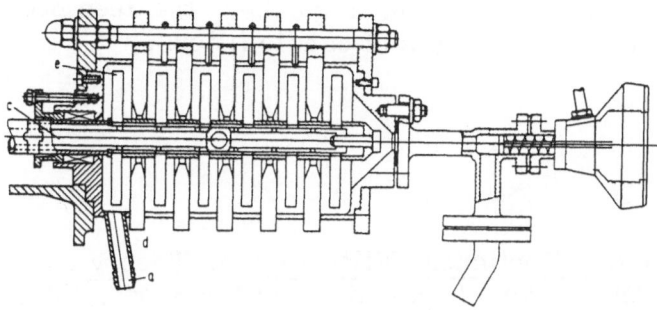

Fig. 17   The Rotary Filter Press

Slurry is fed in at the pipe a, and is moved through the spaces between the adjacent plates not only by the pressure, but also by the centrifugal speed of about 2000 revolutions per minute. The outlet for thickened slurry, i.e. the filter cake which is thixotropic is controlled by an automatic regulating valve b which is actuated by the rise in drive power.

Conclusions

Several filter types are available for a filtration engineer to select the optimal filter for a particular operation. But unfortunately, the selection of the filter and the assessment of its behaviour are still largely based on experience and experiments.

Theory and practice of filtration should make progress by elucidation of the following key problems.

a. Cake compression by flow of suspension, as well as by squeezing,
b. Rate at which the cake is being built up,
   Particular attention is paid to the influence of pumping rates so as to ensure a minimum moisture content, solids concentration, as well as pressure drop,
c. Washing mechanisms,
d. Drainage including cake cracking,
e. Blinding including cake adhesion on filter media,

f. Capillary suction.

It is only to give more importance that the new developments of filtration techniques are pertained to the filtration of pulsating flow, the filtration of moving cakes, the deep bed filtration and interesting designs of candle filters.

References

1. Dahlstrom D.A., Continuous Filtration, in Theory and Practice of Solid-Liquid Separation, Tiller F.M., Ed., Chemical Engineering Department University of Houston, Houston, 1972.

2. Batel W., Vorausberechnung der Restfeuchtigkeit bei der mechanischen Flüssigkeitsabtrennung, Chem.-Ing.-Techn. 27, 497, 1955.

3. Silverblatt C.E. and Dahlstrom D.A., Moisture Content of a Fine-Coal Filter Cake, Ind.& Eng.Chem. 46, 1201, 1954.

4. Rozkydálek J., Eine Methode zur Bestimmung der Entwässerung des Filterkuchens, Chem.Techn. 20, 729, 1968.

5. Wakeman R.J., Prediction of the Washing Performance of Drained Filter Cakes, Filtr.& Separ. 9, 409, 1972.

6. Orlicek A.F., Hackl A.E. and Kindermann P.E., Filtration, Dechema Erfahrungsaustausch, Frankfurt, 1964.

7. Report of Filtration Working Party, The Institution of Chemical Engineers, 1972.

8. Porter H.F., Flood J.E. and Rennie F.W., Filter Selection, Chem.Engng. 78, 39, 1971.

9. Liquid Filtration in the Chemical Industry, Dillier G., Horisberger H., Kaspar J. and Rosch M. Ed., Ciba-Geigy AG, Basel, 1971.